2nd International Conference on Sustainable Technologies and Advances in Automation, Aerospace and Robotics (STAAAR-2023)

MATEC Web of Conferences Volume 393 (2024)

Sehore, India
15-16 December 2023

Editors:

Rohit Sharma
Shiv Manjaree Gopaliya
Ashish Vashishtha

ISBN: 978-1-7138-9397-4

Printed from e-media with permission by:

Curran Associates, Inc.
57 Morehouse Lane
Red Hook, NY 12571

Some format issues inherent in the e-media version may also appear in this print version.

For additional information, please contact EDP Sciences – Web of Conferences
at the address below.

EDP Sciences – Web of Conferences
17, Avenue du Hoggar
Parc d'Activité de Courtabœuf
BP 112
F-91944 Les Ulis Cedex A
France

Phone: +33 (0) 1 69 18 75 75
Fax: +33 (0) 1 69 28 84 91

contact-edps@webofconferences.org

Additional copies of this publication are available from:

Curran Associates, Inc.
57 Morehouse Lane
Red Hook, NY 12571 USA
Phone: 845-758-0400
Fax: 845-758-2633
Email: curran@proceedings.com
Web: www.proceedings.com

TABLE OF CONTENTS

VIT®
BHOPAL
www.vitbhopal.ac.in

2nd INTERNATIONAL CONFERENCE

ON

SUSTAINABLE TECHNOLOGIES AND ADVANCES IN AUTOMATION, AEROSPACE AND ROBOTICS (STAAAR-2023)

15-16 December, 2023

Programme Booklet

School of Mechanical Engineering (SMEC)
VIT Bhopal University, Bhopal 466114,
Madhya Pradesh, India

1. Introduction

About VIT Bhopal University

The VIT legacy, started from Vellore Engineering College 38 years ago under the visionary leadership of **Dr. G. Viswanathan, Founder-Chancellor**, to offer world-class education for the development of society. VIT Bhopal University is ably spearheaded **by Dr. G. Viswanathan, Chancellor, Mr. Sankar Viswanathan, Vice- President, Ms. Kadhambari S. Viswanathan, Assistant Vice-President**. The University is currently led by **Dr. A. Senthil Kumar, Vice-Chancellor**. They share the mission to make VIT a global center of education. Starting from the 2017 academic year, the new Ultra-modern university campus, sprawling over 250 acres, is in the town of Kothrikalan District Sehore, surrounded by green, lush fields and a serene environment. This new and contemporary campus is aiming to become the primary hub of education, with leading-edge and high-tech developments in a wide variety of areas. At present, VIT Bhopal University offers core engineering programs such as Mechanical Engineering, Aerospace Engineering, Computer Science and Engineering, Electronics and Communication Engineering leading to B. Tech degree, M. Tech and allied subject programs such as MSc, MCA, BBA, etc., are also offered. At VIT Bhopal University, the extra-curricular activities and employment needs of the students will be taken care of by various clubs and dedicated career development cells

About School of Mechanical Engineering (SMEC)

The School of Mechanical Engineering (SMEC) at VIT Bhopal offers undergraduate degrees and research programmes in Mechanical Engineering, Mechanical Engineering specialization in Artificial intelligence and Robotics, and Aerospace Engineering. SMEC has been an integral part of the university since its inception. Our faculty members possess diverse expertise, spanning areas such as Machine Design, Thermal & Fluid Sciences, and Manufacturing Engineering within the Mechanical Engineering domain. Our Aerospace Engineering faculty specializes in structures and propulsion, while those in AI and Robotics come from backgrounds in Robotics and Automation. The school offers a comprehensive array of core courses along with a variety of elective courses, carefully designed in consultation with industry experts

About Conference

STAAAR 2023, the 2nd International Conference on Sustainable Technologies and Advances in Automation, Aerospace, and Robotics, in association with the Material Advantage student chapter, offers a global stage for educators, researchers, Ph.D. candidates, research scholars, and industry experts to present, discuss, and share insights on both theoretical and practical research in various aspects of Mechanical, Aerospace, and Robotics engineering. Moreover, STAAAR 2023 strives to establish a leading cross-disciplinary platform to explore emerging innovations, prevailing trends, and the complexities faced in contemporary contexts.

2. Local Organizing Committee

Dr. S. Balaguru, Conference Chair

Dr. Rohit Sharma, Organising Secretary

Dr. Shiv Manjaree Gopaliya, Organising Secretary

Dr. Prashant G K, Convener

Dr. Ashish Vashishtha, Convener

Dr. Vijay Kumar Patidar, Co- Convener

Dr. Pushpdant Jain, Co- Convener

Dr. K Sripriyan, Co- Convener

Dr. Abha Gupta, Co- Convener

Dr. Ankush Tharkar, Co- Convener

Dr. Karthik Rao M.C, Co- Convener

Dr Abhishek Shrivastava, Co- Convener

Dr. Rabindra Prasad, Co- Convener

Mr. Umakanta Meher, Co- Convener

3. Scientific Advisory Committee

Dr. Poonkuntran S, Dean SCSE, VIT Bhopal

Dr. Debashis Adhikari, Dean SEEE, VIT Bhopal

Dr. Hemant Kumar Nashin, Dean SASL, VIT Bhopal

Dr. Sheetal Sharma, Dean SOA, VIT Bhopal

Dr. Prasad Begde, Dean, VIT-BS

Dr. Siddhartha Maiti, SBET, VIT Bhopal

Dr. Dean Callaghan, Director engCORE, SETU, Ireland

Dr. Vincent O'Brien, SETU, Ireland

Dr. Lilibeth A. Zambrano M, SETU, Ireland

VIT® BHOPAL
www.vitbhopal.ac.in

MATERIAL™ ADVANTAGE
VIT BHOPAL UNIVERSITY STUDENT CHAPTER

STAAAR- 2023

Organized by

School of Mechanical Engineering (SMEC)
VIT Bhopal University

Co-sponsored by SERB-DST

SERB INDIA

विज्ञान एवं प्रौद्योगिकी विभाग
DEPARTMENT OF
SCIENCE & TECHNOLOGY

In Joint Collaboration with

SETU IRELAND

2nd International Conference on Sustainable Technologies and Advances in Automation, Aerospace and Robotics (STAAAR)

Hybrid Mode

December 15 – 16, 2023

Important Dates

Revised date for submission of full length paper:	25th Nov. 2023
Comments notification to Authors:	29th Nov. 2023
Final paper approval:	**30th Nov. 2023**
Registration opens:	**1st Dec. 2023**
Last Date for Registration:	**10th Dec. 2023**

Paper submission and presentation is permitted by registered authors only.

About VIT Bhopal University

The VIT legacy, started from Vellore Engineering College 38 years ago under the visionary leadership of Dr. G. Viswanathan, Founder-Chancellor, to offer world-class education for the development of society. VIT Bhopal University is ably spearheaded by Dr. G. Viswanathan, Chancellor, Mr. Sankar Viswanathan, Vice-President, Ms. Kadhambari S. Viswanathan, Assistant Vice-President. The University is currently led by Dr. A. Senthil Kumar, Vice-Chancellor. They share the mission to make VIT a global center of education. Starting from the 2017 academic year, the new Ultra-modern university campus, sprawling over 250 acres, is in the town of Kothrikalan District Sehore, surrounded by green, lush fields and a serene environment. This new and contemporary campus is aiming to become the primary hub of education, with leading-edge and high-tech developments in a wide variety of areas. At present, VIT Bhopal University offers core engineering programs such as Mechanical Engineering, Aerospace Engineering, Computer Science and Engineering, Electronics and Communication Engineering leading to B. Tech degree, M. Tech and allied subject programs such as MSc, MCA, BBA, etc., are also offered. At VIT Bhopal University, the extra-curricular activities and employment needs of the students will be taken care of by various clubs and dedicated career development cells.

About Conference

STAAAR 2023, the 2nd International Conference on Sustainable Technologies and Advances in Automation, Aerospace, and Robotics, in association with the Material Advantage student chapter, offers a global stage for educators, researchers, Ph.D. candidates, research scholars, and industry experts to present, discuss, and share insights on both theoretical and practical research in various aspects of Mechanical, Aerospace, and Robotics engineering. Moreover, STAAAR 2023 strives to establish a leading cross-disciplinary platform to explore emerging innovations, prevailing trends, and the complexities faced in contemporary contexts.

About School of Mechanical Engineering (SMEC)

The School of Mechanical Engineering (SMEC) at VIT Bhopal offers undergraduate degrees and research programmes in Mechanical Engineering, Mechanical Engineering specialization in Artificial intelligence and Robotics, and Aerospace Engineering. SMEC has been an integral part of the university since its inception. Our faculty members possess diverse expertise, spanning areas such as Machine Design, Thermal & Fluid Sciences, and Manufacturing Engineering within the Mechanical Engineering domain. Our Aerospace Engineering faculty specializes in structures and propulsion, while those in AI and Robotics come from backgrounds in Robotics and Automation. The school offers a comprehensive array of core courses along with a variety of elective courses, carefully designed in consultation with industry experts.

CHIEF PATRON

Hon'ble Dr. G. Viswanathan
Founder & Chancellor

PATRONS

Mr. Sankar Viswanathan
Vice President

Ms. Kadhambari S. Viswanathan
Assistant Vice President

Co-PATRONS

Dr. A. Senthil Kumar, Vice-Chancellor

Dr. Pradyumna Yadav, Registrar

CHAIRMAN

Dr. S. Balaguru, Dean, SMEC, VIT Bhopal

ORGANIZING SECRETARIES

Dr. Rohit Sharma, Assistant Professor
Dr. Shiv Manjaree Gopaliya, Assistant Professor

CONVENERS

Dr. Prashant G K, Assistant Professor
Dr. Ashish Vashishtha, SETU, Ireland

Co- CONVENERS

Dr. Vijay Kumar Patidar, Sr. Assistant Professor
Dr. Pushpdant Jain, Sr. Assistant Professor
Dr. K Sripriyan, Assistant Professor
Dr. Abha Gupta, Assistant Professor
Dr. Ankush Tharkar, Assistant Professor
Dr. Karthik Rao M.C, Assistant Professor
Dr Abhishek Shrivastava, Assistant Professor
Dr. Rabindra Prasad, Assistant Professor
Mr. Umakanta Meher, VIT Bhopal

ADVISORY COMMITTEE

Dr. Poonkuntran S, Dean SCSE, VIT Bhopal
Dr. Debashis Adhikari, Dean SEEE, VIT Bhopal
Dr. Hemant Kumar Nashin, Dean SASL, VIT Bhopal
Dr. Sheetal Sharma Dean SOA, VIT Bhopal
Dr. Prasad Begde, Dean, VIT-BS
Dr. Siddhartha Maiti, SBET, VIT Bhopal
Dr. Dean Callaghan, Director engCORE, SETU, Ireland
Dr. Vincent O'Brien, SETU, Ireland
Dr. Lilibeth A. Zambrano M, SETU, Ireland

CONTACT DETAILS

Dr. Rohit Sharma, Assistant Professor, SMEC
Mob: +91–9131960256

Dr. Shiv Manjaree Gopaliya, Assistant Professor, SMEC
Mob: +91–9891354291

Mail: staaar@vitbhopal.ac.in
Website: https://vitbhopal.ac.in/event/staaar-2023/

CALL FOR PAPER

Topics include, but are not limited to:

- ✓ Aerodynamics
- ✓ Applied Mechanics and Finite element analysis
- ✓ Biomaterials and Bio-mechanics
- ✓ Computer-aided engineering design
- ✓ Computational Fluid Dynamics
- ✓ Dynamics and Vibration
- ✓ Experimental Fluid Dynamics
- ✓ Green Manufacturing Engineering Applications of AI
- ✓ Green Manufacturing
- ✓ Internal Combustion Engines
- ✓ Instrumentation and Control
- ✓ Material characterization
- ✓ Manufacturing Techniques
- ✓ Machine Learning in Fluid Dynamics and Heat Transfer
- ✓ Rapid Prototyping
- ✓ Robotics and Automation
- ✓ Renewable Energy
- ✓ Polymers and Composites

REGISTRATION FEE DETAILS

Participants Category	Registration Fee + (Incl. of GST 18%)
Students/ Research Scholars	**Rs. 750**
Academicians/ Faculty	**Rs. 1500**
R&D Organizations/ Industry	**Rs. 2500**
Participants from abroad	**USD 60 / Rs. 4980**

PUBLICATION

After peer review, accepted and presented papers in STAAAR 2023 will be published in **"MATEC Web of Conferences"** indexed in WoS /Scopus with an additional fee of around Rs. 3200. Selected papers from the conference will be published in Technologies (MDPI journal), Impact Factor- 3.6.

technologies

FULL MANUSCRIPT SUBMISSION LINK

https://forms.gle/oMxFm5WYJSWZEo5z6

Scan me to submit the manuscript

REGISTRATION LINK

https://forms.gle/bWSTDz8xfLuBdBtD6

Scan me for Registration

Proceedings of the 1st STAAAR Conference

Scan me for STAAAR 2022 Proceedings

Acknowledgements

The conference organizers extend their gratitude for the financial support provided by SERB-DST, India. Special thanks are extended to our valued collaborators, South East Technological University (SETU), Ireland, and the Material Advantage student chapter of VIT Bhopal University, for their invaluable assistance at crucial junctures.

Additionally, we wish to acknowledge with deep gratitude the cooperation and assistance provided by VIT Bhopal University, where the conference is being organized. Their support has been instrumental in ensuring the success of this significant event.

We are grateful to

our collaborators

Sponsors

2nd International Conference on "Sustainable Technologies and Advances in Automation, Aerospace, and Robotics" (STAAAR 2023)

15-16 December, 2023

Preface

The 2nd International Conference on "Sustainable Technologies and Advances in Automation, Aerospace and Robotics (STAAAR 2023)" was held at VIT Bhopal University on 15-16 December 2023 in hybrid (online and offline) mode. This event strives to establish a leading cross-disciplinary platform to explore emerging innovations, prevailing trends and the complexities faced in the fields of Mechanical, Aerospace and Robotics.

The 2nd edition of STAAAR conference, in collaboration with South East Technological University (SETU), Ireland, and the Material Advantage student chapter of VIT Bhopal University, is a testament to the remarkable progress in Engineering and Technology witnessed in the last decade. This conference provides a crucial platform for researchers and scholars to showcase their pioneering work in various key areas such as Aerospace, Design, Materials and Manufacturing, Robotics, and Thermal Engineering.

Our prestigious Conference showcased a diverse lineup of speakers, reflecting the outstanding quality of the event. Over two days, we had the privilege of hosting 68 paper presentations across various sessions, each chaired by distinguished academic researchers from renowned institutions. It is encouraging to see significant participation from young researchers and scholars across the globe. Through a rigorous screening process, we diligently ensured that only high-quality papers were selected for publication in this proceedings.

The Conference Proceedings consist of 36 scientific papers divided into 4 topics as follows:

1. Materials Science and Manufacturing Processes
2. Design, Development, and Optimization
3. Advancements in Aerothermal Analysis
4. Robotics

We expect this publication, "Proceedings of STAAAR 2023," to be helpful to researchers from both academics and industries.

Best regards,

Dr. Rohit Sharma, Dr. Shiv Manjaree Gopaliya, and Dr. Ashish Vashishtha

Editors

About the editors

Dr. Rohit Sharma is currently serving as an Assistant Professor and Programme Chair (Mechanical Engineering) at VIT Bhopal University, India. He earned his Ph.D. from Politecnico di Milano (POLIMI), Italy in Energy and Nuclear Science & Technology in 2017. His research focused on the near-field region of turbulent swirling jets in the combustion & propulsion laboratory. Notably, Dr. Sharma received an esteemed Erasmus Mundus fellowship sponsored by the European Commission in 2013 to pursue his Ph.D. Prior to his doctorate, Dr. Sharma received his M.Tech in Thermal Engineering from NIT Warangal in 2013 and a B.E. in Mechanical Engineering from LNCT Bhopal in 2010. Over the years, he has amassed extensive experience in the field of research and academia. For three years, he worked as a Senior Project Research Scientist at IIT Bombay. He also served as an Assistant Professor at NIT Jalandhar for some time, and gained invaluable industrial experience as a Graduate Engineer Trainee at Vardhman Yarns for one year.

Dr. Sharma has an impressive record of 23 publications in esteemed journals and conferences and has been honored with best research paper and poster awards. As a core team member, he actively participated in the student solar ambassador workshop organized under the SoULS project at IIT Bombay, which achieved the Guinness World Record on 2nd October 2018 for lighting the most LED lights based on solar energy. He has contributed to other sponsored projects such as the Energy and Economic Growth (EEG) project in collaboration with the University of California, Berkeley, USA, and Lawrence Berkeley National Laboratory, Berkeley, USA. Also, he worked on the TATA Research project in collaboration with Boston College, USA, under the SoULS, IIT Bombay umbrella. He has recently been awarded an International Travel Grant (2022) in the Young Scientist category and a Seminar/Symposia Grant (2023) by SERB, Department of Science and Technology, Government of India, serving as the Principal Investigator. He has traveled to numerous countries, including France, Greece, Italy, Poland, Portugal, Saudi Arabia, Switzerland, and the United Kingdom, to present his research at various international conferences. His diverse research interests encompass Experimental Fluid Dynamics, Optical Diagnostics, Flow Visualization, Internal Combustion Engines, Proper Orthogonal Decomposition, Predictive Modelling, and Sustainable Energy for Rural Development.

Dr. Shiv Manjaree Gopaliya is presently serving as Assistant Professor and Program Chair (Mechanical Engineering with specialization in Artificial Intelligence and Robotics) at VIT Bhopal University. Dr. Gopaliya has served in various academic and administrative positions at various academic Institutes for more than 15 years. Apart from her teaching commitments, she has successfully completed Government and industry-sponsored research projects both. She has one patent published to her credit. She has also served as Audit Coordinator for various Accreditation bodies namely, NBA, NAAC, IQAC to name a few. Dr. Gopaliya holds a Ph.D. in Robotics Engineering and M Tech in Robotics and Automation from Maharishi Dayanand University, Rohtak, Haryana. She is a Mechanical Engineering graduate from Government Engineering College, Jagdalpur, Chattisgarh. She is an active researcher in the area of the application of Artificial Intelligence in Robotics with a good number of publications in journals of repute. Her research areas are Robotics Engineering, Application of Artificial Intelligence, Mechatronics, Instrumentation and Control Engineering. She is a life member of The Robotics Society and life members of SAE and Aeronautical Society of India. Presently. She is also handling the position of Convener, Women Empowerment Cell at VIT Bhopal University.

Dr. Ashish Vashishtha is currently working as PI and Lecturer at the Department of Mechanical and Aerospace Engineering, Faculty of Engineering, South East Technological University Carlow Campus, Ireland. He received his PhD from the Department of Advanced Energy, The University of Tokyo, Japan in the domain of Hypersonic Aerothermodynamics, and completed his postdoctoral research in hypersonic (at Indian Institute of Science Bangalore, India) and gas turbine combustion (University of Galway, Ireland). Prior to his PhD, he had 4.75 years of experience in an automotive industry (Renault-Nissan in India) in the domain of IC Engine combustion modeling and completed his master's research in experimental supersonic flows at IIT Kanpur. After joining SETU Carlow, he expanded his research interest in various applied domains of renewable energy (hydrogen, waste-to-synthetic fuel technologies, offshore wind turbines, IC-Engines/gas-turbine engine combustion, fuel flexibility), high-speed propulsion (hypersonic air breathing engines, space systems, green propellants for rockets as well as emerging coating, repair and additive manufacturing technology of Cold Spray and received funding of €1.5 million in past 4 year. He is actively involved in various professional societies, such as: Senior Member of AIAA, Member of Royal Aeronautical Society, Combustion Institute (Irish Section), EASN (European Aeronautics Association), Interstellar Research Group (IRG), International Shockwave Institute, Cold Spray Club, also member of COST Action 20127, Waste refinery Technologies for sustainable energy processes.

Statement of Peer review

In submitting conference proceedings to *Web of Conferences*, the editors of the proceedings certify to the Publisher that

1. They adhere to its **Policy on Publishing Integrity** in order to safeguard good scientific practice in publishing.
2. All articles have been subjected to peer review administered by the proceedings editors.
3. Reviews have been conducted by expert referees, who have been requested to provide unbiased and constructive comments aimed, whenever possible, at improving the work.
4. Proceedings editors have taken all reasonable steps to ensure the quality of the materials they publish and their decision to accept or reject a paper for publication has been based only on the merits of the work and the relevance to the journal.

Title, date and place of the conference

2nd International Conference on Sustainable Technologies and Advances in Automation, Aerospace and Robotics (STAAAR-2023)

December 15 – 16, 2023

VIT Bhopal Univesity, Sehore, India

Proceedings editor(s):

Dr. Rohit Sharma, VIT Bhopal University, India
Dr. Shiv Manjaree Gopaliya, VIT Bhopal University, India
Dr. Ashish Vashishtha, South East Technological University, Ireland

Date and editor's signature

Rohit Sharma   Shiv Manjaree Gopaliya   Ashish Vashishtha

Date: **29-02-2024**

MATEC Web of Conferences **393**, 01001 (2024) https://doi.org/10.1051/matecconf/202439301001
STAAAR-2023

Phase Transition of Thermally Treated Polyhedral Nano Nickel Oxide with Reduced Band Gap

Peeyush Phogat[1] *, *Shreya*[1], *Ranjana Jha*[1] and *Sukhvir Singh*[1]

[1] Research Systems for Energy Systems, Department of Physics, Netaji Subhas University of Technology, Dwarka, New Delhi, INDIA

Abstract. Nickel oxide (NiO) is a semiconducting material which exhibits a unique electronic structure. Because of its distinctive electronic properties, NiO stands as an intriguing candidate for various applications in optoelectronics, photo catalysis, and energy devices such as solar cells. In the present work efforts have been made to tailor the band gap of NiO. A simple co-precipitation method followed by heat treatment is utilized to synthesize the material. Prior to heat treatment, x-ray diffraction study of the as synthesized material showed the presence of nickel hydroxide [$Ni(OH)_2$]. On calcination for one hour at 1000°C, a single phase NiO was revealed. After heat treatment it was noticed that the particle size was found to be increased. The absorption spectra of [$Ni(OH)_2$] and NiO were recorded using UV-Vis spectroscopy method. Using Tauc plot a band gap of 4.2 eV and 1.8 eV for $Ni(OH)_2$ and NiO was observed respectively. It is observed that a significant decrease in band gap of NiO was noticed. Surface morphological study was carried out by using FESEM, which revealed the transformation of sheet like structures of [$Ni(OH)_2$] to polyhedral shaped NiO on calcination. The presence of nickel and oxygen was confirmed by the energy dispersive spectrometry analysis.

1 Introduction

Substantial progress has been achieved in the expansion of renewable and sustainable energy sources during the last decade. The advancement of renewable energy sources such as solar cells and electrochemical technologies such as capacitors/supercapacitors and batteries (lithium ion) is critical for fostering a cleaner world [1–3]. To advance these technologies, a primary requirement is the study of material properties. Extensive research has been conducted on anode materials for solar cell applications, whether involving II–VI semiconductors or transition metals in general [4, 5]. Nevertheless, there remains a need to explore and investigate novel cathode materials for solar cell applications.

Nickel oxide (NiO) is a well-known cathode material with a band gap of 3.5 to 4 eV [6]. The unique characteristics of nickel oxide nanoparticles make this material highly promising for a variety of applications like energy storage, optoelectronics, etc. [7–9].

* Corresponding author: peeyush.phogat@gmail.com

NiO has gained a prominence in the field of solar cells primarily due to its tunable band gap, which enables the material to absorb more sunlight and convert it efficiently into electric energy. NiO-based solar cells hold particular appeal due to their potential cost-effectiveness and sustainability. While NiO is inherently a wide band gap material, the reduction in its band gap can present intriguing and innovative avenues in material science and its applications. Conventionally, the wide band gap of NiO has limited its utility in optoelectronic and energy-related applications [10]. However, through the techniques presented in this manuscript, namely nanostructuring, and thermal treatment, the authors have successfully engineered NiO with a reduced band gap. This reduction in the band gap holds immense promise, facilitating improved light absorption, enhanced charge transport, and expanded opportunities in photovoltaics. Increased absorption of light results in improved photovoltaic performance, enhancing sensitivity in photodetectors, wider range of emission spectra, will provide better performance in infrared applications and facilitates tuning electronic properties and carrier transport properties.
An effort has been undertaken in the present research work to synthesize NiO with a reduced band gap. Co-precipitation technique was used to easily complete the synthesis in one step. The phase of the nickel oxide and nickel hydroxide materials was identified using the X-ray powder diffraction (XRD) method. The crystallite size and microstrain were determined using the Debyes-Sherrer equation and the Williamson Hall (W-H) plot, respectively. UV-Vis Spectroscopy was utilized to investigate the optical properties and band gap of the materials. The morphological traits associated with them were studied using a field emission scanning electron microscope (FESEM). Elemental compositional analysis was performed utilizing energy dispersive spectrometer (EDS) as an attachment of FESEM.

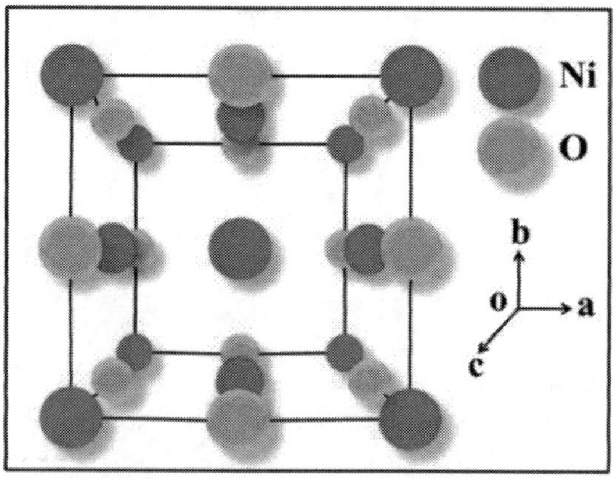

Fig. 1. 3D model of cubic NiO.

2 Experimental Procedure

2.1 Chemicals Used

15Mega-Ohm resistivity distilled water and pure alcohol were used for washing and synthesis. DI water was used to make all of the solutions. Table-1 lists the other compounds that were utilized.

Table 1. Chemicals Used.

Chemicals Name	Company

Nickel (II) nitrate hexahydrate [$Ni(NO_3)_2 .6H_2O$]	M/s Thomas Baker
Sodium Hydroxide (NaOH)	M/s Thomas Baker

2.2 Synthesis

Nickel nitrate hexahydrate was diffused in DI water to prepare a 0.96 M solution. Concurrently, a 1.25 M solution of sodium hydroxide was prepared. Magnetic stirrers were used to mix both solutions for one hour. The addition of the NaOH solution to the $Ni(NO_3)_2 .6H_2O$ solution for $Ni(OH)_2$ formation was carried out using a burette, and a controlled dropwise addition was crucial for optimal nanoparticles development. The subsequent blend was further diffused for four hours at room temperature. Subsequently, the liquid was then subjected to filtration. Filtration and washing steps were repeated three times, involving rinsing with distilled water and then with 100% ethanol. Afterward, the sample was dried in a vacuum oven at pressure of 600 mm of Hg and a temperature of 120 °C. Once dried, it was allowed to cool overnight before being finely powdered. The fine powder was finally calcinated in oxygen atmosphere at 1000°C for 1 hour.

2.3 Characterization Details

To understand the structure and phase of the material, the fine powder of the as synthesized sample was studied with an X-Ray Diffractometer (make:Panalytical and model X'pert PRO). Copper (Cu) K(alpha)-radiation (=1.5406A^o) was used to record the x-ray diffraction pattern. For the morphological and elemental compositional analysis, a FESEM (model:7610FPlus, make:JEOL) with an EDS attachment was employed. The optical response of the as produced material was measured using a SHIMADZU UV-2600i UV Vis spectrometer.

3 Results and Discussion

3.1 X-Ray Diffraction

The x-ray diffraction pattern for the as synthesized powder samples was recorded over a 2θ range of 10° to 80°. The diffraction peaks were analyzed and indexed to the standard JCPDS file: ***00-001-1047,*** which exhibited a good match with nickel hydroxide [$Ni(OH)_2$]. The as synthesized $Ni(OH)_2$ displayed a hexagonal crystal system, with identified peaks corresponding to the hkl planes (100), (101), (102), (110) and (111). $Ni(OH)_2$ showed low crystallinity with high noise, specifically in the region of 10° to 30° as shown in figure-2(a). However, following the thermal treatment, XRD measurement was again performed for calcinated sample in the same range of 2θ as mentioned above. The observed pattern demonstrate high crystallinity with well resolved and distinct peaks, which were subsequently analyzed and indexed to JCPDS file: ***01-078-0423***. The calcinated sample showed a strong correlation with nickel oxide (NiO) and exhibited a cubic crystal system when compared with standard JCPDS file ***01-078-0423.*** The reflection peaks corresponded to the hkl planes (111), (200), (220), (311) and (222) belongs to the cubic structure having polycrystalline nature as shown in figure-2(b).

The JCPDS file indicated the crystal structure of NiO to be in a space group Fm3m. The "Fm3m" crystal system refers to the face-centered cubic (fcc) crystal system. In an fcc crystal system, each lattice point is shared by eight adjacent unit cells, which results in a highly symmetric and dense packing of atoms or ions. This crystal system is commonly encountered in various materials, including metals like aluminium, copper, and gold, as well as some ionic compounds. Figure-1 depicts the cubic crystal system of NiO.

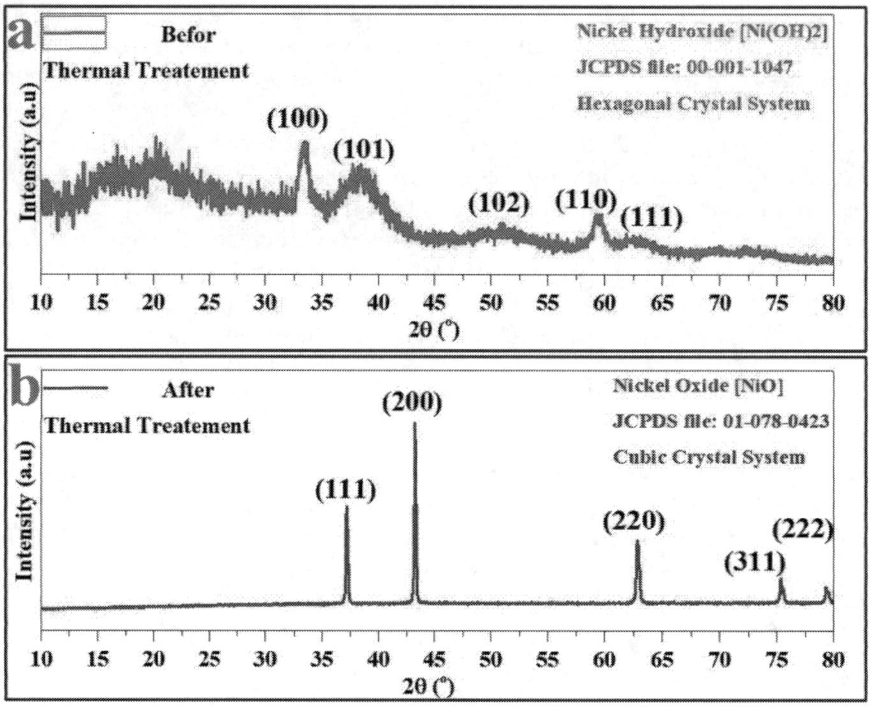

Fig. 2.(a) XRD pattern for as synthesized $Ni(OH)_2$, **(b)** XRD pattern for calcined $Ni(OH)_2$ revealing the formation of single phase cubic NiO with polycrystalline nature.

3.2 Lattice Parameters, Crystallite Size and Strain

Nickel nitrate Crystallite size refers to the size of individual crystal grain within a material. This unique property of any nanomaterial is determined through a well-known equation known as Debye Scherrer formula given in equation-1 as [11],

$$D = \frac{K\lambda}{\beta \mathrm{Cos}\theta} \tag{1}$$

Here, D associates to the crystal grain size (crystallite size), K is a constant of 0.9 known as Scherrer constant, λ is the wavelength of x-ray source given by 0.154nm, β is full with at half of maxima (FWHM) and θ is half the position XRD pattern. The calculation of crystallite size was performed by taking an average of all the analyzed peaks. The resulting crystallite size for $Ni(OH)_2$ and NiO was approximately 12 nm and 78 nm respectively. The crystallite size for $Ni(OH)_2$ lies in quantum dot region, which signifies quantum confinement and existence of quantum electronical states [12]. Whereas after the thermal treatment of $Ni(OH)_2$, the crystallite size increases due to several factors such as recrystallization, annealing, energy redistribution and grain growth.

However, Debye-Scherrer analysis is based on the assumption that there is no strain present in the material. To verify the presence of strain, lattice parameters were calculated. $Ni(OH)_2$

and NiO showed an hexagonal and cubic crystal system respectively. Hence to calculate the deviation in lattice parameters the formula presented in equation-2(a&b) has been used [13, 14],

$$\frac{1}{d^2} = \frac{4}{3}\left(\frac{h^2+k^2+hk}{a^2}\right) + \frac{l^2}{c^2} \quad \text{(for Hexagonal)} \tag{2a}$$

$$\frac{1}{d^2} = \frac{h^2}{a^2} + \frac{k^2}{b^2} + \frac{l^2}{c^2} \quad \text{(for Cubic)} \tag{2b}$$

The interplaner spacing is d, the miller indices are h,k,l, and the lattice parameters are a,b,c. Table-2 shows the comparison of lattice parameters between JCPDS file and calculated values.

Table 2. Lattice Parameters.

Parameters	**As synthesized Ni(OH)2**	**$Ni(OH)_2$ in JCPDS file: *00-001-1047***	**As synthesized NiO**	**NiO in JCPDS file: *01-078-0423***
a (Å)	3.08	3.07	4.18	4.17
b (Å)	3.08	3.07	4.18	4.17
c (Å)	4.71	4.60	4.18	4.17

The table above illustrate that the calculated value deviates from the standard once due to the presence of micro strain present in the materials. To calculate the strain a well-known method known as Williamson Hall plot was applied. Williamson Hall plot (W-H plot) states that broadening of XRD peaks not only from crystallite size but also from strain. The W-H equation is given as equation-3 [1, 15, 16],

$$\beta_T Cos\theta = (\varepsilon \times 4Sin\theta) + \frac{K\lambda}{D} \tag{3}$$

Here in equation-3, ε is the micro strain present in the as synthesized material. Plotting the graph between 4Sinθ versus βCosθ and comparing the value of slope depicted in figure-3(a&b) reveals the micro strain to be -0.0087 and -1.02×10^{-4} for $Ni(OH)_2$ and NiO respectively. The negative strain present in the material reveals compression present in the material. The compression in $Ni(OH)_2$ causes the confinement. The comparison of intercept revealed the crystallite size of approximately 6 nm and 66 nm for $Ni(OH)_2$ and NiO respectively. In the Debye-Scherrer equation, the assumption was made that there was no strain present in the material. However, when calculating the Williamson-Hall Plot, microstrain was taken into account, leading to differences in the inferred particle size. The variability in the Williamson-Hall plot was attributed to the measured Full Width Half Maxima (FWHMs) resulting from instrumental broadening. Also, due to the presence of strain, the crystallite size of the material could not be confined to a single value; instead, it always existed within a range. Table-3 presents all the values calculated from the Debye-Scherrer equation and the W-H plot.

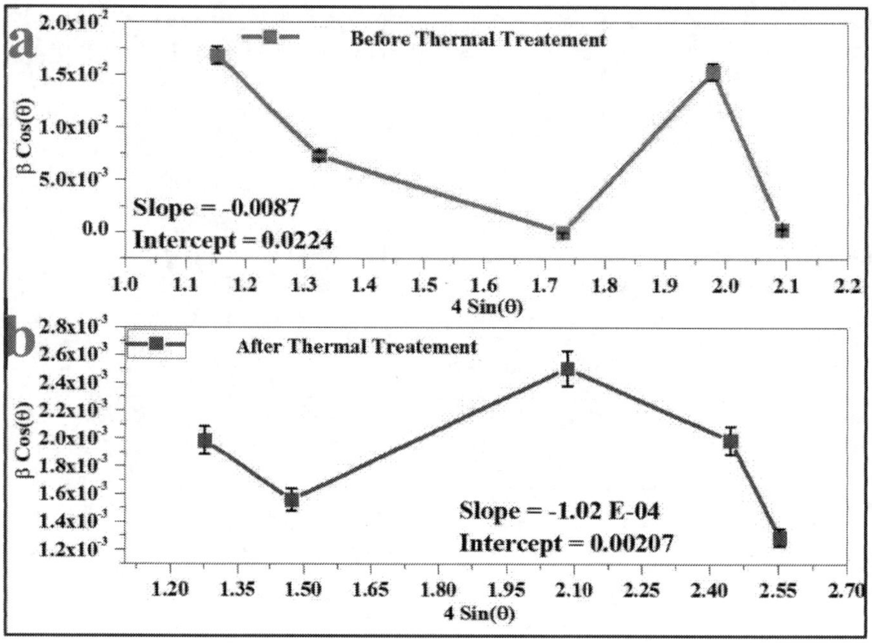

Fig. 3.(a) Williamson Hall plot for $Ni(OH)_2$, **(b)** Williamson Hall plot for NiO.

Table 3. Crystallite Size and strain.

Parameters	Debye Scherrer Size	Williamson Hall Size	Williamson Hall Strain	W-H Plot Slope	W-H Plot Intercept
$Ni(OH)_2$	12 nm	6 nm	-0.0087	-0.0087	0.0224
NiO	78 nm	66 nm	-1.02×10^{-4}	-1.02×10^{-4}	0..00207

3.3 UV-Vis Spectroscopy

The optical response of as synthesized powder was recorded in a range of 190 nm to 850 nm using diffusion reflectance spectroscopy (DRS). The absorbance curve reveals that $Ni(OH)_2$ exhibited absorbance solely in the UV region, specifically between 193 nm to 235 nm with small absorbance at 390 nm and from 665 nm to 671 nm (as shown in figure-4(a)). The maximum absorbance was observed at a wavelength of 231 nm. However, after the thermal treatment, the absorbance peaks exhibit a red shift towards visible region. The as synthesized NiO showed absorbance from 191 nm to 850 nm, covering the entire absorption spectrum, indicating that the material is well suited for solar cell applications. The maximum absorbance for NiO was observed at 722 nm, 724 nm and 725 nm as shown in figure-4(b). The material exhibits an intriguing absorbance phenomenon, demonstrating absorbance in both the UV and the entire visible spectrum. To obtain the band gap of both samples, Tauc's plot was plotted as shown in figure-4(c&d). Tauc plot was obtained using equation-4, given as [17],

$$\alpha h\upsilon = A[h\upsilon - E_g]^{\frac{1}{n}} \quad (4)$$

Where α is absorbance coefficient, hυ is the input energy of photon, A is the constant for energy, Eg is the bandgap, and n is the exponent reflecting the type of the transition. The band gap for $Ni(OH)_2$ and NiO was computed using the aforementioned equation and establish to be 4.2 eV and 1.8 eV, respectively. NiO shows a band gap of 3.5 eV to 4 eV, however due to quantum confinement in $Ni(OH)_2$ the transition from larger band gap to smaller band gap has caused an extreme reduction in NiO band gap.

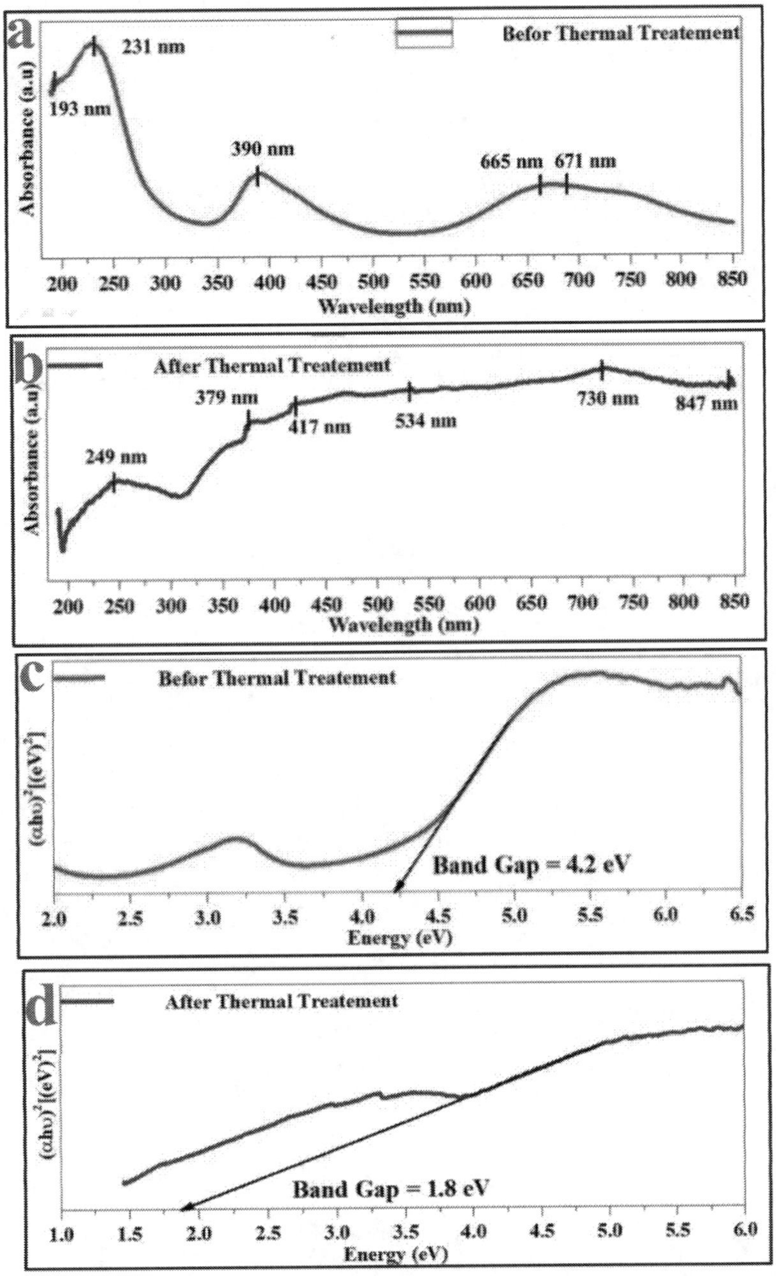

Fig. 4.(a) Absorption spectra for $Ni(OH)_2$, **(b)** absorption spectra for NiO, **(c)** Tauc's plot for $Ni(OH)_2$, **(d)** Tauc's plot for NiO revealing band gap.

3.4 Refractive Index

Refractive index (RI) plays an important role in solar cell application of the sample. The RI can affect how efficiently light interacts with the layer of solar cell. When the refractive indices of the material and the medium are matched, more light can penetrate the medium and reach the absorbing layer, increasing the productivity of solar cell. For calculating the refractive index of the samples the equation (5) has been taken in consideration [18-20],

$$\frac{n^2-1}{n^2+2} = 1 - \sqrt{\frac{E_g}{20}} \quad (5)$$

Here, n represents the refractive index, whereas E_g is the respective band gap of $Ni(OH)_2$ and NiO (i.e., 4.2 eV and 1.8 eV). Using the values of band gap the calculated refractive index came out to be 2.18 and 2.82 for $Ni(OH)_2$ and NiO respectively. In general, materials with refractive indices greater than 1.0 indicate that light travels more slowly in that medium compared to a vacuum. The higher the refractive index, the more a material can bend light and more suitable for solar applications. Table 4. Summarizes the max. absorbance, band gap and refractive index of $Ni(OH)_2$ and NiO.

Table 4. Band Gaps and Refractive Index.

Parameters	Max Absorbance	Band Gap	Refractive Index
Ni(OH)2	231 nm	4.2 eV	2.18
NiO	722-725 nm	1.8 eV	2.82

3.5 FESEM

Figure 5(a) shows FESEM image of nickel hydroxide recorded at suitable magnifications. Nanosheets stacked together are clearly seen in the SEM image along with some scattered particles. Thermal treatment of $Ni(OH)_2$ underwent morphological changes throughout the calcination process, which led to an increase in particle size as depicted in figure 5(b) having polyhedral shape attached together. Average size of these polyhedral particles are found to be of the order of 500 nm. As indicated in fig. 5(c&d), elemental compositional analysis of the relevant sections revealed the existence of just oxygen and nickel

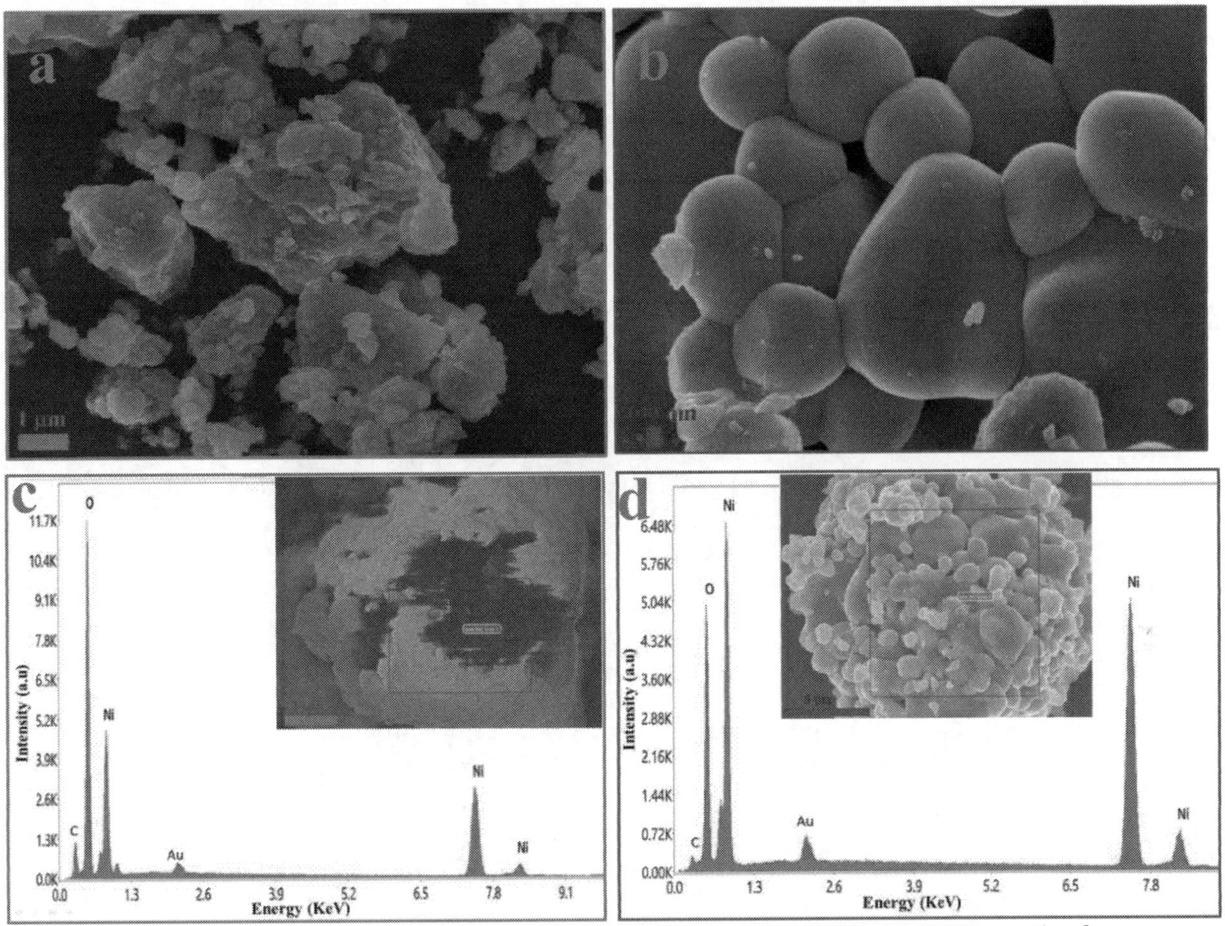

Fig. 5.(a) FESEM images for $Ni(OH)_2$, **(b)** FESEM images for NiO, **(c)** EDX spectra for $Ni(OH)_2$ with inserted selected area, **(d)** EDX spectra for NiO with inserted selected area.

4 Conclusion

The present work is devoted to the synthesis of reduced band gap NiO particles for suitability in solar cell applications. The NiO was synthesized using facile co-precipitation synthesis route. Structural study confirms the presence of both $Ni(OH)_2$ before thermal treatment and NiO phase after thermal treatment. Detailed morphological analysis depicts the stacked sheet like morphology for $Ni(OH)_2$ whereas after thermal treatment the morphology shows a round and polyhedral like morphology with significant increase in the particle size. The novelty of the present work is the noteworthy reduction in band gap. $Ni(OH)_2$ showed a large band gap of 4.2 eV which lies in insulating region while NiO showed an extreme reduction in energy band gap with 1.8 eV which is highly suitable for solar cell applications.

The contributors warmly credit Prof. Anand Srivastava, Hon. Vice Chancellor of Netaji Subhas University of Technology, for providing infrastructural resources.
Peeyush Phogat: Conceptualization, Writing - original draft and Formal analysis. Shreya: Formal analysis and software. Ranjana Jha: Supervision. Sukhvir Singh: Supervision, Writing – review & editing.
All data is presented in the mansucript
There are no conflicts of interest among the writers.

References

1 Dipti, P. Phogat, Shreya, D. Kumari, S. Singh, Fabrication of tunable band gap carbon based zinc nanocomposites for enhanced capacitive behaviour, *Phys. Scr.*, **98**, 9, 95030, (2023). doi: 10.1088/1402-4896/acf07b.

2 S. Sharma, P. Phogat, R. Jha, S. Singh, Electrochemical and Optical Properties of Microwave Assisted MoS_2 Nanospheres for Solar Cell Application, **12**, 3, 66–72, (2023). doi: 10.12720/sgce.12.3.66-72.

3 P. PHOGAT, Shreya, R. JHA, S. Singh, Diffusion Controlled Features of Microwave Assisted ZnS/ZnO Nanocomposite with Reduced Band Gap, *ECS J. Solid State Sci. Technol.*, **12**, 3, 034004, (2023). doi: 10.1149/2162-8777/acc426.

4 T. Kumar, Shreya, P. Phogat, V. Sahgal, R. JHA, Surfactant-Mediated Modulation of Morphology and Charge Transfer Dynamics in Tungsten Oxide Nanoparticles, *Phys. Scr.*, **98**, 8, 085936, (2023). doi: 10.1088/1402-4896/ace566.

5 P. Phogat, Shreya, R. Jha, S. Singh, Optical and Microstructural Study of Wide Band Gap ZnO@ZnS Core--Shell Nanorods to be Used as Solar Cell Applications, in *Recent Advances in Mechanical Engineering*, (2023), 419–429.

6 R. Goel, R. Jha, C. Ravikant, Nickel oxide (NiO) nano-triangles with enhanced electrochromic and photovoltaics properties, *Chem. Pap.*, **77**, 5, 2885–2903,(2023). doi: 10.1007/s11696-023-02673-y.

7 R. Goel, R. Jha, C. Ravikant, Solution-processed spin coated multilayer structured nickel oxide thin films for anodic electrochromism, *J. Appl. Electrochem.*, **53**, 4, 713–735, (2023). doi: 10.1007/s10800-022-01807-6.

8 R. Goel, R. Jha, C. Ravikant, Synergistic effect of Urea and Potassium Sulphate during hydrothermal synthesis of NiO nanospheres with reduced crystallite size and enhanced electrical conductivity, *Inorg. Chem. Commun.*, **141**, 109563,(2022). doi: https://doi.org/10.1016/j.inoche.2022.109563.

9 R. Goel, R. Jha, M. Bhushan, R. Bhardwaj, C. Ravikant, Hydrothermally synthesized nickel oxide (NiO) nano petals, *Mater. Today Proc.*, **48**, 687–689, (2022). doi: https://doi.org/10.1016/j.matpr.2021.08.094.

10 R. Goel, R. Jha, C. Ravikant, Investigating the structural, electrochemical, and optical properties of p-type spherical nickel oxide (NiO) nanoparticles, *J. Phys. Chem. Solids*, **144**, 109488, (2020), doi: https://doi.org/10.1016/j.jpcs.2020.109488.

11 Shreya, P. Phogat, R. Jha, S. Singh, Elevated Refractive Index of MoS2 Amorphous Nanoparticles with a Reduced Band Gap Applicable for Optoelectronics BT - Recent Advances in Mechanical Engineering, in *Recent Advances in Mechanical Engineering*, B. Sethuraman, P. Jain, and M. Gupta, Eds. Singapore: Springer Nature Singapore, (2023), 431–439.

12 P. Phogat, Shreya, R. Jha, S. Singh, Impedance Study of Zinc Sulphide Quantum Dots via One Step Green Synthesis, *Mater. Sci. Forum*, **1099**, 119–125, (2023). doi: 10.4028/p-G1CCxq.

13 Shreya, P. Phogat, R. Jha, S. Singh, Microwave-synthesized γ-WO3 nanorods exhibiting high current density and diffusion characteristics, *Transit. Met. Chem.*, **48**, 3, 167–183, (2023), doi: 10.1007/s11243-023-00533-y.

14 P. Phogat, Shreya, R. Jha, S. Singh, Electrochemical analysis of thermally treated two dimensional zinc sulphide hexagonal nano-sheets with reduced band gap, *Phys. Scr.*, **98**, 12, 125962, (2023), doi: 10.1088/1402-4896/ad0d93.

15 Shreya, A. Yadav, R. Khatri, N. Jain, A. Bhandari, N. K. Puri, "Double Zone

Thermal CVD and Plasma Enhanced CVD Systems for Deposition of Films/Coatings with Eminent Conformal Coverage BT - Advances in Manufacturing Technology and Management," in *Advances in Manufacturing Technology and Management*, R. M. Singari, P. K. Jain, and H. Kumar, Eds. Singapore: Springer Nature Singapore, (2023), 273–283.

16 A. Yadav, Shreya, and N. K. Puri, "Preliminary Observations of Synthesized WS2 and Various Synthesis Techniques for Preparation of Nanomaterials," in *Advances in Manufacturing Technology and Management*, Springer Science and Business Media Deutschland GmbH, (2023), 546–556.

17 R. Bhardwaj, R. Jha, M. Bhushan, R. Sharma, Comparative study of the electrochemical properties of mesoporous 1-D and 3-D nano- structured rhombohedral nickel sulfide in alkaline electrolytes, *J. Phys. Chem. Solids*, **144**, 109503, (2020). doi: 10.1016/j.jpcs.2020.109503.

18 M. Bhushan, R. Jha, Surface activity correlations of mesoporous 3-D hierarchical ZnS nanostructures for enhanced photo and electro catalytic performance, *Appl. Surf. Sci.*, **528**, 146988, (2020), doi: 10.1016/j.apsusc.2020.146988.

19 J. Dahiya, P. Phogat, A. Hooda, and S. Khasa, Investigations of Praseodymium doped LiF-ZnO-Bi2O3-B2O3 glass matrix for photonic applications, *AIP Conf. Proc.*, **2995**, 1, 20065, (2024). doi: 10.1063/5.0178197.

20 D. Kumari, Shreya, P. Phogat, Dipti, S. Singh, R. Jha, Enhanced electrochemical behavior of C@CdS Core-Shell heterostructures, *Mater. Sci. Eng. B*, **301**, 117212, (2024), doi: https://doi.org/10.1016/j.mseb.2024.117212.

Experimental Investigation of Al-6061 Metal Matrix Composite Reinforced with Selenium

Sri Harsha Kandala Tirumala[1*], *Sai Sujith* Kavili[1], *Venugopal Naidu* Manubolu[1] and *Diwakar Reddy* V[1]

[1]Department of Mechanical Engineering, SVU College of Engineering, Tirupati, India, 517502.

Abstract. The creation of innovative materials with essential properties is required due to the industry's rapid advancements. The most desirable properties of metals used in the aerospace industry, automotive industry, and marine industry are hardness, compression, shear, etc. Aluminium metal matrix composites (AMMCs) are effective and meet industrial standards. This study focuses on the influence of reinforcement particles (Selenium) on Al-6061 by using the stir-casting fabrication method and the composites are prepared. Mechanical characteristics are studied by varying Selenium weight percentages such as 0.5%, 1%, and 1.5%. Based on the Mechanical tests it is noticed that there is an improvement in Mechanical strength such as tensile strength, flexural strength, and compression strength by 32%, 29.5%, and 25% at 1 Wt% of Selenium. By implying the Powder XRD test on the prepared sample which is a significant parameter drawn from mechanical tests, based on the Standard XRD data it is noticed that a Hexagonal Crystal Structure is formed.

Keywords: Al-6061, Selenium, Stir casting, Mechanical Tests, XRD test.

1 Introduction

Metal matrix composites (MMCs) emerged as a strong material class with a distinctive set of qualities that appeal to a variety of applications. These composites consist of a metal matrix reinforced with one or more secondary materials, such as ceramics, fibres, or particles. One of the primary motivations for investigating MMCs[1] is their improved mechanical properties. The addition of reinforcements within the metal matrix enhances its strength, stiffness, and wear resistance. The selection of reinforcement material depends on the desired properties such as (Hardness, Strength, etc), with popular options including ceramic particles, Fibers (such as carbon, silicon carbide, or alumina), and whiskers.

Aluminium alloy 6061, also known as Al-6061, is a widely used and highly versatile material in various industries. It is an Aluminium alloy that comes from the 6xxx series, [2] which is

* Corresponding author: ktsriharsha1476@gmail.com

recognized for its exceptional strength-to-weight ratio and good formability. Al-6061 offers a unique combination of mechanical properties, suitable for applications [3] in aerospace, automotive, construction, etc. The composition of Al-6061 primarily consists of aluminium, magnesium, and silicon with negligible amounts of other elements. This alloy exhibits exceptional strength and stiffness, enabling it to withstand high loads and stresses while maintaining structural integrity. Its strength-to-weight ratio makes it particularly advantageous [4] for applications where weight reduction is crucial without compromising performance. The presence of magnesium and silicon enhances [5] its ability to withstand corrosive environments, including atmospheric conditions and certain chemicals. This corrosion resistance [6] contributes to the longevity and durability of components made from Al-6061, ensuring their reliability in a wide range of operating conditions.

The current work is an inquiry into the influence of Selenium particles in Al- 6061. The Specific reason behind choosing Selenium as a reinforcement is some similar minerals like Magnesium, etc are to be found which may affect strength in an incremental way. The stir-casting process is used to fabricate three samples by varying weight percentages of Selenium (0.5, 1 & 1.5). These samples are tested to study mechanical characteristics such as tensile strength, flexural test, compression strength, and hardness. Later the Powder XRD test is performed on the Al-Se composite to identify its crystal structure.

2 Materials & Fabrication Method

2.1 Materials

The Al-6061 Alloy bars are acquired from the Sachin Steel Center in Mumbai, India. Sigma Aldrich in Bangalore, Karnataka, India provided the selenium particles. Table 1. [7] lists the chemical composition of the matrix Aluminium alloy 6061.

Table 1. Composition of matrix material

Constituent	Al	Mg	Si	Fe	Cu	Zn	Mn	Cr	Other
Wt. %	96.5	0.956	0.562	0.532	0.236	0.202	0.102	0.046	0.864

2.2 Fabrication Method

In this work, Al-6061 material is considered as a matrix, and the micro-size Selenium particles are considered as reinforcement. The samples are fabricated using a stir-casting process. Initially, the Al-6061 is heated up to 850-900°C (Fig. 1.) for complete melting. After that, the Selenium particles are added and stirred to get a uniform mixture. A metallic die made up of cast iron gets filled with the molten melt and it is kept in an open environment to cool down for easy removal from the mold on its own. The entire process is repeated to prepare new samples by varying weight percentages. The samples are:

S1: 99.5 wt.% of Al6061 + 0.5 wt.% of Selenium
S2: 99 wt.% of Al6061 + 1 wt.% of Selenium
S3: 98.5 wt.% of Al6061 + 1.5 wt.% of Selenium

 https://doi.org/10.1051/matecconf/202439301002

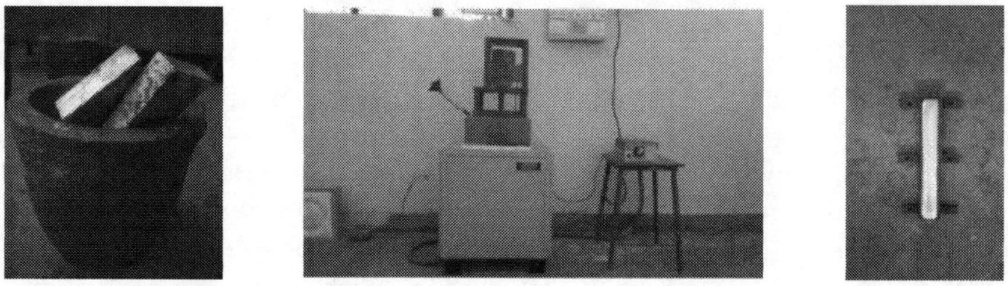

(a) (b) (c)

Fig. 1. (a)Al-6061 material in Graphite crucible (b)Experimental Setup (c)Al6061- Se composite Mold.

3 Results and Discussion

Mechanical tests have been carried out and detailed below to study the mechanical properties of fabricated Al-Se composite material.

3.1 Hardness test

The specimens' hardness is determined by performing Brinell hardness test. A Brinell hardness test instrument that has an indenter radius of five millimeters is employed in this investigation to determine the hardness of composite materials. The samples are fabricated with ASTM E10 guidelines. Each specimen receives an average of three readings (Table 2). The Al 6061-Se composites outperformed the aluminium alloy according to the findings of the hardness test (Fig. 2). The reinforcement (Se) presence in aluminium matrix increases the hybrid composites' resistance to plastic deformation.

Table 2. Composites' hardness values in HBW

Specimen number	Composition	Reading 1	Reading 2	Reading 3	Average
S1	99.5 wt.% of Al6061 + 0.5 wt.% of Selenium	56	59	61	58.67
S2	99 wt.% of Al6061 + 1 wt.% of Selenium	67	65	69	67
S3	98.5 wt.% of Al6061 + 1.5 wt.% of Selenium	66	63	70	66.33

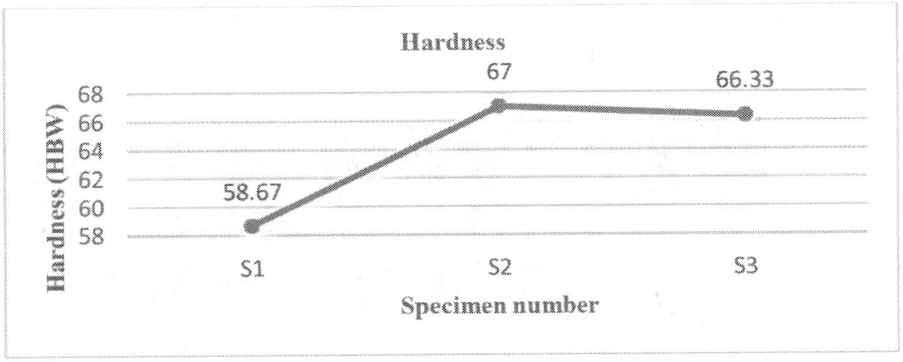

Fig. 2. Hardness Vs Specimen number

3.2 Flexural test

The rupture modulus test, generally known as the flexural strength test, it is a fundamental mechanical test used [8] to determine a material's capacity to endure bending or flexural loads. The material flexural strength is determined by subjecting a test specimen to a bending moment or a transverse load. This test gauges how composite materials respond to simple bending. The samples are fabricated in conformity with ASTM A370 guidelines. The samples were subjected to a flexural test using UTM. The samples' flexural strength is then compared with the base material's flexural strength, and it is observed that Sample 2 (S2) is found to have the highest breaking load and flexural strength of an increase of 29.5% than the other samples (Fig. 3). Also, in the Fig. 3 let $(\sigma_s)/\sigma_o$ % be equal to F.

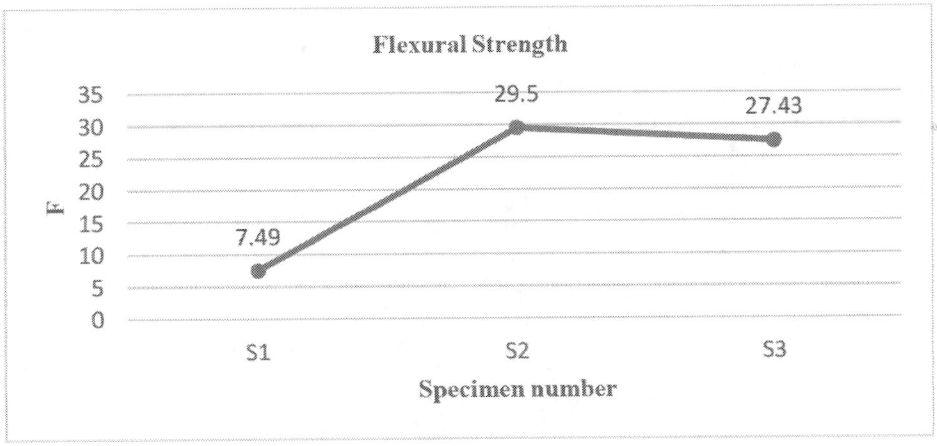

Fig. 3. Flexural Strength Vs Specimen number

3.3 Tensile Strength

Materials often get selected for engineering applications [9] based on characteristics including modulus of elasticity, yield strength, and ultimate tensile strength. In this investigation, a test for tensile strength is done on a UTM (Universal Testing Machine), and the composite specimens S1, S2, and S3 are [10] manufactured in accordance with ASTM specifications . The material tensile behavior is varied by the varying levels of compositions. The Three specimens have different Young's modulus percentage elongation, percentage reduction, yield strength, and ultimate strength which are all measured by the universal testing machine (Fig.4 Samples before and after the tensile test). Fig. 5 shows the load-deformation behavior of individual specimens and compositions, where it can also be observed that the tensile strength is increased by up to 32%. In fig 5 where $(\sigma_s)/\sigma_o$ % is represented as T.

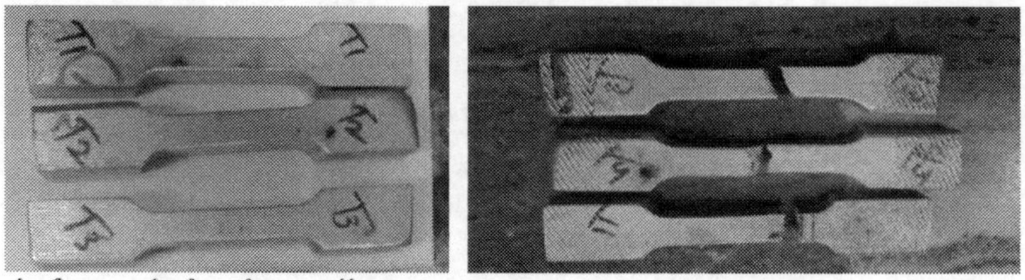

Fig. 4. Samples before and after the tensile test

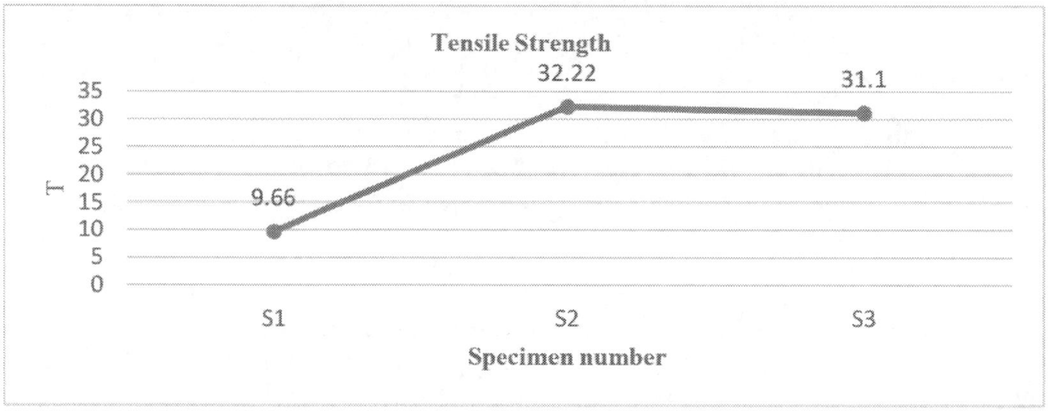

Fig. 5. Tensile Strength Vs Specimen number

3.4 Compression Test

The compression test is a fundamental mechanical test used to evaluate the behavior and strength of materials under compressive forces. Compression testing involves [11] applying a compressive force or load on a test specimen in a controlled manner until it undergoes a significant deformation. To study the compressive strength of the material, a compressive test was performed on the prepared samples. The Compressive Strength of the materials varied by varying the levels of compositions as shown in Fig. 6. Among the three specimens, we can also observe that the compression strength was increased from 10% to 25% by an increase in the wt% of Selenium. In Fig 6, C denotes $(\sigma_s)/\sigma_o$ %

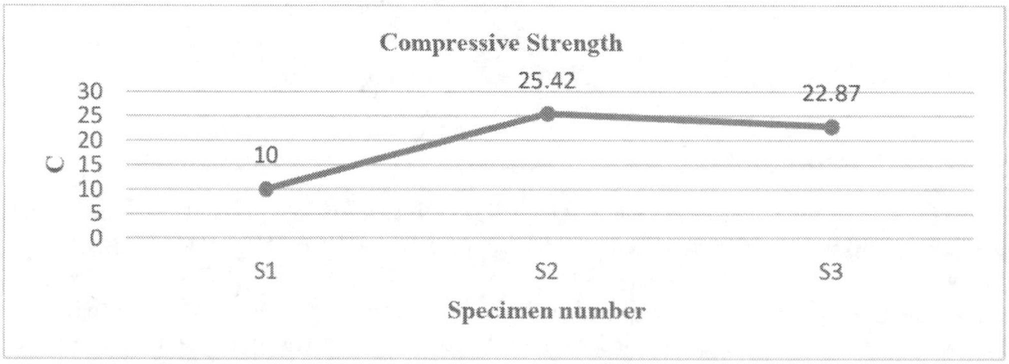

Fig. 6. Compressive Strength Vs Specimen number

3.5 Powder XRD Test

The XRD technique is one of the key phase analysis methods used [12] in the MMC to assess the reaction between the alloy and ceramic components The Powder XRD Test is performed to know more about the characteristics of crystalline materials. It is a non-destructive testing procedure that offers information on the material's structure. As the XRD test is performed on the Al-Se composite sample.

Now, the Al-Se composite sample is placed in a wire EDM machine to cut the sample into the required size of a 5mm, shape cube. Later, the specimen obtained after the wire EDM process is filed thoroughly on the surfaces of the cube. After filing the sample, it is placed in the Powder XRD machine, and a test is done on it. The output responses of the XRD test are obtained as shown in Table 3. Also, by using the raw data the graph is drawn and the peak points are found (Fig. 7).

Table 3. Experimental values of XRD experiment

S.No.	2θ°	Phase name	Chemical Formula	DB Card No	d-spacing, Å	Int I, cps°
1	39.047	Selenium: 1 0 1	Se	01-077-3596	2.30494	11983
2	39.139	Aluminium Titanium: 1 1 1	Al+x Til-x	00-42-1137	2.29970	30890
3	45.438	Selenium: 1 1 0, Aluminium Zinc: 0 1 2	Se, Al0.71, Zn0.29	01-077-3596, 00-019-0057	1.99448	31349
4	65.765	Aluminium Titanium: 2 0 2	Al+x Til-x	00-42-1137	1.418807	23272
5	78.817	Aluminium Zinc: 0 2 1	Al0.71, Zn0.29	00-019-0057	1.213350	10224
6	79.915	Selenium: 2 1 1	Se	01-077-3596	1.1995	309

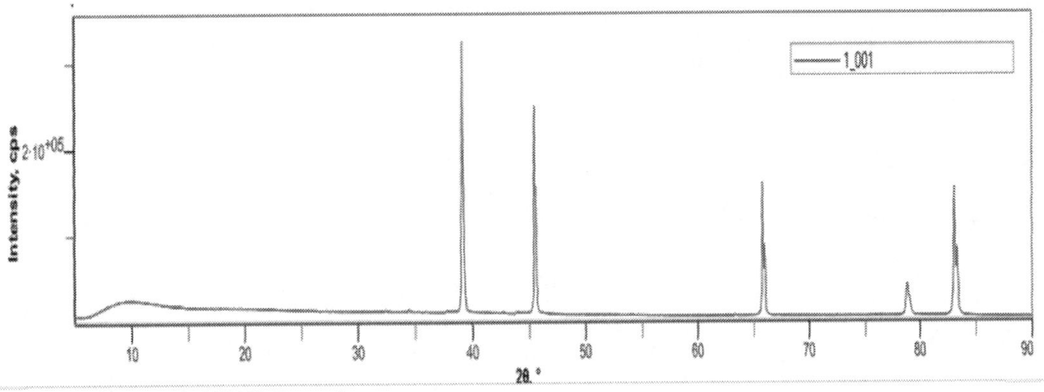

Fig. 7. XRD graph 2θ° Vs Intensity, cps

As, from the above experimental data it can be observed that Selenium is identified at three 2θ° and those are 39.047°, 45.4383° and 79.915°. Also, it can be seen that traces of Aluminium Titanium and Aluminium Zinc are also identified in the XRD test.

The Crystalline size(D) depends on the Full-width Half Maximum (FWHM) (β) and 2 Theta (2θ) and the values of it are given below in Table 4. The values of dislocation density (δ) in nm-2 and microstrain (ε) are given in Table 5 and by using the Braggs equation the d-Spacing values are predicted. The experimental & predicted values of the d-Spacing are given in Table 6. Also, From Table 6, it can be observed that the values of d-Spacing values of both experimental and predicted ones match each other.

Table 4. Crystalline size (D)

2θ (degrees)	θ (degrees)	θ (radians)	FWHM (degrees)	FWHM (β) (radians)	D (nm)
39.047	19.5235	0.3407	0.100	0.001745	82.9908
45.4383	22.71915	0.3965	0.1029	0.001796	80.6346
79.915	39.9575	0.6974	0.87	0.01518	9.5406
Crystalline Size					173.166

Table 5. Dislocation density (δ) and Microstrain (ε)

D (nm)	δ(nm^{-2})	ε	ε *10^3
82.9908	1.45191*10^{-4}	0.073363	73.363
80.6346	1.53800*10^{-4}	0.064881	64.881
9.5406	0.01098622	0.311767	311.767

Table 6. Experimental and Predicted values of d-Spacing

d-Spacing Experimental	d-Spacing Predicted
2.30494	2.30495
1.99448	1.99448
1.1995	1.19943

The Lattice parameters of Selenium from the experimental data are a = 3.98884 Å, b = 3.98884 Å, c = 3.09448 Å, $\alpha = 90°$, $\beta = 90°$ and $\gamma = 120°$. The Crystal System of Selenium is a Hexagonal Crystal where $a = b \neq c$, and $\alpha = \beta = 90°$, $\gamma = 120°$. So, from the XRD Test, it is concluded the presence of Selenium particles in the Al-Se composite with hexagonal crystal structure. It self- defined that the Selenium particles are uniformly distributed in specified areas during fabrication which effects the considerable strength improvement when compared with Al-6061 material.

4 Conclusion

By observing all the output results from various tests, the following conclusions are drawn.

- Al6061 - Se composite is successfully fabricated using stir casting process, by varying wt.% of Se reinforcement such as 0.5, 1, and 1.5 wt.% of Se particles.
- The mechanical parameters of the selenium-reinforced Al 6061 metal matrix composite, such as tensile strength, flexural load, compression load, and Brinell hardness, have significantly improved.
- The Brinell hardness of the fabricated composites shows an increase in hardness.
- The Mechanical Properties of the Al6061-Se composite material such as Tensile Strength, Flexural & Compression strength values are increased by 32%, 29.5%, 25% respectively at 1 wt% of Selenium.
- At 1.5 wt% of Selenium, it is observed that the mechanical properties of the composite material such as the Tensile, Flexural, and compressive strength were decreased. From this, we can observe that in between 1-1.5 wt% of Selenium, the maximum strength can be achieved.

- The Selenium is also identified at three 2θ° (39.047°, 45.4383° and 79.915°) by using the Powder XRD test. It also confirms that the Selenium is in hexagonal crystal structure.
- Further research could explore the composite's behavior weight percentage between 1-1.5 of Selenium. As it can understand the trend in change of mechanical properties by an increase in wt.% of Selenium to the MMC.
- Overall, the results of this study suggest that the Al 6061 metal matrix composite reinforced with selenium is a promising material as it could improve the properties of the base material (Al-6061).

References

1. Mahendra Boopathi M., K. P. Arulshri , N. Iyandurai " Evaluation of Mechanical Properties of Aluminium alloy 2024 reinforced with Silicon Carbide and Fly ash Hybrid Metal Matrix Composites". American Journal of Applied Sciences, 10(3):219-229, (2013), ISSN: 1546-9239,
   doi: https://doi.org/10.3844/ajassp.2013.219.229
2. Bhargavi Rebba, N. Ramanaiah " Evaluation of Mechanical Properties of Aluminium alloy (Al-2024) reinforced with Molybdenum Disulphide (MOS_2) Metal Matrix Composites". 3rd International Conference on Materials Processing and Characterisation (ICMPC 2014), Procedia Materials Science 6 (2014) 1161-1169,
   doi: https://doi.org/10.1016/j.mspro.2014.07.189
3. B. Sai Jagadish " Synthesis and Characterisation of Aluminum 2024 and Graphene Metal Matrix Composites by Powder Metallurgy means". SSRG International Journal of Mechanical Engineering (SSRG-IJME), Volume 2 Issue 7-July (2015),
   doi : 10.14445/23488360/IJME-V2I7P103
4. Prathipa R, Sivakumar C, Shanmugasundaram B "Mechanical Characteristics Of Aluminum 5083 Metal Matrix Strengthen With Fly Ash". Turkish Journal of Computer and Mathematics Education Vol.12 No.10 (2021), 2153-2164,
   doi: https://doi.org/10.17762/turcomat.v12i10.4736
5. Shinoop Khan S, M A Joseph, K Sekar "Study of Impact Strength, Hardness, Microstructure and Tribological Properties of Al5052 Composite with SiC and Graphite as Reinforcements". International Journal of Innovative Technology and Exploring Engineering (IJITEE) ISSN: 2278-3075, Volume-9 Issue-9S, July (2020),
   doi: 10.35940/ijitee.I1002.0799S20
6. Ch Hima Gireesh, K. G. Durga Prasad, Koona Ramji "Experimental investigation on mechanical properties of an Al6061 Hybrid Metal Matrix Composite". Journal of Composite science, J.Compos.Sci. (2018),2,49;
   doi: https://doi.org/10.3390/jcs2030049
7. Dharmesh M. Patoliya, Sunil Sharma "Preparation and Characterization of Zirconium Dioxide Reinforced Aluminium Metal Matrix Composites". International Journal of Innovative Technology and Exploring Engineering (IJITEE) ISSN (Online): 2319 - 8753 ISSN (Print): 2347 - 6710 Vol. 4, Issue 5, May (2015),
   doi: 10.15680/IJIRSET.2015.0405051

8. Sam Daniel Fenny. A Seenivasakumar.M "Characterization of Mechanical Properties of Aluminum based Metal Matrix Composite AA6061/ZrO2/SiC using Stir casting method". International Journal of Advances in Engineering and Management (IJAEM) Volume 2, Issue 11, pp: 258-262, (2020), doi: 10.35629/5252- 0211258262
9. S. Sivananthan, K. Ravi, C. Samson Jerold Samuel " Effect of SiC particles reinforcement on mechanical properties of aluminium 6061 alloy processed using stir casting route". Materials Today : Proceedings (2019)
doi: https://doi.org/10.1016/j.jallcom.2017.06.309
10. Manoj Singla, D. Deepak Dwivedi, Lakhvir Singh, Vikas Chawla "Development of Aluminum Based Silicon Carbide Particulate Metal Matrix Composite". Journal of Minerals & Materials Characterization & Engineering, Vol. 8, No.6, pp 455-467, (2009) , doi: 10.4236/jmmce.2009.86040
11. Adeolu Adesoji Adediran, Abayomi Adewale Akinwande, Oluwatosin Abiodun Balogun, Olanrewaju Seun Adesina, Adeniyi Olayanju, Tajudeen Mojisola" Evaluation of the properties of Al-6061 alloy reinforced with particulate waste glass". Scientific African, Scientific African 12(2021) e00812, doi: https://doi.org/10.1016/j.sciaf.2021.e00812
12. Nagendra Kumar Maurya, Manish Maurya, Ashish Kumar Srivastava, Shashi Prakash Dwivedi, Abdhesh Kumar, Sandeep Chauhan " Investigation of mechanical properties of Al-6061/SiC composite prepared through stir casting technique". Materials Today: Proceedings, (2020), doi: https://doi.org/10.1016/j.matpr.2019.09.003

Tool path planning for hole-flanging process using single point incremental forming

Shashi Ranjan Mohan[1], *Yogesh Dewang*[1,*], *and Vipin Sharma*[2]

[1]Department of Mechanical Engineering, LNCT, Bhopal, India
[2]Department of Mechanical Engineering, Medicaps University, Indore India

Abstract. Hole-flanging is a process used to create flanges around holes in sheet metal parts. It traditionally relies on dies and punches, posing cost and time challenges for small quantities or prototypes. While Incremental forming does not require dedicated sets of tools and dies, and it provides more flexibility and adaptability with enhanced formability, and emerged as a better alternative to conventional press forming. However, precise and efficient tool path planning is essential to achieve high-quality hole-flanged components. Precise toolpath planning is one of the crucial and important part in designing of hole flanging process for both through numerical simulations and experiments. The objective of the present work is to generate tool paths as well as associated G-codes using MATLAB code for different geometries of incremental hole flanging process using SPIF. Six different geometries of hole flanges were considered to generate tool path using MATLAB code. The proposed methodology was successfully applied for different shapes of hole flanges. Since the proposed methodology provides unified programming technique in the current MATLAB environment, which performs dual duties, creating toolpath coordinates against time data and outputting a G-code file and it saves computational time and programming efforts. Hence, proposed methodology is found to be better for incremental hole flanging in contrast to previously existed methodology for development of toolpath for incremental forming cases. Thus, it was concluded that this proposed methodology can be effective for capturing any intricate shapes for incremental hole flanging. You should leave 8 mm of space above the abstract and 10 mm after the abstract. The heading Abstract should be typed in bold 9-point Arial. The body of the abstract should be typed in normal 9-point Times in a single paragraph, immediately following the heading. The text should be set to 1 line spacing. The abstract should be centred across the page, indented 17 mm from the left and right page margins and justified. It should not normally exceed 200 words.

Keywords: Hole flanging, Tool path, MATLAB, Incremental forming,

* Corresponding author: dewang.yogesh3@gmail.com

1 Introduction

Flanging is one of the most versatile sheet metal forming processes [1-2]. It is a contoured bending operation which finds its application in manufacturing of automobiles and aircraft parts, home appliances etc. [3-4]. Flanges appeared in variety of geometries namely straight, stretch, shrink, hole and joggled flange [5-7]. Out of these different variations of flanges, hole flanging is most widely ap-plied form of flanges in industrial applications [8-9].

In the context of hole flanging through incremental forming different investigations on the basis of parametric studies were made by researchers in the recent past. Hole flanges can be made both through conventional forming as well as incremental forming [10]. Higher formability limit and reduced tooling cost, uniform wall thickness, and improved neck height were major benefits of incremental forming in contrast to conventional forming for forming of hole flanges [8,11]. Cui and Gao [11] found multistage diameter deformation strategy as one of the effective strategies for imparting improved formability of hole flanges through single point incremental forming. Makwana et al. [12] noticed increment in form-ability of square hole flanging upon increment in pre-cut square hole radii and on the contrary, there is a reduction in formability due to increment in tool rotation. Mushudi et al. [13] found flange angle as the most influential process parameter which aids in achieving greater flange height. Makwana et al. [14] observed grain refinement upon microscopic examination during square hole flanging of AA 5052 H-32 sheets, as deformation imparted by rotating tool. Fernandez et al. [15] evaluated the formability of stretch flanges in terms of principal strain by the medium of forming limit diagram. In another attempt, Dewang et al. [16] found through review work that localized necking is minimized in hole flanging through incremental forming as in contrast to conventional hole flanging.

In the realm of manufacturing of hole flanges forming that it can be realized through experimental procedures via CNC machines, robotic arms and amino machines [17-21]. Before actual execution of experimental formation of hole flanges, an integrated approach of CAD and CAM is required [17-21]. In the first step, initially a 3D CAD model has been prepared via any CAD modeling software [17-21]. In next step, formulation has been done by already prepared CAD geometry into coding software such as MATLAB in order to get tool paths [17-21]. Afterwards X, Y and Z coordinates of tool path are extracted and further used for generation of G –codes of hole flanges [17-21]. Now these extracted G-codes can be fed to CNC machines. On the other hand, same coordinates can be integrated with time which can be utilized as input for FEM simulation of hole flanging process [17-21].

On the other hand, in the context of tool path definitions in incremental forming, investigations were conducted in the recent past. It was realized that in incremental forming that both toolpath as well as intrusion depth had a considerable effect on geometrical accuracy of formed parts, since higher intrusion and lower depth lead to springforward as well as springback respectively [22]. Besides this, new methodology was devised for toolpath preparation via incremental forming approach by stretching angle through employment of C++, VC++, and OpenGL library [23]. In another attempt, two stage incremental forming investigation through SPIF, optimization of tool path via design of experiments (DoE) statistical approach resulted in reduction of time of processing as well as inaccuracy of formed parts [24]. In another attempt of double-sided incremental forming, tool-path was successfully generated and employed for manufacturing of multi-feature parts by integrating toolpaths in correct order for forming.

It was noticed that majority of researchers worked on generation of tool paths using G-code generation for experimental and numerical simulation of incremental forming process by employing CAD, CAM and CAE and other software tools for arbitrary as a well for complex shapes [17-21]. It was also reported that tool-path planning was attempted for simple geometries of hole flanges through single point incremental forming (SPIF) and exclusive study for toolpath planning and generation for entire set of geometries of hole flanges using MATLAB program as well by integrated approach of CAD and CAM was not reported yet. The objective of the present work is to generate tool paths as well as associated G-codes using MATLAB code for different geometries of hole flanges which can be fur-ther utilized for experimental realization and numerical simulation.

2 Methodology

The process for generating a helical toolpath for single point incremental forming (SPIF) for hole flanging process using MATLAB involves several steps as shown in figure 1. Initially, the dimensions of the model or part geometry is created and then provided as input to MATLAB code. Based on this input, a three-dimensional part is formulated. After that again MATLAB code is utilized to create the toolpath for incremental hole flanging. Following that, the X, Y, and Z coordinates of the tool path are retrieved and utilized to generate G -codes for hole flanges. These derived G-codes may now be used by CNC machines. On the other hand, the same coordinates may be integrated over time and used as input for a FEM simulation of the hole flanging process. Figure 1 shows the flowchart to generate the tool path in terms of coordinate for numerical simulations and G-Code files for CNC machine.

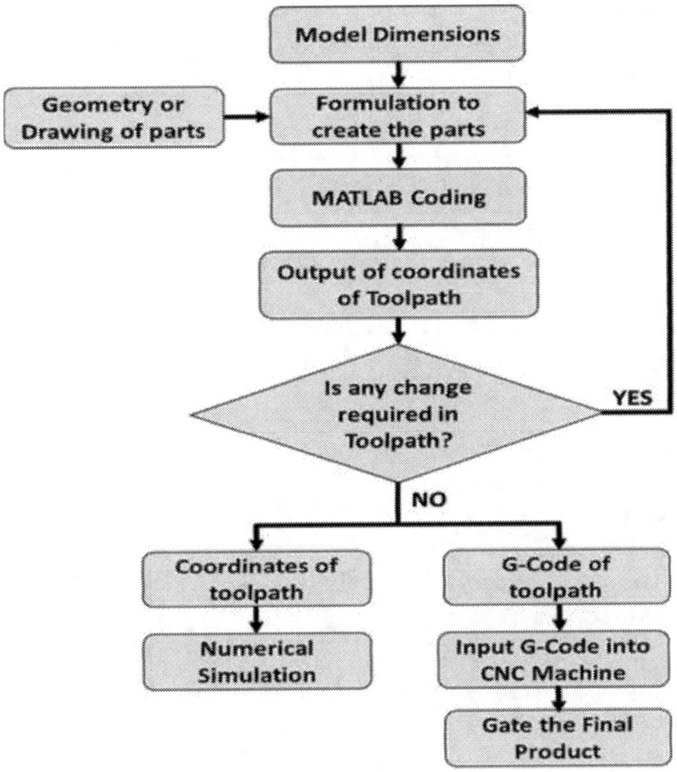

Fig. 1. Method to generate tool-path and G-code for incremental hole flanging.

https://doi.org/10.1051/matecconf/202439301003

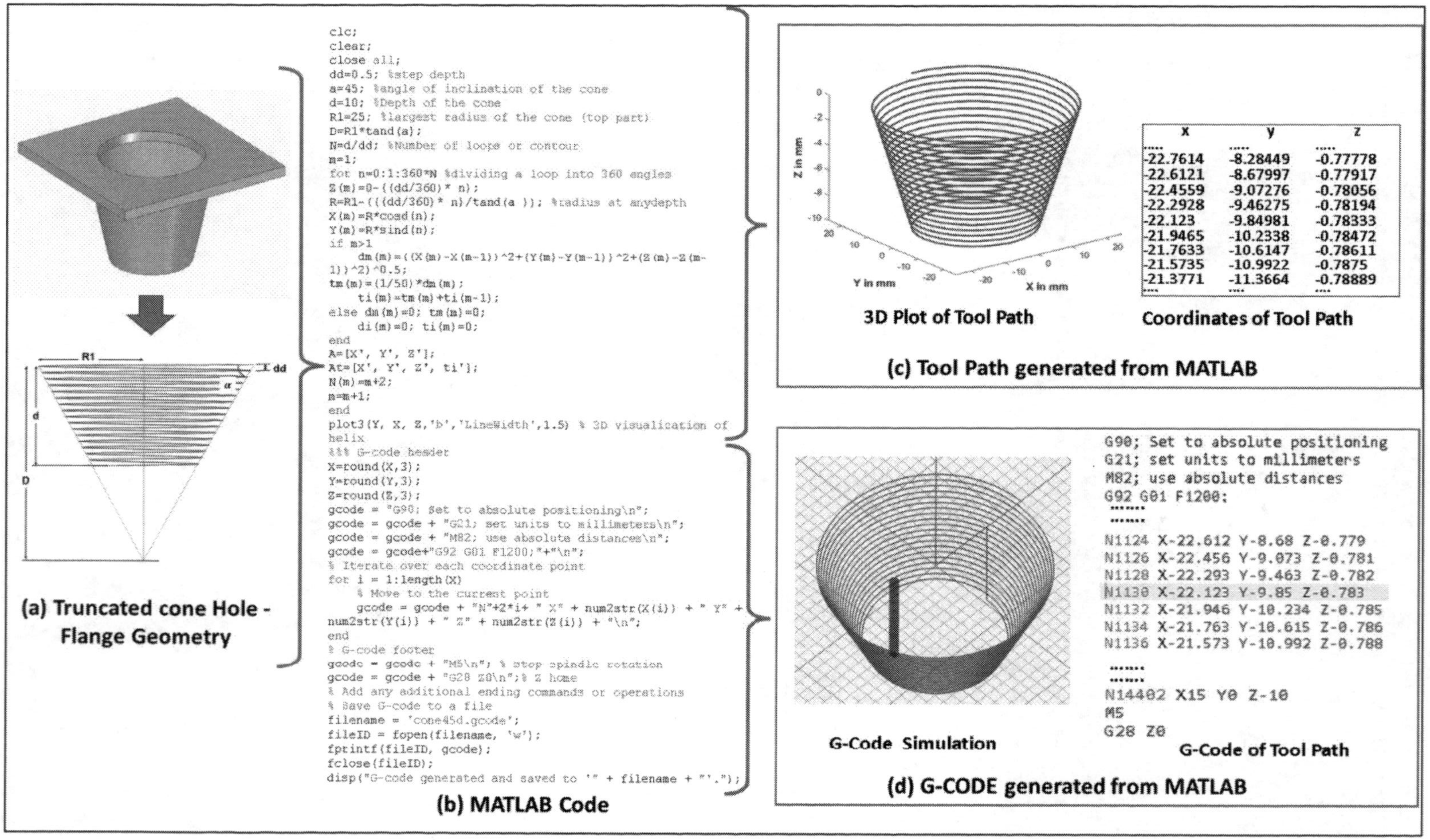

Fig 2: Steps to generate incremental tool path and G-code for conic hole flange

This procedure is illustrated through an example, where the CAD geometry of the tapered hole-flange parts is being formulated using input parameters, depicted in figure 2 (a). Table 1 provides input parameters to formulate toolpath in MATLAB programming. This CAD geometry is encoded in MATLAB to generate a helical toolpath for incremental hole flanging, detailed in the first half segment of figure 2 (b). The MATLAB code yields X, Y, and Z-coordinates for the tool-path, displayed in figure 2 (c) which has been visualized in three-dimensional graph using function "plot3". Subsequently, the second half segment of MATLAB code is given in figure 2 (b) which produces G-Code using the toolpath coordinates, as shown in figure 2 (d). MATLAB programming used to generate G-Code has been divided in to three parts, start, intermediate and end. This entire process is crucial for understanding and implementing the sequential steps involved in the creation and implementation of the helical toolpath for incremental hole flanging.

Table 1. Input parameters of tapered hole-flange.

Geometrical Parameters	Notation	Values
Flange radius	R1	25mm
Depth of cone	d	10mm
Step depth	δd	0.5mm
Inclination angle of cone	α	45^0

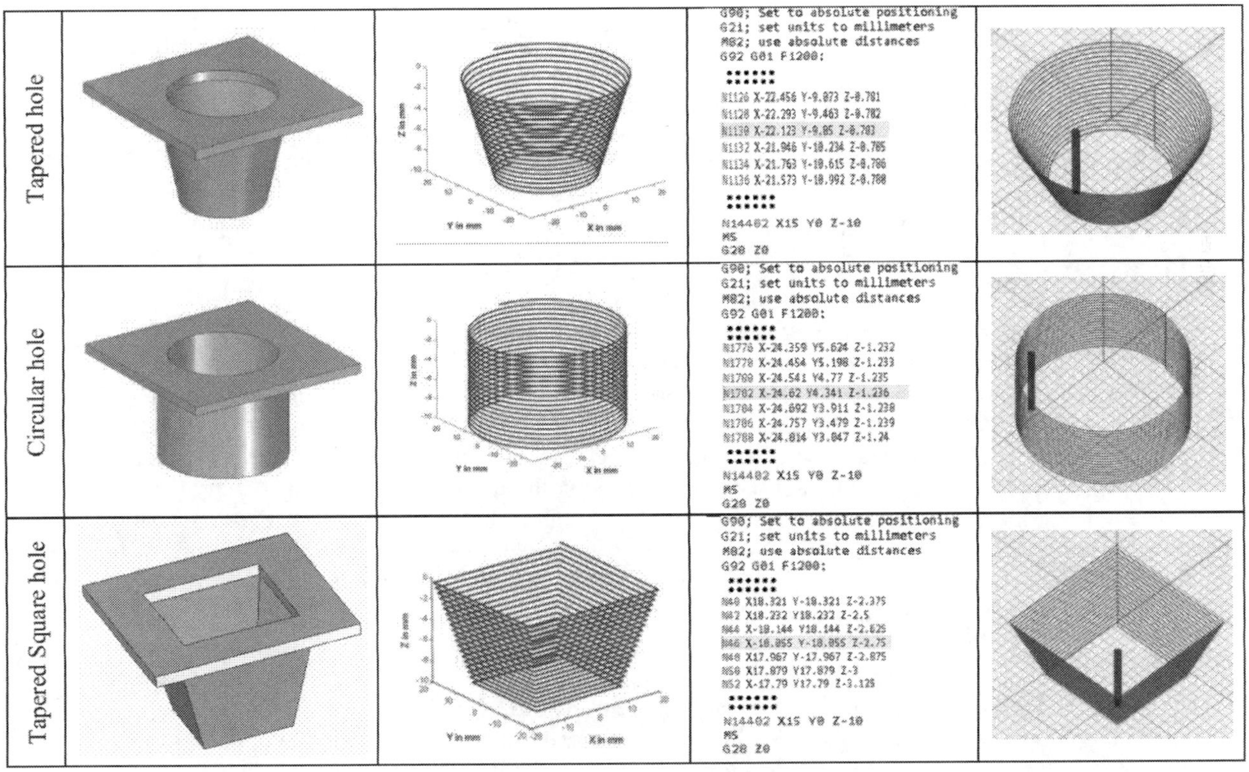

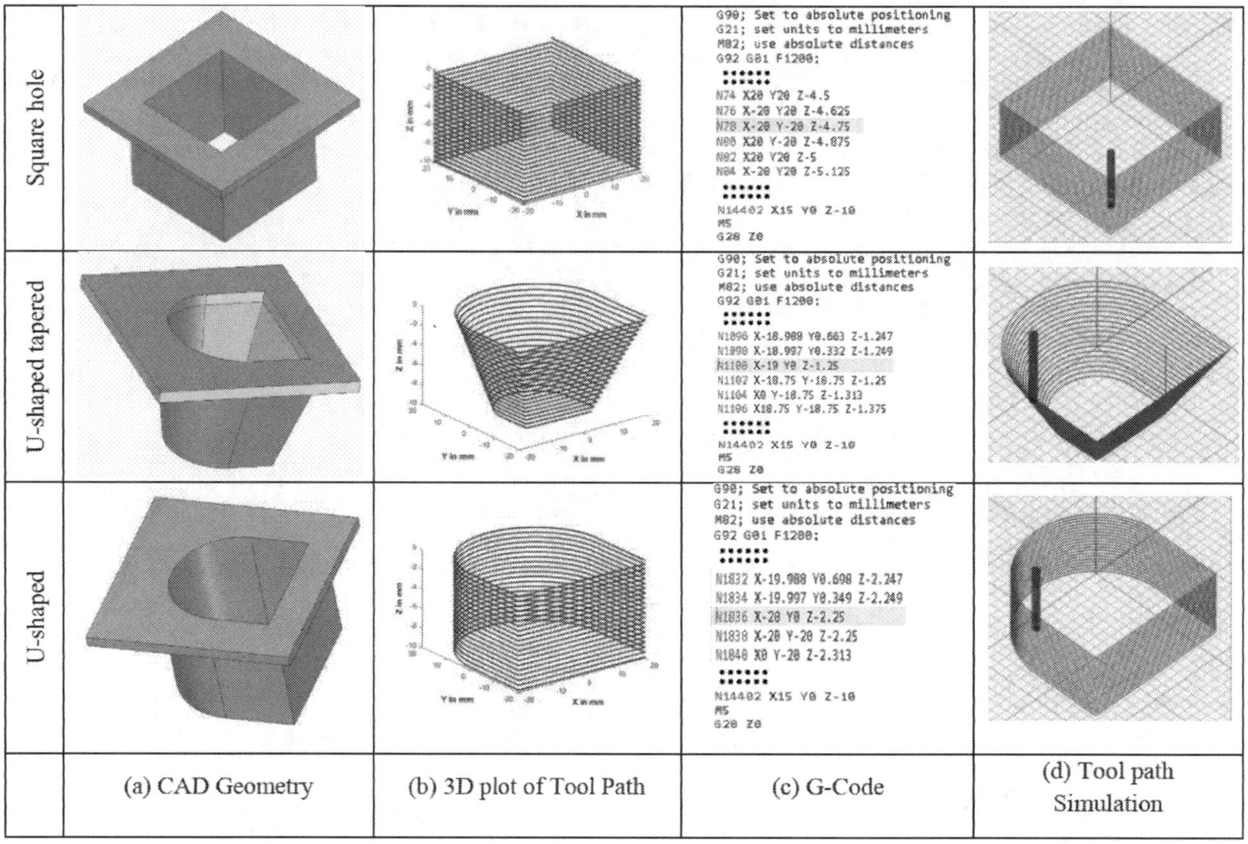

Fig. 3. (a) CAD geometry (b) 3D plot of Tool Path (c) G-Code, (d) Tool path Simulation.

3 Results

In the present work, six exemplary instances of unique hole flanging geometries were investigated to verify the current methodology. 3D CAD models of hole flanges were displayed in figure 3(a). The first one is a tapered hole (cone) flange part with a curved surface, the second one is a circular hole flange part with a zero-inclination angle, the third one is a tapered square hole (pyramid) flange part with flat surfaces, and the fourth one is a square hole flange. The fifth and sixth are asymmetric U-shaped hole flanges with tapered and without tapered holes, respectively.

The toolpath for incremental hole flanging was generated by formulating the geometry of holes using input data in MATLAB code. Figure 3 (b) shows the X, Y, and Z coordinates of the created toolpath shown on 3-dimensional axes. G-Code has also been created using MATLAB code based on the toolpath coordinates. Figure 3 (c) depicts the three sections of G-Code: the start or header, the intermediate or tool movement, and the end or footer. Header section consists G90 to represent absolute location, G21 to represent millimeter units, and M28 to represent absolute distance. While the intermediate segment includes CNC tool X, Y, and Z movements. The end part features M5 to stop spindle rotation and G28 for tool home position. To test or visualize the toolpath motions, the resulting G-code was simulated using an open-source software "NC viewer v1.1.3" which is shown in Figure 3 (d).

The current MATLAB code performs dual functions, generating toolpath coordinates versus time data and producing a G-code file through a unified programing approach. This process entails input geometry dimensions, formulating parts, utilizing MATLAB to create a toolpath, and subsequently employing the generated toolpath for both simulation and experimental fabrication purposes.

4 Discussions

Tool path definition is one of the crucial parts of design and fabrication of sheet metal parts formed by SPIF [10,16]. It was notice that proper definition as well as generation of tool path is significant process because it influences the geometric accuracy of formed parts [26]. Besides this it was also gathered that proper generation of tool path in incremental forming also influences resultant force acting on tool as well as on forming limit [26].

In the present work, tool path for various geometries of hole flanges were presented and proper selection of tool path is quite interrelated with the three different output parameters namely geometry accuracy, resultant forces, and forming limit [10,11,26]. Therefore, tool path for hole flanging should be selected on the basis of such tool path which yield geometric accuracy, low resultant force on forming tool and higher forming limit [10,11,26]. Another aspect of selection of optimized tool path depends upon step size of tool path and speed of tool which in turn reduces the surface roughness of hole flanges [26].

5 Conclusion

The objective of the present work is to generate tool paths as well as associated G-codes using MATLAB code for different geometries of incremental hole flanging process using SPIF. Six different geometries of hole flanges were considered to generate tool path using MATLAB code. It is worth noting that the majority of researches emphasized the importance and use of tool path creation and G-code generation for experimental and numerical modelling of the incremental forming process utilizing various CAD, CAM, and CAE software tools. It was also observed that relatively few research focused solely on the creation of tool paths and G-Code for various hole flange shapes. The proposed methodology was successfully applied for different shapes of hole flanges. Since the proposed methodology provides unified programming technique in the current MATLAB environment, which performs dual duties, creating toolpath coordinates against time data and outputting a G-code file and it saves computational time and programming efforts. The current work concludes selecting tool paths for various hole flange geometries, emphasizing the correlation between tool path selection and geometry accuracy, resultant forces, and forming limit. Optimizing the tool path also considers step size and speed to reduce surface roughness. Hence, proposed methodology is found to be better for incremental hole flanging in contrast to previously existed methodology for development of toolpath for incremental forming cases. Thus, it was concluded that this proposed methodology can be effective for capturing any intricate shapes for incremental hole flanging.

References

1. Dewang, Y., Hora, M.S. and Panthi, S.K., Prediction of crack location and propagation in stretch flanging process of aluminum alloy AA-5052 sheet using FEM simulation. Transactions of Nonferrous Metals Society of China, **25**(7), pp.2308-2320 (2015).

2. Dewang, Y., Hora, M.S. and Panthi, S.K., Finite element analysis of non-axisymmetric stretch flanging process for prediction of location of failure. Procedia Materials Science, **5**, pp.2054-2062 (2014).
3. Dewang, Y., Panthi, S.K. and Hora, M.S., Binder force effect on stretch flange forming of aluminum alloy. Materials and Manufacturing Processes, **34**(13), pp.1516-1527 (2019).
4. Dewang, Y., Panthi, S.K. and Hora, M.S., Some aspects of blank holding force in stretch flanging process. Materials Today: Proceedings, **5**(2), pp.6789-6798 (2018).
5. Dewang, Y., Hora, M.S. and Panthi, S.K., Effect of process parameters on deformation behavior of AA 5052 sheets in stretch flanging process. Materials Today: Proceedings, 4(8), pp.9316-9326 (2017).
6. Dewang, Y., Hora, M.S. and Panthi, S.K., Influence of blank holding force on stretch flange forming of aluminum alloy. Materials Today: Proceedings, **2**(4-5), pp.1934-1941 (2015).
7. Dewang, Y., Hora, M.S. and Panthi, S.K., FEM Simulation of Non-Axisymmetric Stretch Flange Forming of Aluminum Alloy 5052 Based on Shell Type Elements. Iranian Journal of Materials Science and Engineering, **14**(4), pp.69-80 (2017).
8. Dewang, Y. and Sharma, V., Sheet metal shrink flanging process: a critical review of current scenario and future prospects. Materials and Manufacturing Processes, **38**(6), pp.629-658 (2023).
9. Kumar, S., Soni, M., Panthi, S.K., Ahmed, M. and Dewang, Y., Analysis the Influence of Various Step Punch Shape and Clearance in Stretch Flanging Process Using FEM Simulation. NanoWorld J, **9**(S1), pp.S390-S396 (2023).
10. Dewang, Y., Purohit, R. and Tenguria, N., A study on sheet metal hole-flanging process. Materials Today: Proceedings, **4**(4), pp.5421-5428 (2017).
11. Cui, Z. and Gao, L., Studies on hole-flanging process using multistage incremental forming. CIRP Journal of Manufacturing Science and Technology, **2**(2), pp.124-128 (2010).
12. Makwana, R., Modi, B. and Patel, K., Single-stage single point incremental square hole flanging of AA5052 material. Materials and Manufacturing Processes, **38**(6), pp.680-691 (2023).
13. Mashudi, I., Fakhruddin, M. and Hardjito, A., April. Forming parameters effect toward flange height on single-step incremental backward hole flanging process. In AIP Conference Proceedings, **2531**, No. 1) (2023).
14. Makwana, R., Modi, B. and Patel, K., In-process grain refinement on AA5052-H32 sheet metal in single-stage single point square hole flanging. Manufacturing Letters, **35**, pp.232-238 (2023).
15. Fernández, J.A.L., Borrego Puche, M., Centeno Báez, G. and Vallellano, C., Formability Analysis of Stretch and Shrink Flanging by Single Point Incremental Forming Based on Stresses. Key Engineering Materials, **955**, pp.111-119 (2023).
16. Dewang, Y., Tenguria, N., Sharma, V. and Dubey, M.K., Formation of Hole Flanges Through Incremental Forming: A Review. Advances in Engineering Design: Select Proceedings of FLAME 2018, pp.547-561 (2019).
17. K. Suresh, A. Khan, and S. P. Regalla, "Tool path definition for numerical simulation of single point incremental forming," Procedia Eng. **64**, 536–545 (2013).
18. Azaouzi, M., & Lebaal, N., Tool path optimization for single point incremental sheet forming using response surface method. Simulation Modelling Practice and Theory, **24**, p.49-58. (2012)
19. Z.Yeshiwas, and A. Krishniah,"Spiral Toolpath Definition and G-code Generation for Single Point Incremental Forming," Journal of Mechanical Engineering, **17**(1), 91-102, (2020)

20. R. Malhotra, N. V. Reddy, and J. Cao, "Automatic 3D Spiral Tool path Generation for Single Point Incremental Forming," J. Manuf. Sci. Eng.**132** (6), 061003 (2010).
21. Skjoedt, M., Hancock, M. H., and Bay, N., Creating Helical Tool Paths for Single Point Incremental Forming, Key Eng. Mater., **344**, p. 583–590. (2007).
22. Schmitz, R.U.C., Bremen, T., Bailly, D.B. and Hirt, G.K.P., On the influence of the tool path and intrusion depth on the geometrical accuracy in incremental sheet forming. Metals, **10**(5), p.661 (2020).
23. Zhu, H., Cheng, G. and Jung, D., Toolpath planning and generation for multi-stage incremental forming based on stretching angle. Materials, **14**(17), p.4818 (2021).
24. Yan, Z., Hassanin, H., El-Sayed, M.A., Eldessouky, H.M., Djuansjah, J.R.P., A. Alsaleh, N., Essa, K. and Ahmadein, M., Multistage tool path optimisation of single-point incremental forming process. Materials, **14**(22), p.6794 (2021).
25. Jiang, Z., Liao, S., Slocum, A.H., Leem, D., Ehmann, K.F. and Cao, J., Toolpath Planning for Manufacturing of Complex Parts Through Incremental Sheet Forming. ASME Open Journal of Engineering, (2022). doi: https://doi.org/10.1115/1.4053751
26. Besong, L.I., Buhl, J., Ünsal, I., Bambach, M., Polte, M., Blumberg, J. and Uhlmann, E., Development of tool paths for multi-axis single stage incremental hole-flanging. Procedia Manufacturing, **47**, pp.1392-1398 (2020).

MATEC Web of Conferences **393**, 01004 (2024) https://doi.org/10.1051/matecconf/202439301004
STAAAR-2023

Reduction mechanism of hydrothermally synthesized wide band gap $ZnWO_4$ nanorods for HER application

Shreya[1*], *Peeyush* Phogat[1], *Sukhvir* Singh[1] and *Ranjana* Jha[1]

[1]Research Lab for Energy Systems, Department of Physics, Netaji Subhas University of Technology, New Delhi, India

Abstract. This manuscript presents a detailed investigation of the synthesis of zinc tungstate ($ZnWO_4$) nanoparticles and various characterizations of the as-synthesized sample to reveal its potential for hydrogen evolution reaction (HER). The study focuses on a simple and efficient hydrothermal method that facilitates the production of $ZnWO_4$ nanoparticles, which involves the controlled reaction between zinc and tungstate ions in a specific solution. The resulting $ZnWO_4$ nanoparticles were characterized by various characterization techniques which include XRD, UV-vis spectroscopy, TEM and electrochemical study to get insights into their size, structure, properties and electrochemical behaviour. The characterization includes the analysis of the nanoparticles' structural features and optical properties. The material's electrochemical properties were also investigated by employing cyclic voltammetry (CV) and potentio electrochemical impedance spectroscopy (PEIS). The charge transfer process was studied for the material revealing its diffusion controlled behaviour and reduction peaks in the cathodic region. These properties suggest that the material is a potential candidate for HER catalysis.

1 Introduction

Zinc tungstate ($ZnWO_4$) has received wide attention in the materials research because of its intriguing ceramic characteristics and potential applications. It is a crystalline compound which exhibits excellent optical and luminescent characteristics as observed in various studies [1]. It is an optically transparent material which makes it suitable for various optical applications [2]. This material possesses a high chemical stability, thermal resistance, and a wide bandgap, which make it suitable for applications in radiation detection, scintillation detectors, solar cells and phosphors for white light-emitting diodes (LEDs) [3], [4].

A variety of methods can be used to synthesize $ZnWO_4$ nanoparticles like hydrothermal, chemical vapour deposition, solvothermal, etc. [5], [6] Researchers have synthesized $ZnWO_4$ nanoparticles by a variety of physicals and chemical techniques and confirmed their

* Corresponding author: shreyasharma.aug15@gmail.com

suitability for different applications. G. Bhaskar Kumar et. al. used solid-state route to synthesize $ZnWO_4$ ceramic powder presented its structural, optical, morphological and dielectric properties [1]. Severo et. al. synthesized $ZnWO_4$ nanoparticles by solvo-hydrothermal technique and revealed their application for inulinase immobilisation [7]. Another research by Abubakar et. al. presented review of zinc tungstate as a photocatalyst [8].

The synthesis is a key component in the development of advanced radiation detectors used in medical imaging, homeland security, and nuclear physics research. Its scintillating properties enable the accurate detection of ionizing radiation, making it crucial in healthcare diagnostics and safety measures. Zinc tungstate has also shown promising applications in the field of LED technology, contributing to the production of energy-efficient lighting sources [9]. As the world transitions towards sustainable energy solutions, materials like zinc tungstate play a vital role in reducing energy consumption and environmental impact.

Despite the significant progress made in understanding zinc tungstate, there remain research gaps that need to be addressed. One notable gap is the exploration of novel synthesis methods and the optimization of existing ones. The research gap also includes the study of $ZnWO_4$ for hydrogen evolution reaction (HER). This manuscript has reported the synthesis of $ZnWO_4$ by one step facile hydrothermal route. Another research by Siriwong et. al. [10] reported the synthesis of $ZnWO_4$ by hydrothermal method using zinc nitrate and sodium tungstate as precursors and keeping it in autoclave at 200°C for 24 hours. They studied the optical and microstructural properties of $ZnWo_4$ nanorods. Another group, Mashkani et. al. [11] have used precipitation process to synthesize $ZnWO_4$ nanorods and studied its optical, morphological and magnetic properties along with studying its photocatalytic applications.

The present work delves into the characterization of the material required to analyze the material's structure, composition, and properties comprehensively. The electrochemical properties of the as-synthesized zinc tungstate material are also investigated which show the diffusive behaviour of the material and reduction process involved revealing its possible usage for HER. It provides a holistic overview of this remarkable material's properties. This work is pivotal in advancing our knowledge and harnessing the full potential of zinc tungstate for technological advancements. This manuscript presents a comprehensive exploration of the synthesis and characterization of zinc tungstate, shedding light on its remarkable attributes, the importance of studying it, and addressing pertinent research gaps.

2 Experimental Section

2.1 Synthesis of $ZnWO_4$ nanorods

ZnO nanoparticles of average particle size 10-30 nm and Sodium tungstate dihydrate AR grade ($Na_2WO_4.2H_2O$; >99) were procured from M/s Nanoshel LLC and M/s Thomas Baker Pvt. Ltd. respectively. HNO_3 (70 %) of M/s CDH ltd. was also used during the synthesis. $ZnWO_4$ nanoparticles were synthesized by hydrothermal method. 100 mg ZnO powder was ultrasonically dispersed in 20 ml DI water. Subsequently, 0.2 mmol Na_2WO_4 was dissolved in 30 ml DI water. The former solution was mixed dropwise into the later one and further stirred. Later few drops of 5M HNO_3 was added into the solution to make pH acidic. Later, the solution was kept in autoclave at 180°C for 24 hours. It was then left to cool down naturally followed by washing and drying at 60°C.

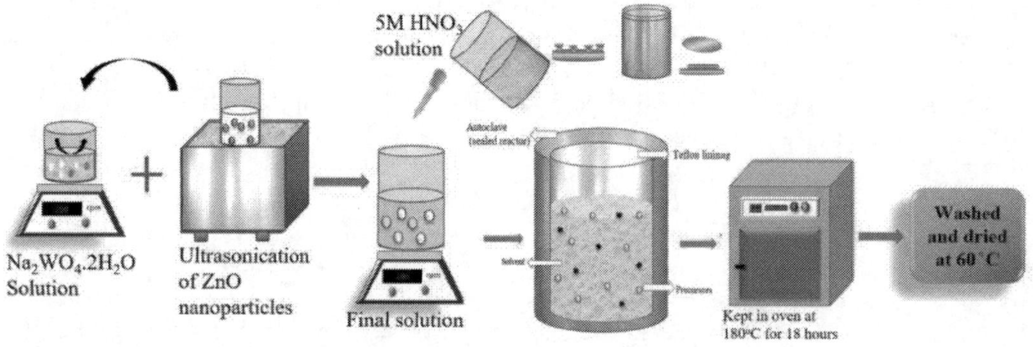

Fig. 1. Schematic diagram depicting the synthesis process of ZnWO4 nanorods.

2.2 Deposition of thin film and electrochemical study

The thin film of the as-synthesized $ZnWO_4$ sample was deposited on a fluorine doped tin oxide (FTO) glass substrate. A uniformly dispersed solution of 50 mg/ml concentration was prepared in ethanol and dispensed onto the substrate once, spinning at 1500 rpm for a minute to prepare a uniformly deposited thin film. The electrochemical behavior of $ZnWO_4$ nanorods to study its usage for HER has been studied by using three electrode electrochemical workstation (Biologic make, SP-240 model). The setup includes three electrodes which includes working, reference and counter electrode. Ag|AgCl was used as the reference electrode in the present study while Pt was used as the counter electrode. The deposited film of the sample was used as the working electrode.

3 Results and Discussion

3.1 X-ray Diffraction

The structural properties of the as-synthesized sample were investigated by using X-ray diffraction technique. The obtained XRD spectra (fig. 2 (a)) revealed polycrystalline material indexed to zinc tungstate as matched with JCPDS file 98-008-7933. The diffraction peaks corresponding to the (hkl) planes as depicted in fig. 2. were found to match with the monoclinic crystal structure of $ZnWO_4$ with the space group P12/c1 and space group number 13. The lattice parameters of the as-synthesized $ZnWO_4$ material were calculated by well-known Bragg's law [12] and were found to be a=4.64 A, b=5.60 A and c=4.93 A The values were compared with the standard values of JCPDS file as shown in table 1 which revealed little deviation from the standard values inferring negligible strain in the synthesized $ZnWO_4$ material.

The crystallite size and strain of $ZnWO_4$ were estimated by two different approaches which include Williamson-Hall (W-H) plot [13] (fig. 2 (b)) and size-strain (S-S) plot [14](fig. 2 (c)). The crystallite size calculated using intercept of W-H plot and slope of S-S plot was found out to be 31.86 nm and 24 nm respectively. Alternatively, the strain was found out to be 0.00136 and 0.0109 by W-H and S-S plot respectively. The crystallite size and strain values are summarized in table 2.

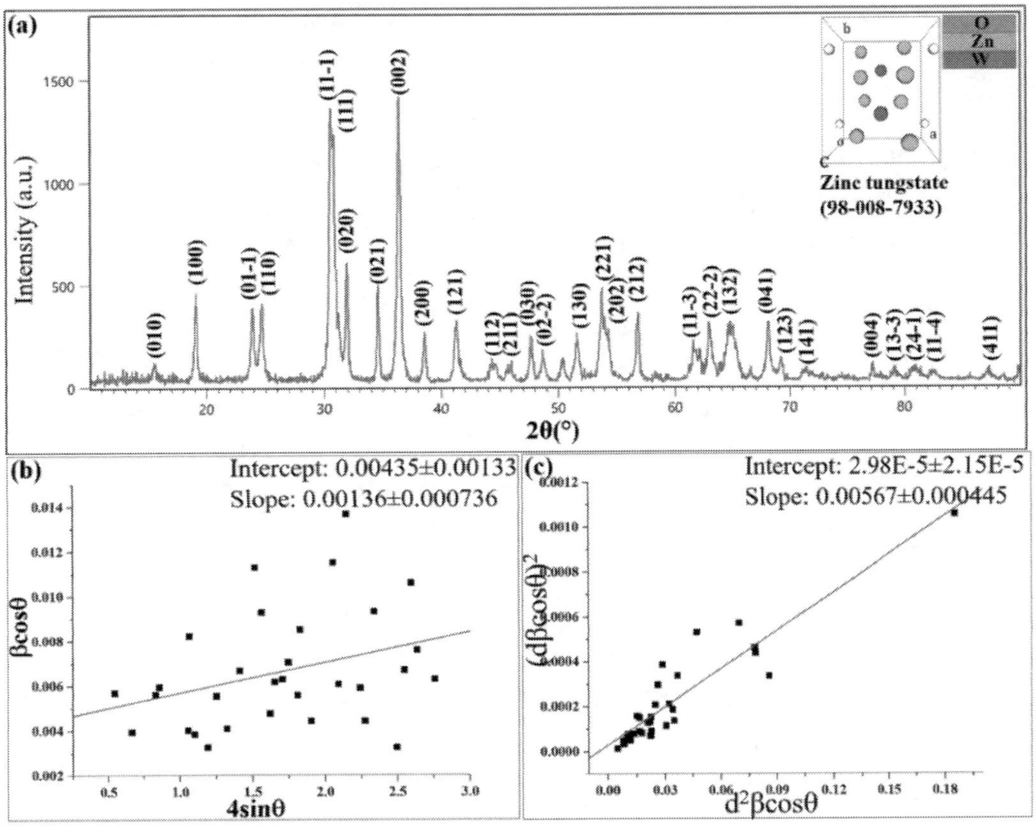

Fig. 2. (a) XRD spectra of as-synthesized $ZnWO_4$, **(b)** W-H plot and **(c)** S-S plot.

Table 1. Lattice parameters of the as-synthesized $ZnWO_4$ as matched with JCPDS data.

Lattice parameter	JCPDS Value	Calculated value
a (Å)	4.69	4.64
b (Å)	5.73	5.60
c (Å)	4.93	4.93

Table 2. Crystallite size and strain values of as-synthesized $ZnWO_4$ using W-H and S-S plot.

Method	Crystallite size	Strain
Williamson-Hall plot	31.86 nm	0.00136
Size-Strain plot	24 nm	0.00109

3.2 UV-vis Spectroscopy

UV-visible spectroscopy was utilised to study the optical properties of the as-obtained $ZnWO_4$ sample. The absorbance spectrum (fig. 3 (a)) was obtained using Shimadzu make 2600i model UV-visible Spectrophotometer. The spectrum revealed the absorbance by the material in UV region with maximum prominent absorbance peaks at 220, 266 and 360 nm. The spectrum was further used to estimate the optical band gap of the material by employing Tauc plot method [15], [16]. The linear part of the Tauc plot was extrapolated as shown in fig. 3 (b) to get the value of optical band gap to be 3.23 eV which makes it a wide band gap semiconductor. The refractive index of the material was also investigated based on the optical band gap and found out to be 2.34 [17].

Furthermore, HOMO and LUMO levels of the material were identified using the obtained band gap value and 6.31 eV as absolute electronegativity of $ZnWO_4$. [12]. Using these values, the valence and conduction bands were observed to be 3.426 eV and 0.194 eV respectively. The whole data is summarized in table 3.

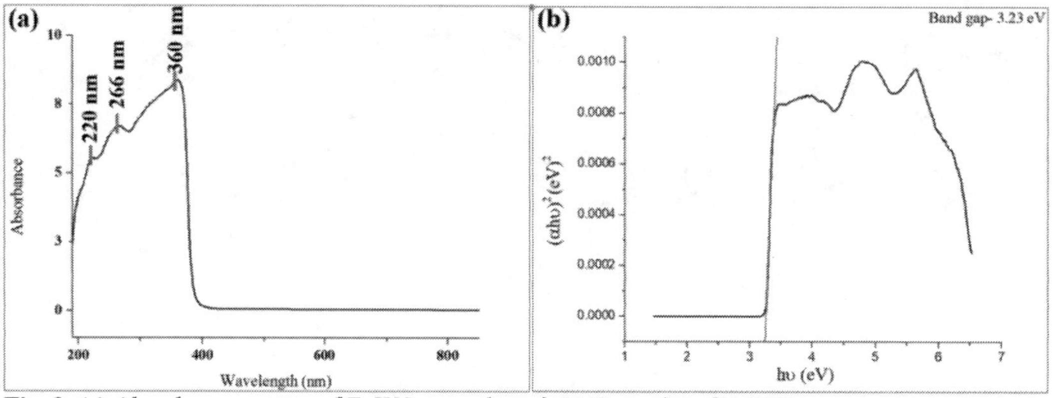

Fig. 3. (a) Absorbance spectra of $ZnWO_4$ sample and **(b)** Tauc plot of $ZnWO_4$.

Table 3. Optical properties of as-synthesized $ZnWO_4$ sample.

Property	Value
Absorbance peaks	220 nm, 266 nm, 360 nm
Band gap	3.23 eV
Refractive index	2.34
HOMO level	3.426 eV,
LUMO level	0.194 eV

3.3 Transmission Electron Microscopy

The microstructural characteristics of the $ZnWO_4$ sample were studied by TEM of JEOL make JEM-1400 model microscope. TEM image (fig. 4 (a)) reveals the formation of nanorods of varied lengths from 50 nm to 200 nm. The corresponding area's selected area electron diffraction pattern (SAEDP) was captured, as seen in fig. 3(b). The different spots of varying intensities forming the circles were observed revealing the polycrystalline nature of the material with fine crystallites. To study the d-spacing and further determine the appropriate (hkl) planes, one of the spots was investigated. It was indexed to the (100) plane, which is well in line with the XRD spectra displayed in fig. 2(a).

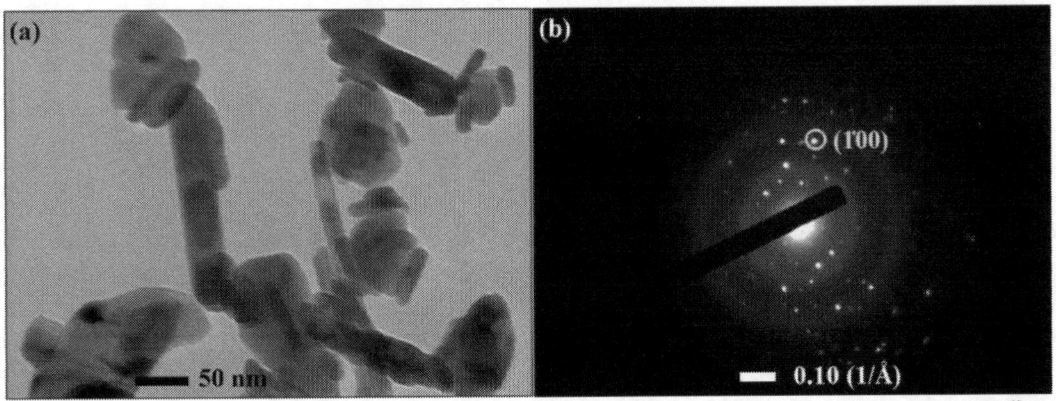

Fig. 4. (a) TEM image at 50 nm scale showing nanorods and **(b)** SAED pattern of the corresponding area.

3.4 Electrochemical Studies

CV and PEIS techniques were used to study the electrochemical characteristics of the as-synthesised ZnWO4 nanomaterial. Fig. 5 (a) depicts the cyclic voltammograms at varying scan rates ranging from 40 mV/s to 140 mV/s. The significant reduction peaks were observed in the voltammograms. The peak current was observed to be increasing with increased scan rate which is in good agreement with Randle Sevick equation. Fig. 5 (c) shows the variation of peak current with scan rate. It showed the linear relationship between the two revealing diffusion controlled behaviour of the working electrode which is thin film of $ZnWO_4$. Another plot utilized to verify the diffusion controlled behaviour is log (peak current) vs log (scan rate) plot (fig. 5 (d)). The slope of this plot was obtained as 0.39 which verifies the diffusive behaviour of the working electrode [18]. The diffusive behaviour of $ZnWO_4$ makes it a potential candidate to be used as a catalyst in hydrogen evolution reaction. Another research group of Shad et. al. [19] have also studied the cyclic voltammetry study of $ZnWO_4$ and observed no oxidation. Alternatively, the reduction peaks were observed at approximately 1-2 μA which is quite less than the value obtained in the present work.

PEIS technique was analysed to look into the redox process of the working electrode and thus resistances and capacitances involved. Fig. 5 (b) shows nyquist plots at varying potential values ranging from 0 mV to -400 mV. The potential applied to an electrochemical system affected the kinetics of electrode reactions [20]. As the potential changed, the rate of charge transfer processes at the electrode-electrolyte interface also changed. This, in turn, affected the impedance response observed in the Nyquist plot. At small |V| values, large impedance was observed leading to larger radii semicircles. Alternatively, at larger |V| values, impedance was observed to be decreasing as the radii of semicircles decreased. The Nyquist plots at -350 mV and -400 mV were of same radii suggesting the saturation. The smaller radii semicircles showed the presence of Warburg impedance also suggested by a slant line followed by the semicircle which shows the formation of infinite layer of ions at the interface [21]. It shows good diffusion of charge carriers in the system [22], [23].

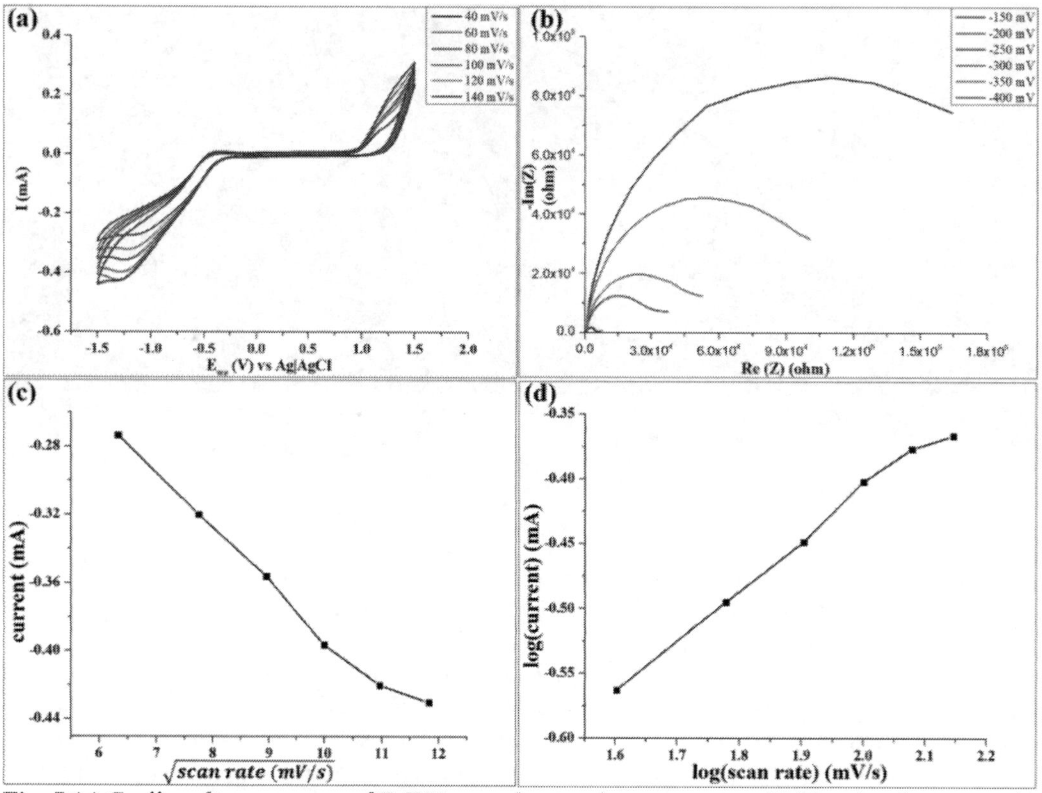

Fig. 5 (a) Cyclic voltammograms of $ZnWO_4$ sample at various scan rates, **(b)** Nyquist plots at varying potential values, **(c)** peak current vs $\sqrt{scan\ rate}$ and **(d)** log (peak current) vs log (scan rate).

4 Conclusion

A comprehensive study on the synthesis and characterization of zinc tungstate nanoparticles using the hydrothermal route is presented in the manuscript. The findings from various characterization techniques shed light on the material's properties and potential applications. XRD confirmed the polycrystalline nature of the synthesized $ZnWO_4$, matching its monoclinic crystal structure. UV-Visible spectroscopy revealed significant absorbance in UV region, indicating material's potential for applications in UV-sensitive devices. The estimated band gap of 3.23 eV and refractive index of 2.34 indicates the material's optical properties. The determination of the HOMO - LUMO levels at 3.323 and 0.195 eV, respectively, provides valuable information for potential electronic and photonic applications. TEM demonstrated the formation of nanorods, showcasing the control over morphology achieved through the hydrothermal synthesis method. SAED pattern analysis confirmed the presence of $ZnWO_4$ crystals, with one of the dots corresponding to the (100) plane, further revealing polycrystalline material. Cyclic voltammetry revealed the presence of reduction peaks, indicating the material's electrochemical activity. The diffusion-controlled behaviour observed suggests its potential in HER catalysis. PEIS at various potential values demonstrated the presence of Warburg impedance, highlighting the material's suitability for photocatalytic activity.

The authors acknowledge the support of Prof. Anand Srivastava, Vice Chancellor, NSUT, Delhi for providing the essential resources required for the research work.

All the data used for this research work is included within the manuscript.

Shreya did the Conceptualization, formal analysis and wrote the original draft. Peeyush Phogat helped in the formal analysis and with software. Ranjana Jha supervised the research work and arranged resources for the work. Sukhvir Singh reviewed the edited the manuscript.

No conflicts of interest exist among the authors.

References

1 G. B. Kumar, K. Sivaiah, S. Buddhudu, Synthesis and characterization of ZnWO4 ceramic powder, *Ceram Int*, **36**, 1, 199–202 (2010) https://doi.org/10.1016/j.ceramint.2009.07.005

2 K. M. Garadkar, L. A. Ghule, K. B. Sapnar, S. D. Dhole, A facile synthesis of ZnWO4 nanoparticles by microwave assisted technique and its application in photocatalysis, *Mater Res Bull*, **48**, 3, 1105–1109, (2013). https://doi.org/10.1016/j.materresbull.2012.12.002

3 U. K. Ghorui, P. Mondal, J. Satra, B. Adhikary, and A. Mondal, In situ metallic copper incorporation into novel g-C3N4/ZnWO4 nanocomposite semiconductor for efficient thin film solar cell application, *Mater Sci Semicond Process*, **143**, 106559 (2022). https://doi.org/10.1016/j.mssp.2022.106559

4 W. Ran, Q. Wang, Y. Zhou, S. Ding, J. Shi, J. H. Jeong, Fabrication of ZnWO4:Sm3+, Bi3+, Li+ with tunable white light-emitting properties for W-LEDs, *Mater Res Bull*, **64**, 146–150 (2015). https://doi.org/10.1016/j.materresbull.2014.12.050

5 A. Yadav, Shreya, N. K. Puri, Preliminary Observations of Synthesized WS2 and Various Synthesis Techniques for Preparation of Nanomaterials, In: Singari, R.M., Jain, P.K., Kumar, H. (eds) Advances in Manufacturing Technology and Management. Lecture Notes in Mechanical Engineering. Springer, Singapore, 546–556 (2023). https://doi.org/10.1007/978-981-16-9523-0_61

6 Shreya, A. Yadav, R. Khatri, N. Jain, A. Bhandari, N. K. Puri, Double Zone Thermal CVD and Plasma Enhanced CVD Systems for Deposition of Films/Coatings with Eminent Conformal Coverage, In: Singari, R.M., Jain, P.K., Kumar, H. (eds) Advances in Manufacturing Technology and Management. Lecture Notes in Mechanical Engineering. Springer, Singapore, 273–283 (2023). https://doi.org/10.1007/978-981-16-9523-0_31

7 E. da C. Severo, E. R. Abaide, C. G. Anchieta, V. S. Foletto, C. T. Weber, T. B. Garlet, G. C. Collazzo, M. A. Mazutti, A. Gundel, R. C. Kuhn, E. L. Foletto, Preparation of Zinc Tungstate ($ZnWO_4$) Particles by Solvo-hydrothermal Technique and their Application as Support for Inulinase Immobilization, *Materials Research*, **19,** 4, 781–785 (2016) https://doi.org/10.1590/1980-5373-MR-2015-0100

8 H. L. Abubakar, J. O. Tijani, S. A. Abdulkareem, A. Mann, S. Mustapha, A review on the applications of zinc tungstate ($ZnWO_4$) photocatalyst for wastewater treatment, *Heliyon*, **8**, 7, e09964, (2022). https://doi.org/10.1016/j.heliyon.2022.e09964

9 D. Sivaganesh, S. Saravanakumar, V. Sivakumar, S. Sasikumar, J. NandhaGopal, R. Ramanathan, "$ZnWO_4$:Eu^{3+} phosphor with intense blue LED excitation: photoluminescence and electron density distribution analysis.," *Luminescence*, **36**, 1, 99–109 (2020). https://doi.org/10.1002/bio.3920

10 P. Siriwong, T. Thongtem, A. Phuruangrat, and S. Thongtem, Hydrothermal synthesis, characterization, and optical properties of wolframite ZnWO4 nanorods, *CrystEngComm*, **13**, 5, 1564–1569 (2011) https://doi.org/10.1039/C0CE00402B

11 S. M. Hosseinpour-Mashkani, M. Maddahfar, A. Sobhani-Nasab, "Precipitation Synthesis, Characterization, Morphological Control, and Photocatalyst Application of ZnWO4 Nanoparticles," *J Electron Mater*, **45**, 7, 3612–3620 (2016). https://doi.org/10.1007/s11664-016-4532-3

12 Shreya, P. Phogat, R. Jha, S. Singh, Microwave-synthesized γ-WO3 nanorods exhibiting high current density and diffusion characteristics, *Transit Met Chem*, **48**, 167-183 (2023). http://dx.doi.org/10.1007/s11243-023-00533-y

13 Shreya, P. Phogat, R. Jha, S. Singh, Elevated Refractive Index of MoS2 Amorphous Nanoparticles with a Reduced Band Gap Applicable for Optoelectronics, In: Sethuraman, B., Jain, P., Gupta, M. (eds) Recent Advances in Mechanical Engineering. STAAAR 2022. Lecture Notes in Mechanical Engineering. Springer, Singapore., 431–439 (2023). http://dx.doi.org/10.1007/978-981-99-2349-6_39

14 P. Phogat, Shreya, R. Jha, and S. Singh, Optical and Microstructural Study of Wide Band Gap ZnO@ZnS Core–Shell Nanorods to be Used as Solar Cell Applications, In: Sethuraman, B., Jain, P., Gupta, M. (eds) Recent Advances in Mechanical Engineering. STAAAR 2022. Lecture Notes in Mechanical Engineering. Springer, Singapore., 419–429 (2023). http://dx.doi.org/10.1007/978-981-99-2349-6_38

15 P. PHOGAT, S. Shreya, R. JHA, and S. Singh, Diffusion Controlled Features of Microwave Assisted ZnS/ZnO Nanocomposite with Reduced Band Gap, *ECS J. Solid State Sci. Technol.*, **12**, 3, 034004 (2023). http://dx.doi.org/10.1149/2162-8777/acc426

16 J. Dahiya, P. Phogat, A. Hooda, S. Khasa, Investigations of Praseodymium doped LiF-ZnO-Bi_2O_3-B_2O_3 glass matrix for photonic applications. *AIP Conf Proc*, **2995**, 020065 (2024). http://dx.doi.org/10.1063/5.0178197

17 T. Kumar, Shreya, P. Phogat, V. Sahgal, and R. Jha, Surfactant-mediated modulation of morphology and charge transfer dynamics in tungsten oxide nanoparticles. *Phys Scr*, **98**, 8, 085936 (2023). http://dx.doi.org/10.1088/1402-4896/ace566

18 S. Sharma, P. Phogat, R. Jha, S. Singh, Electrochemical and Optical Properties of Microwave Assisted MoS2 Nanospheres for Solar Cell Application. *International Journal of Smart Grid and Clean Energy*, **12**, 3, 66–72 (2023). http://dx.doi.org/10.12720/sgce.12.3.66-72

19 N. A. Shad, S. Z. Bajwa, N. Amin, A. Taj, S. Hameed, Y. Khan, Z. Dai, C. Cao, W. S. Khan, Solution growth of 1D zinc tungstate ($ZnWO_4$) nanowires; design, morphology, and electrochemical sensor fabrication for selective detection of chloramphenicol. *J Hazard Mater*, **367**, 205–214 (2019). https://doi.org/10.1016/j.jhazmat.2018.12.072

20 P. Phogat, Shreya, R. Jha, S. Singh, Electrochemical analysis of thermally treated two dimensional zinc sulphide hexagonal nano-sheets with reduced band gap. *Phys Scr*, **98**, 12, 125962 (2023). http://dx.doi.org/10.1088/1402-4896/ad0d93

21 Dipti, P. Phogat, Shreya, D. Kumari, S. Singh, Fabrication of tunable band gap carbon based zinc nanocomposites for enhanced capacitive behaviour. *Phys Scr*, **98**, 9, 095030 (2023). http://dx.doi.org/10.1088/1402-4896/acf07b

22 P. Phogat, S. Shreya, R. Jha, S. Singh, Impedance Study of Zinc Sulphide Quantum Dots via One Step Green Synthesis. *Materials Science Forum*, **1099**, 119–125, (2023). http://dx.doi.org/10.4028/p-G1CCxq

23 D. Kumari, Shreya, P. Phogat, Dipti, S. Singh, R. Jha, Enhanced electrochemical behavior of C@CdS Core-Shell heterostructures. Materials Science and Engineering: B, **301**, 117212, (2024). https://doi.org/10.1016/j.mseb.2024.117212

Optimization of Manufacturing Plant Layout Using Systematic Layout Planning (SLP) Method

E Badharinath[1*], *V Diwakar Reddy*[2], *P Sravan Kumar*[3], *and A K Damodaram*[4]

[1&2]Department of Mechanical Engineering, SVU College of Engineering, Tirupati, India, 517502.
[3]Quality Manager ZF Groups, USA.
[4]SV College of Engineering&Technology, Chittoor, India, 517127.

Abstract. The plant shop floor layout of a company manufactures gear bush is analysed in this study along with the application of the systematic layout planning (SLP) technique as a means of resolving the issue.The interaction between machines, operational tasks, and material flow is taken into consideration when deciding where each machine should be placed. The relationship between machines and plant area has looked into in detail, together with the activity relationship chart, process chart and plant layout. The proposed production shop floor plan was designed using the SLP technique, and the performance of the proposed layout in terms of material travelling distance, travel time, and labour cost was compared to that of the existing layout. The measures included the layout's actual sizes, the activities between machines, the separation between machines, and the material travelling between machines in the plant. The total distance travelled by the material is reduced to 799ft. from 1227ft., production is doubled and ideal time is eliminated.

Keywords: Systematic Layout Planning (SLP) method, Plant Floor, Comprehensive Activity relationship chart, Existing Layout, proposed Layout.

1 Introduction

Systematic Layout Planningis a methodical planning process that enables users to recognise, plot, and rank the different tasks, connections, and options associated with a layout project. The technique's three main components are relationships, space, and adjustment. The aim of the work is to improving factory floor layout is to reduce expansion costs by maximising space and flow within the available limits. In order to reduce operating costs and increase process efficiencies, the ideal factory floor layout method entails allotting space and positioning physical assets. At the micro-level layout changes can and

*Corresponding Author: badharinath555@gmail.com

should be made as and when a need arises. The layout of a factory's equipment, workspaces, and amenities for workers all have an impact on how effectively goods are produced. The smooth and quick transfer of materials from the raw material to the end product can only be guaranteed by a correctly designed plant. It can be characterised as a method of placing equipment, procedures, and plant services inside the factory to produce the desired amount and quality of output at the lowest possible production cost. It requires careful planning of manufacturing facilities to ensure direct workflow.

2 Literature Review

Richard Muther [1] , Lee Hales [2] described about the SLP, it was first being developed in the 1950s, their commitment to solid fundamental concepts, attention on addressing actual issues, and personal encouragement of Richard Muther were all greatly valued and beneficial. In text book Systematic Layout Planning longer than in 50 years, their effective use of SLP on countless projects has led to add helpful updates and explanations to both this version and earlier ones. Carlo et. al. [3] explains, Maximising a manufacturing system's productivity is one of its primary objectives. This depends on a number of variables, including the nature and complexity of the manufactured good, calibre of the raw materials, complexity of the manufacturing process, and configuration of the workstations that make up the manufacturing process. While some of these criteria are fixed and cannot be changed since they are determined by the product, others are variable and can be improved. Subham Barnwal, Prasad Dharamadhikari [4] in their journal describes, To create a new layout using the notion of the Systematic Layout Planning pattern. The proposed paradigm, which is based on the Systematic Layout Planning Pattern Theory (SLP), has been proven successful. The overall distance that the material went decreased, the production rate increased, and the production time decreased. Suhardini, Septiani, and Fauziah [5] describe design simulations and four layout alternatives, each of which is assessed according to two criteria: The processing time by simulation and the cost of material handling as determined by material handling evaluation sheets (MHES). Xiaorong Xu [6] describes the program's entire process is simulated using Flexsim modelling, and the laborious reduction of the entire process is confirmed. Gómez et. al.[7] describes the resolve layout problems in facilities where there are aisles using genetic algorithms.

3 Methodology

3.1 Systematic Layout Planning

The process of arranging a plant's workspace so, that areas with high frequency to be close to one another is known as systematic layout planning. SLP approach used to maximise the current configuration. It is anticipated that the application will create fastest material flow with least amount of material handling and lowest cost. SLP method procedures are represented in Fig.1.
The four phases of SLP are as follows:
Step I: Choose the site which required modifications.
Stage II: Create the general layout of the building.
Stage III: Completely design the layout of the facility.
Stage IV: Getting the design outcomes ready and installed.

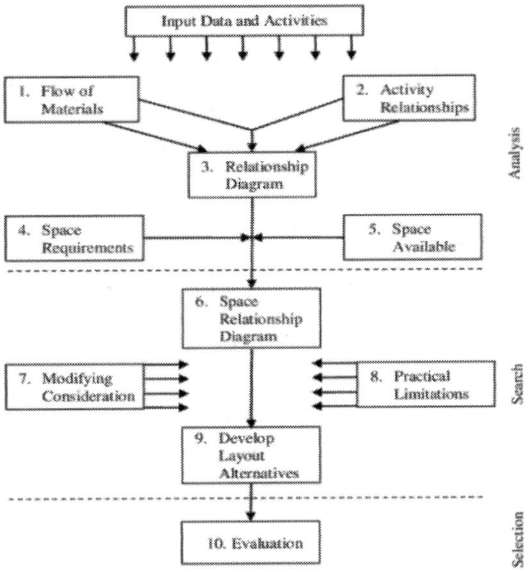

Fig.1. SLP method procedures [7]

3.2 Demand status of gear bush

The work centers are not situated in the Gear Bush manufacturing facility's product flow order. As a result, the current layout has multiple cross movements and extended movements. Additionally, the current layout arrangement has a lot of time-consuming tasks. The status of demand of Gear bush is given in the following Table 1. [5], which gives the complete details of demand for last six months i.e., October 2022 to March 2023.

The gear bush manufacturing company could not meet the demand for production in last six months (as shown in Table 1.). The cause of the problem is the layout which is not in proper alignment the production floor.

Table 1. Demand Status Of Gear Bush

Month	Demand	Capacity	Deviation	Status
October	15,000	12,000	-3,000	Not Fulfilled
November	11,000	12,000	1,000	Fulfilled
December	13,000	12,000	-1,000	Not Fulfilled
January	16,000	12,000	-4,000	Not Fulfilled
February	12,500	12,000	-500	Not Fulfilled
March	11,500	12,000	500	Fulfilled

3.3 Analysis of Present Layout

To enable improvement, a thorough analysis of the current layout is conducted in this part. The initial step that needs to be done as part of the work is gathering departmental details. In order to obtain a fundamental comprehension of the material flow and to detect the primary activities, their successors, and their predecessors, a sequence diagram is also produced.

Table 2. Total Department Areas (sq. ft)

Name of the Department	Total Area	Name of the Department	Total Area
Maintenance Room	3,000	Gauges Inspection (Machining completed parts)	200
Storage Area (Casting parts)	600	Storage Area (Machining completed parts)	600
General Machine Shop	9,200	Final Inspection	2,500
Calibration Room	1,500	Storage Area (FI Completed)	600
CMM	2,000	Pack and Dispatch	3,200
Standard Room	2,000	Rejection Area	600

The Table 2. explain about all the areas of departments, the main changes are need to be done at General Machine Shop because of inefficient utilization of shop floor and long flow of material, Time consuming at setups to new operations every time.

3.4 Analysis of General Machine Shop

In the General machine shop the machines which are used for the production of Gear Bush are LMW, SPINNER 400, DMG MORI. In the company mainly focused on manufacturing of gear bush only. In the below Fig. 2. every possible flow of process is represented, from the understanding of diagram the machine LMW is out of the line, Quality test is important after every step of machining, Gauge inspection at surface table is also required, DMG MORI Machine used for two setups one for Operation-20 and Operation-40, no changes are required in the FI and Dispatch section[6].

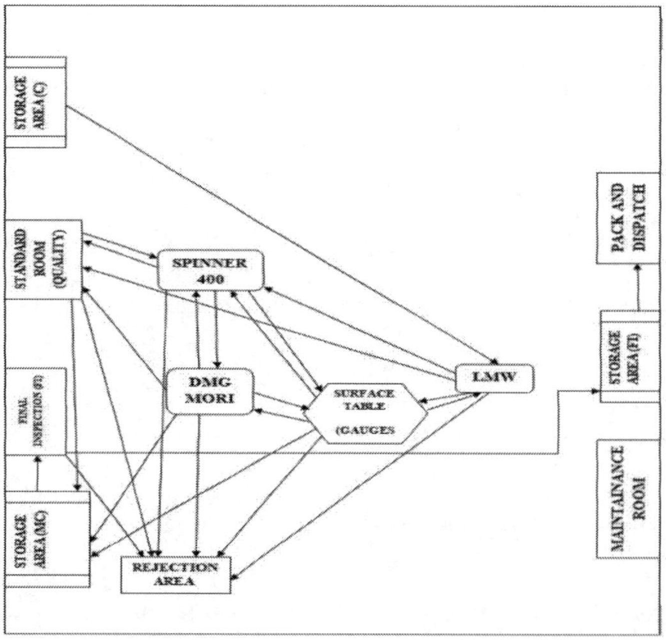

Fig. 2. Present Flow Process and Space Diagram

3.5 Analysis of the Operating Units Relationships

Here, focused primarily on seven operational areas that are directly related to the entire process which are Storage, Machining, Quality, Gauge Inspection, Rejection area, Final Inspection, Pack, and Dispatch from these areas Comprehensive relationship diagram [1] [2] is drawn (as shown in Fig. 2.). The SLP approach was used to arrange the crucial manufacturing sequence in order to continuously optimise the plant's workflow.

The primary issue causes with the current layout are as follows:

- There is a huge distance between the LMW Machine and Quality.
- DMG MORI allows for two operations, but each time a new setup is needed.
- The setup process is primarily to blame for part rejection.
- The storage location for final inspection items is far from final inspection.
- The total manufacturing time is longer per component.

The work unit relationship classification are :
A-Absolutely Important,
E-Very Important,
I-Important,
O-General,
U-Unimportant,
X-Don't Approach.

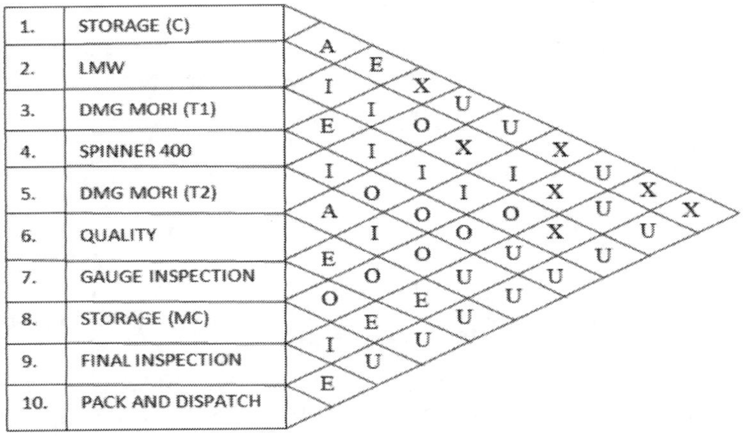

Fig. 3. Comprehensive relationship diagram

The new layout is prepared and implemented (as shown in Fig. 3.). on the shop floor.As the production of the gear bush did not meet the customer requirement a new machine is introduced i.e., TSUGAMI TURNING MACHINE. There is only one crosswalk in the new layout, this cannot be avoided since in the future, other machines occupy the available space [5].

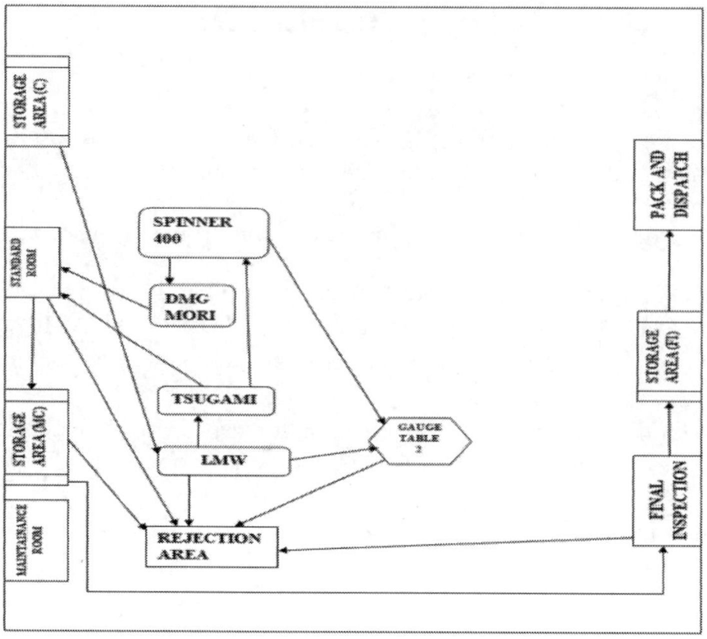

Fig. 4. New Flow Process and Space Diagram

3.6 Proposed Layout Design

Analysis of the plant layout using SLP to examination of manufacturing process revealed that wasted space could be eliminated and long distance decreased. Analysis of the current layout and the altered layout can be done in terms of

- Total Distance (sq. ft).
- Increase in Number of Parts Produced (Total Production).
- Elimination of Ideal Time

- **Total Distance :** The material has travelled a total of 1227 ft. [4] in the current arrangement. However, the altered plan is 799 ft. Table 3. gives the details of distances between departments.

Table 3. Modified Distance Table

S.No	Operation	Distance
1.	Storage Area (Casting)	-
2.	Storage Area (Casting) To LMW Machine	500 (150)
3.	LMW Machine To Gauge Inspection	30
4.	LMW Machine To Quality	200 (40)
5.	LMW Machine To Rejection Area	10
6.	LMW Machine To TSUGAMI Machine	0 (2)
7.	TSUGAMI Machine To Quality	0 (32)
8.	TSUGAMI Machine To Rejection Area	0 (8)
9.	TSUGAMI Machine To Spinner 400	0 (10)
10.	Spinner 400 Machine To Gauge Inspection	10
11.	Spinner 400 Machine To Quality	10 (20)
12.	Spinner 400 Machine To Rejection Area	20
13.	Spinner 400 Machine To DMG MORI	2

14.	DMG MORI Machine To Quality	5
15.	DMG MORI Machine To Rejection Area	10
16.	DMG MORI To Storage Area (M.C)	10
17.	Gauge Inspection To Rejection Area	30
18.	Quality To Rejection Area	100
19.	Storage Area (M.C) To Final Inspection	10 (200)
20.	Final Inspection To Rejection Area	20 (80)
21.	Final Inspection To Storage Area (F.I.)	250 (10)
22.	Final Inspection To Pack and Dispatch	10
	TOTAL DISTANCE	1227 (799)

- **Increase in Number of Parts Produced (Total Production)** : From, Table 4. around 12,000 parts are produced monthly on average. As a result, 400 pieces are created overall in 24 hours. Therefore, a total of 8 hours are needed for the fabrication of 150 pieces. Now that production time [3] has been cut by 4 hours per day, 24,000 more parts may be produced monthly. This will raise output from 12,000 to 24,000 pieces per month, enhancing the company's overall production[10].

Table 4. Total Production Comparison Table

Parts Produced In	Before (Qty)	After (Qty)
Month	12,000	24,000
Day (3 Shifts)	400	800
Shift (8 Hrs)	150	250

- **Elimination of Ideal Time**

A new setup is required at DMG MORI Machine and Setup Approval is obtained from the Quality once LMW Machine operation is complete and the manufactured components are kept in the storage area (Machined parts). The waiting period for the new setup of the machine and Quality Approval is not necessary because the new machine (The Tsugami Turning Machine) has been installed[9]. Total ideal time after new layout is represented in Table 5.

Table 5. Ideal Time Representation Table

Machine Operation	Ideal Time Before	Ideal Time After
LMW Machine	8 Hrs	0
Spinner 300	8 Hrs	0
DMG Mori	8 Hrs	0

3.7 Results And Discussions

The outcomes were examined and contrasted in terms of a number of parameters, which are displayed in Table 6. The total distance, production time, elimination of ideal time were some of the parameters used to compare the original layout and the changed configuration[11]. The total distance, production rate are the primary improvement parameters from the comparison of the present layout and the improved layout.

Table 6. Results of Plant Layout

Conditions	Present Layout	New Layout
Total Distance (Sq. Ft)	1227	799
Number Of Parts Produced(Total Production)	12,000	24,000
Elimination of Ideal Time (Hrs)	8	0

4 Conclusion

Through the usage of a real-world work study, the proposed strategy is demonstrated to be a workable method for resolving production shop floor layout design problems. The manufacturing flow area may be cut from 1227 feet to 799 feet thanks to the more cost-effective alternative arrangement which means that the total distance that materials went decreased by 34%, number of parts produced is doubled and ideal time is zero. The problem with the current plan is relative distance between various departments, which forces them to go a long distance, obstructs the efficient flow of materials, and raises costs[8]. Based on an activity relationship map, it changed the placement of several departments in relation to other departments in suggested plan.

References

Text Books

1. Richard Muther (1913 – 2014) Founder, Richard Muther & Associates. Founder, Muther International. Founder, Institute of High Performance Planners. Co-author of the books and videos Production Line Technique, Practical Plant Layout, Systematic Handling Analysis (SHA).
2. Lee Hales President, Richard Muther & Associates. Founder and President, High Performance Concepts, Inc. Co-author of the books and videos Systematic Planning of Industrial Facilities (SPIF), Volumes I and II. DOI:https://richardmuther.com/wp-content/uploads/2016/07/Systematic-Layout-Planning-SLP-4th-edition-soft-copy.pdf

Journals

3. Carlo, F. D., Antonietta, M., Borgia, O., & Tucci, M. (2013). Layout design for a low capacity manufacturing line: a case study. International Journal of Engineering Business Management: Special Issue on Innovations in Industry. DOI:10.5772/56883
4. Shubham Barnwal, Prasad Dharmadhikari. Optimization of plant layout using SLP method. IJIRSET /Vol. V/ Issue III/March 2016.
   DOI:http://www.ijirset.com/upload/2016/march/46_Optimization.pdf
5. D Suhardini, W Septiani and S Fauziah Paper : Design simulation plant layout using systematic layout planning.
   DOI:https://iopscience.iop.org/article/10.1088/1757-899X/277/1/012051
6. Xiaorong Xu, Paper : SLP-based Technical Plant Layout Planning and Simulation Analysis.

DOI:https://iopscience.iop.org/article/10.1088/1757-899X/772/1/012020

7. A. Gomez, Q.I. Fernandez, D. De la Fuente Garcia, P.J. Garcia. Paper: Using generic algorithms to resolvelayout probles in facility where there are aisles DOI :https://doi.org/10.1016/S0925-5273(02)00468-1
8. Shashank Singh Pawar, Dr. Devendra Singh Verma Paper : Design Cost Engineering Through Quality Function Deployment.
   DOI :https://www.ijrte.org/wp-content/uploads/papers/v8i2/B2679078219.pdf
9. Dr. Devendra S. Verma, Rajdeep Chaurasia. A study to identify the factors affecting employee turnover in small scale industries.
   DOI:10.5281/zenodo.57056
10. W. Wiyaratn and A. Watanapa, 2010.Improvement Plant Layout Using Systematic Layout Planning (SLP) for Increased Productivity World Academy of Science, Engineering and Technology.
    DOI:10.7763/IJET.2013.V5.515
11. Gaurav Goyal, Devendra S Verma. Paper : Optimization of plant layout in manufacturing industry.
    DOI:https://www.ijrte.org/wp-content/uploads/papers/v8i2/B2679078219.pdf

Ranking of epoxy/Kota stone dust composite by MCDM approach using hybrid AHP-MOORA methods

Virendra Rajput[1*], *Rohit* Soni[1], *Alok* Jha[2], and *Alok* Agrawal[3]

[1]Mechanical Engineering Department, Trinity Institute of Technology & Research, Bhopal-462021, India
[2]Mechanical Engineering Department, Shri Ramdeobaba College of Engg. & Management Nagpur-440013, India
[3]Mechanical Engineering Department, Sagar Institute of Research & Technology, Bhopal-462041, India

Abstract. The selection of appropriate materials is crucial in mechanical and structural design processes. This research study focuses on the fabrication of epoxy-based composites reinforced with Kota stone dust using a manual hand layup process. To simplify the material selection process, various Multi-Criteria Decision Making (MCDM) approaches were utilized. In this research, hybrid MCDM approaches, such as the analytical hierarchy process (AHP) and multi-objective optimization based on ratio analysis (MOORA), were employed to rank the epoxy composites filled with Kota stone dust. The AHP method was utilized to determine the weightage of each criterion, while the MOORA technique was used to rank the composites. Manual hand layup method was used for fabrication of composites. The selection of criteria was based on the physical and mechanical properties of the fabricated composites. The results indicate that epoxy composites containing 20% Kota stone dust are the most efficient for structural applications.

1 Introduction

Polymers offer numerous advantages due to their unique combination of hydrogen and carbon atoms. They possess excellent resistance to microorganisms and have a long lifespan. While petroleum-based plastics have found extensive use in applications where strength is not a primary concern, their environmental impact cannot be overlooked. Recycling has gained paramount importance in various sectors today [1]. Industries produce significant amounts of waste, which contributes to pollution in the environment. The stone industry, including the well-known Kota Stone Industries in India, generates industrial waste in the form of Kota stone powder or dust. Since Kota stones are extensively utilized in civil construction for various purposes such as exteriors, roadways, corridors, driveways, and balconies, the stone industry generates a substantial amount of Kota stone dust [2]. However, indiscriminate disposal of waste in open spaces can lead to environmental issues. Therefore, employing such waste as a filler in polymer matrix composites not only helps protect the environment but also reduces overall product costs [3].

*Corresponding Author: virendra.rajput89@gmail.com

In the present era, multiple Multi-Criteria Decision Making (MCDM) methods are widely employed to select optimal materials and composites. Material selection plays a crucial role in the fabrication of epoxy composites for structural applications. Numerous researchers evaluate different properties of fabricated composites and employ various MCDM and Multiple Attribute Decision Making (MADM) techniques to rank the composites [4]. One of the extensively used decision-making tools is the Analytic Hierarchy Process (AHP) introduced by T.L. Saaty [5]. AHP combines mathematics and psychology to address complex decisions. In this study, the AHP method is utilized to calculate the weightage of criteria, which is then employed in the MOORA method to obtain the final ranking of epoxy composites.The AHP-MOORA hybrid model has proven to be effective in enhancing the quality of public transportation services [6].

Kalpesh and Mania [7] utilized the AHP-MOORA approach to select the optimal wire cut electrical discharge machining process parameters for cutting EN31 alloy steel with brass wire. Their results indicated the satisfactory and practical applicability of the hybrid AHP-MOORA method. Vendor selection, which is crucial for meeting customer satisfaction, has also benefited from the AHP-MOORA technique [8]. Utama et al. [9] employed an integrated AHP-MOORA approach in the textile industry for the selection of green suppliers, utilizing eight criteria and 15 sub-criteria to assess and choose the most suitable suppliers. Sonawane and Wangikar [10] stated that the integrated AHP-MOORA method can be utilized for optimizing Wire Electrical Discharge Machining (WEDM) parameters. The AHP method determines the relative weights, while the MOORA method determines the most optimal cutting conditions. Senthil et al. [11] utilized the AHP-MOORA approach to select wire-cut electrical discharge machining process parameters. Additionally, the AHP-MOORA approach can be applied to identify the best set of machining variables that enhance overall machining performance, surface quality, and minimize tool wear [12].

Kumar et al. [13] proposed a hybrid AHP-MOORA method for optimizing brake friction composites. The AHP determined the weighted priorities of selected attributes, while the MOORA approach ranked the friction composites. Taski and Gorener [14] implemented the AHP-MOORA technique in the medical tourism sector, where the AHP approach determined the relevance of various factors and the multi-objective optimization based on ratio analysis (MOORA) was utilized for ranking alternative solutions. Satoglu and Turkekul [15] employed the AHP-MOORA approach for selecting material handling equipment. Gadakh et al. [16] applied the MOORA method to solve multiple function optimization problems in welding. Kumar and Nazir [17] provides a comprehensive understanding of the role of AI-based smart sensors in smart cities, with a specific focus on their assessment using Analytic Hierarchy Process (AHP) and Multi-Objective Optimization on the Basis of Ratio Analysis (MOORA). The integration of these methodologies presents a robust approach to evaluating the effectiveness of smart sensors in contributing to the development of intelligent urban environments.

Singh et al.[18] conducted a thorough investigation into the material selection process for AA7075/SiC composite foam, employing the COPRAS (Complex Proportional Assessment) technique. The AA7075/SiC composite is renowned for its favorable mechanical properties and widespread use in aerospace and automotive industries. The COPRAS technique, classified as a Multi-Criteria Decision-Making (MCDM) method, is strategically employed to address the intricate nature of material selection, taking into account various criteria such as strength, weight, and cost. Lokhande et al.[19] focus on the numerical simulation and experimentation of an endodontic file, a critical component in dentistry. The study utilizes Taguchi Design of Experiments (DoE) in conjunction with numerical simulations to optimize the design parameters of the endodontic file.

2 Methodology

2.1 AHP Method

Here are the step-wise instructions for applying the Analytic Hierarchy Process (AHP) methodology:

1. Define the decision problem: Clearly define the decision problem and identify the criteria and alternatives that will be considered in the decision-making process.

2. Develop a decision hierarchy: Create a decision hierarchy that shows the relationship between the decision problem, criteria, sub-criteria, and alternatives. The decision hierarchy should be structured in a way that allows for easy evaluation of the alternatives based on the criteria [17].

3. Pair-wise comparisons: Perform pair-wise comparisons to evaluate the relative importance of each criterion and sub-criterion in relation to the others. This is done by comparing each element in the hierarchy with every other element using a scale ranging from 1 to 9, with 1 meaning the two elements are equally important, and 9 meaning one element is much more important than the other. [20]. The table is a square matrix where the elements represent the pairwise comparisons between criteria or alternatives. The entries in the matrix are numerical values that reflect the relative importance or preference of one element over another. The scale used for these comparisons is typically a series of values such as 1, 3, 5, 7, and 9, where:

1: Equal importance

3: Slightly more important

5: Moderately more important

7: Strongly more important

9: Extremely more important

4. Construct a priority matrix: Use the pairwise comparison data to construct a priority matrix that shows the relative weights of each element in the hierarchy.

5. Calculate consistency: Check the consistency of the pairwise comparisons by calculating the consistency index (CI) and the consistency ratio (CR) of the priority matrix. If the CR is greater than 0.1, the pairwise comparisons need to be revised [21].

6. Calculate final priorities: Calculate the final priorities for each element in the hierarchy by multiplying the priority weights from the priority matrix by the priority weights of the higher-level criteria.

7. Sensitivity analysis: Conduct a sensitivity analysis to evaluate the robustness of the results by varying the weights of the criteria and sub-criteria.

8. Make a decision: Use the final priorities to make a decision based on the decision problem and alternatives.

By following these steps, the AHP methodology provides a structured approach to decision-making that can help decision makers to systematically evaluate complex decision problems and make more informed decisions.

2.2 MOORA Method

The following are the steps of the MOORA method.

Step 1: The construction of a paired comparison matrix

$$Y_{ij} = \begin{bmatrix} y_{11} & y_{12} & \cdots & y_{1n} \\ y_{21} & y_{22} & \cdots & y_{2n} \\ \vdots & \vdots & \vdots & \vdots \\ y_{m1} & y_{m2} & \cdots & y_{mn} \end{bmatrix}$$

Step 2: Normalization of the comparison matrix.

MOORA is a ratio-based system where the alternative value on a criterion is compared to the denominator, which represents all alternative criteria. The denominator for each alternative's value for each criterion is the sum of the squares of all alternatives.

Step 3: Multi-objective optimization value calculation

$$y_i = \sum_{j=l}^{g} x_{ij} - \sum_{i=g+1}^{n} x_{ij}$$

Step 4: Determine the alternative rank based on the MOORA calculation findings.

The suitable option in the rank with the highest value is ranked first, and the alternative with the lowest value is ranked last.

3 Results and Discussion

The AHP (Analytic Hierarchy Process) algorithm is used to establish the importance or priority order of the criteria that determine performance. Table 1 presents the pair wise comparison matrix between the selected criteria, which includes Hardness, Tensile strength

(T.S.), Flexural strength (F.S.), Compressive strength (C.S.), density, and Water absorption (W.A.) [4]. Table 2 displays the normalized matrix and the priority values assigned to each criterion.

Table 1 Pair-wise comparison matrix among parameters

Criteria	F.S.	C.S.	T.S.	Density	W.A.	Hardness
F.S.	1	2	1/3	7	1/3	3
C.S.	1/2	1	1/5	3	1/3	1/3
T.S.	3	5	1	6	2	4
Density	1/7	1/3	1/6	1	1/5	1/3
W.A.	3	3	1/2	5	1	3
Hardness	1/3	3	1/4	3	1/3	1

Table 2 Normalized matrix

Criteria	F.S.	C.S.	T.S.	Density	W.A.	Hardness	Priority values
F.S.	1	2	0.33	7	0.33	3	0.167
C.S.	0.5	1	0.2	3	0.33	0.33	0.071
T.S.	3	5	1	6	2	4	0.368
Hardness	0.33	3	0.25	3	0.33	1	0.102
Density	0.14	0.33	0.17	1	0.2	0.33	0.036
W.A.	3	3	0.5	5	1	3	0.256
			CR= 6.6%				

Table 2 presents the pairwise comparison matrix for the selected parameters, where the priority values are displayed in the eighth column. Additionally, priority values can be calculated for all parameters by formulating the pairwise comparison matrix between the different alternatives, using other alternatives. According to the eighth column of Table 4, the order of relative weights is Tensile strength (T.S.) > Water absorption (W.A.) > Flexural strength (F.S.) > Hardness > Compressive strength (C.S.) > Density. The Consistency Ratio (CR) is calculated to be 0.066, which is less than 0.1.

Table 3 represents the results of the prepared composites. After calculating the individual weights for the responses, the weighted normalized values will be calculated using equation 1. The normalized matrix is shown in Table 4, and the weighted normalized matrix is presented in Table 5.

Table 3 Results of prepared composites

Criteria	F.S. (MPa)	C.S. (MPa)	T.S. (MPa)	Density	W.A.	Hardness
EPO+40KSD	40.24	108.78	28.34	1.44	0.81	94
EPO+30KSD	46.96	103.63	29.97	1.37	0.56	87
EPO+20KSD	60.55	99.63	31.52	1.3	0.49	82
EPO+10KSD	49.22	91.45	28.9	1.23	0.45	80

Table 4 Normalized matrix

Critiria	F.S.	C.S.	T.S.	Density	W.A	Hardness
EPO+40KSD	0.40	0.53	0.47	0.54	0.68	0.54
EPO+30KSD	0.47	0.51	0.5	0.51	0.47	0.50
EPO+20KSD	0.60	0.49	0.53	0.48	0.41	0.48
EPO+10KSD	0.49	0.45	0.48	0.46	0.37	0.46

Table 5 Weighted normalized decision matrix

Criteria	F.S.	C.S.	T.S.	Density	W.A	Hardness
EPO+40KSD	0.067	0.038	0.175	0.019	0.174	0.055
EPO+30KSD	0.078	0.036	0.185	0.018	0.121	0.051
EPO+20KSD	0.101	0.035	0.195	0.017	0.105	0.049
EPO+10KSD	0.082	0.032	0.179	0.016	0.097	0.047

The result is displayed in Table 6.

Table 6 Multi-objective optimization value (***yi***)

Critiria	Max	Min	yi = Max - Min
EPO+40KSD	0.354	0.174	0.18
EPO+30KSD	0.368	0.121	0.247
EPO+20KSD	0.397	0.105	0.292
EPO+10KSD	0.356	0.097	0.259

The Ranking results of composite are shown in Table 7.

Table 7 Alternative raking

S.No.	Critiria	y_i	Rank
4	**EPO+40KSD**	0.18	**4**
3	**EPO+30KSD**	0.247	**3**
2	**EPO+20KSD**	0.292	**1**
1	**EPO+10KSD**	0.259	**2**

4 Conclusion

This research article presents a study on the selection of an epoxy /KSD composite for structural applications using a hybrid AHP-MOORA approach. The aim was to rank six alternatives by first weighting the criteria using the AHP method and then evaluating the alternative experiments using the MOORA technique. The results obtained through the hybrid AHP-MOORA method indicate that the alternative sequence 2-1-3-4 was determined, with alternative 2 (EPO+20KSD) receiving the highest score (0.292) and being identified as the optimal choice for structural applications. The research highlights the effectiveness of the MCDM approach as a valuable tool for addressing material selection decision problems. This study is the first to explore the use of this technique for this matrix and filler material combination. Furthermore, the developed model can be applied to other decision-making scenarios involving different numbers of alternatives and criteria. This research introduces a novel opportunity for the repurposing of industrial by-products such as Kota stone dust. It provides an expansive platform for future researchers to delve into various facets of similar composite materials. Potential applications involve exploring alternative industrial waste materials as fillers, considering polymeric resins beyond epoxy, and incorporating natural fibers to create innovative hybrid composites, wherein Kota stone dust serves as the primary filler material.

References

1. Shejkar, S. K., Agrawal, A., & Agrawal, B. (2020). Walnut Shell Particulates as Filler Material in Polymeric Matrix: A Review. International Journal of Engineering Research in Current Trends. 2(3). 41-43.

2. V. Makwana, S. Ahmad Qureshi, IJSRD - International Journal for Scientific Research & Development. 2018 | ISSN (online): 2321-0613.

3. H. Awad, R El-gamasy, A. A. Abd El-Wahab, M. H. Abdellatif, Construction and Building Materials, 228 (2019) 116766.

4. Virendra Rajput, Neelesh Kumar Sahu, Alok Agrawal, Optimization of waste kota stone dust filled epoxy composite by analytical hierarchy process (AHP) approach, Materials Today: Proceedings, 2021,,ISSN 2214-7853, https://doi.org/10.1016/j.matpr.2021.09.168.

5. V. Rajput, D. kumar, A. sharma, S. Singh , Rambhagat, lasted journal for advanced research in applied sciences ,5 2018 , 2394-8442.

6. Moslem, S., Çelikbilek, Y. An integrated grey AHP-MOORA model for ameliorating public transport service quality. Eur. Transp. Res. Rev. 12, 68 (2020). https://doi.org/10.1186/s12544-020-00455-1.

7. Jaksan D. Patel, Kalpesh D. Maniya, Application of AHP/MOORA Method to Select Wire Cut Electrical Discharge Machining Process Parameter to Cut EN31 Alloys Steel with Brasswire, Materials Today: Proceedings, Volume 2, Issues 4–5, 2015, Pages 2496-2503, ISSN 2214-7853, https://doi.org/10.1016/j.matpr.2015.07.193.

8. Khoiry, I'tishom & Gernowo, Rahmat & Surarso, Bayu. (2021). Fuzzy-AHP MOORA approach for vendor selection applications. Register: Jurnal Ilmiah Teknologi Sistem Informasi. 8. 24. 10.26594/register.v8i1.2356.

9. Utama, D. M., Asrofi, M. S., & Amallynda, I. (2021, June). Integration of AHP-MOORA algorithm in green supplier selection in the Indonesian textile industry. In *Journal of Physics: Conference Series* (Vol. 1933, No. 1, p. 012058). IOP Publishing.

10. Sonawane, S. & Wangikar, Sandeep. (2021). Optimization of WEDM Parameters During Machining of Ni-75 Using AHP-MOORA Method. 10.1007/978-3-030-69921-5_102.

11. Senthil, P., Vinodh, S., and Singh, A. K. (2014). Parametric optimisation of EDM on Al-Cu/TiB2 in-situ metal matrix composites using TOPSIS method, International Journal of Machining and Machinability of Materials, 16(1), 80-94.

12. Debnath, S., & Ghosh, S. (2021). Experimental investigation of Electro discharge machining process by AHP-MOORA technique. Journal of Industrial Engineering and Decision Making, 2(1), 1-7. https://doi.org/10.31181/jiedm200201001d.

13. Kumar, Naresh & Fekete, Gusztáv & Singh, Tej & Grewal, J.S.. (2019). Natural Fiber Reinforced Brake Friction Composites: Optimization Using Hybrid AHP-MOORA Approach. AIP Conference Proceedings. 2142. 150015. 10.1063/1.5122564.

14. Görener, Ali & Taşçı, Emre. (2016). Evaluating the strategies of medical tourism sector through AHP and MOORA in a SWOT framework. Journal of Global Strategic Management. 2. 15-15. 10.20460/JGSM.20161024353.]

15. Satoglu, S. I., & Türkekul, İ. (2021). Selection of Material Handling Equipment using the AHP and MOORA. *Jurnal Teknik Industri*, *22*(1), 113-124.

16. V. Gadakh, V. B. Shinde, N. S. Khemnar (2013) Suitable welding parameters selection indifferent weldingProcesseswithMOORAmethod.InternationalJournalofAdvancedTechnology69:9-12.

17. Perez-Vega, S., et al., Analytical hierarchy processes (AHP) for the selection of solvents in early stages of pharmaceutical process development. Process Safety and Environmental Protection, 2011. 89(4): p. 261-267.

18. Singh, N. K., S. Balaguru, R. K. Rathore, A. K. Namdeo, and A. Kaimkuriya. "Multi-Criteria Decision-Making Technique for Optimal Material Selection of AA7075/SiC Composite Foam Using COPRAS Technique". *Journal of Mines, Metals and Fuels*, Nov. 2023, https://www.informaticsjournals.com/index.php/jmmf/article/view/34005.

19. Pravin R. Lokhande, Sachin S. Salunkhe, Balaguru Sethuraman, Corrigendum to: Numerical Simulation and Experimentation of Endodontic File using Taguchi DoE, Int. J. Simul. Multidisci. Des. Optim. 13, 11 (2022).

20. Dağdeviren, M., S. Yavuz, and N. Kılınç, Weapon selection using the AHP and TOPSIS methods under fuzzy environment. Expert Systems with Applications, 2009. 36(4): p. 8143-8151.

21. Albayrak, E. and Y.C. Erensal, Using analytic hierarchy process (AHP) to improve human performance. An application of multiple criteria decision making problem. Journal of Intelligent Manufacturing, 2004. 15: p. 491-503.

MATEC Web of Conferences **393**, 01007 (2024) https://doi.org/10.1051/matecconf/202439301007
STAAAR-2023

Fabrication and Characterization of a Stir Casting-Based Aluminium Hybrid MMC Reinforced with SiC, TiC, and MoS_2

M Arun [1*], *K* Ragupathy[1], *T* Anand[1] and Vishvanathperumal *S*[2]
[1]Agni college of Technology, Chennai, Tamilnadu, India
[2]S.A. Engineering College, Chennai, Tamilnadu, India

Abstract. The need for strong, lightweight materials has prompted the creation of innovative metal matrix composites based on aluminum. The properties of metal matrix composites that are uniformly dispersed with nanoparticles are much superior to those of monolithic alloy and microparticle-reinforced composites. The objective of this work was to create and evaluate a metal matrix composite reinforced with MoS_2, SiC, and TiC that is a hybrid aluminum alloy, Al6061. It was also investigated how the weight percentages (3, 6, 9, and 12%) of MoS_2, SiC, and TiC reinforcement affected the mechanical, morphological, tribological, and physical characteristics of the metal matrix composite. The addition of SiC and MoS_2 increased the density of the reinforced Al6061 composite when compared to as-cast non-reinforced Al6061. It was found that the hybrid composite Al6061/12% SiC/4% MoS_2 had the maximum density. The hybrid metal matrix composite's toughness increased as the proportion of TiC weight increased. The composite made of Al6061, 12% TiC, and 4% MoS_2 had the maximum hardness, measuring 114.03 HV. The composite Al6061/12% TiC/4% MoS_2 has the most ultimate tensile strength. The tribology analysis revealed that when applied stress increased from 10 to 50 N, mass loss increased dramatically. Because of the solid MoS_2 lubricant and the development of the TiC layer at the contact zone, Double- and triple-reinforced specimens had less wear loss than non-reinforced specimens, as shown by the wear performance of hybrid composites. The main wear mechanisms of the composites were delamination wear and wear debris.

Keywords: Silicon Carbide, Molybdenum disulphide, wear behaviour , stir casting method

1 Introduction

High wear resistance, good thermal conductivity and high strength to weight ratio are desirable properties that make MMCs hybrid MMCs are popular in various industries including medical, electronics, automotive, aerospace and marine. MMCs have both the metallic characteristics of matrix alloys, like toughness and ductility, and the ceramic features of reinforcements, including high modulus and high strength. Greater shear, tensile, and compression resistance as well as service temperature capabilities result from the combination of these qualities. Compared to microparticle-reinforced composites and monolithic alloys, uniformly dispersed nanoparticle-reinforced MMCs show significantly superior characteristics. However, the manufacturing of nanocomposite materials is complicated and costly due to a number of fundamental challenges, including dispersion, particle-matrix interface debonding and wettability. Among the various types of metal alloys,

* M. Arun: amarun100@gmail.com

aluminum alloys are the most favored matrix materials due to their high corrosion resistance, good thermal and electrical conductivity, lightweight nature, and effective conductivity. Novel aluminum-based MMCs have been developed through a variety of processes in response to the need for lightweight materials with high strength. In their assessment of the growth and production of MMCs in the foundry sector over the last 50 years, Kumar et al.17 covered a variety of reinforcement kinds and processing methods. In general, the reinforcing phase, also known as reinforcement, has the following properties: low density, high thermal stability, high Young's modulus, compression, and tensile strength, as well as good mechanical and chemical compatibility.18. When it comes to metal matrix composites, stir casting is the most widely used production method because it is inexpensive and suitable for mass producing machine components. This is in contrast to other fabrication techniques like powder metallurgy, which necessitate considerable handling of powder. The process of stir casting is typically employed to create MMCs that are reinforced with particles. Researcher after researcher has proposed that adding SiC, TiC, and MoS_2 to the alloy as single reinforcement or hybrid reinforcement is one method of enhancing the mechanical and physical properties of monolithic Al6061 alloy. In general, the reinforcing phase, also known as reinforcement, has the following properties: low density, high thermal stability, high Young's modulus, compression, and tensile strength, as well as good mechanical and chemical compatibility.18. When it comes to metal matrix composites, stir casting is the most widely used production method because it is inexpensive and suitable for mass producing machine components. This is in contrast to other fabrication techniques like powder metallurgy, which necessitate considerable handling of powder. The process of stir casting is typically employed to create MMCs that are reinforced with particles. Researcher after research has proposed that adding SiC, TiC, and MoS_2 to the alloy as single reinforcement or hybrid reinforcement is one method of enhancing the mechanical and physical properties of monolithic Al6061 alloy by putting two or more reinforcements together.20, 19. These substances provide outstanding wettability and improve the matrix structure. The tensile and compression behavior of Al6061 MMCs reinforced with TiC after being processed through the number of studies have also examined the wear characteristics of Al6061 MMCs with different types of reinforcements. Mohammad et al.'s study from 24 examined the wear characteristics of aluminum MMCs reinforced with press- and sinter-synthesized SiC and MoS_2. Al6061 10% SiC 2% MoS_2 hybrid MMCs had the coefficient of friction (COF) and lowest wear rate. The friction and wear characteristics of stir-cast Al6061 MMCs reinforced with TiC were examined by Dou et al. (25). The findings demonstrated that when the sliding period, applied load, and sliding velocity approached critical values, the mass loss and COF rose noticeably. Powder metallurgy-produced SiC and TiC were used in hybrid reinforcement of Al6061 to enhance the composite's mechanical and tribological characteristics.26. The wear characteristics of aluminum MMCs reinforced with MoS_2 and SiC, which were produced by press and sintering, were examined by Rouhi et al. 24. What was the lowest COF and wear rate for Al6061? 10% SiC 2% hybrid MoS_2. MMCCs. The primary cause of wear was found to be mild delamination. Shinde and colleagues observed the impact of different TiC levels (0.5, 1, 1.5, and 2 weight percent) on the scratch resistance and hardness of aluminum metal micron composites. 27. The hardness of the composites increased by 26.59% at 2 weight percent TiC, and at a scratch load of 5 N, the scratch quality improved by 73%. The wear resistance of the aluminum MMCs was shown to increase with an increase in TiC concentration.28. By integrating TiC, MoS_2, and SiC reinforcements, which have good physical, chemical, mechanical, and tribological properties, the qualities of the base alloy, aluminum Al6061, can be improved. A comprehensive examination of the previously mentioned literature clearly identifies a number of investigations on the mechanical, tribological, and physical characteristics of composites made of Al6061 with a single reinforcement. Nevertheless, little to no research was done to comprehend TiC/SiC/

MoS_2-reinforced Al6061-based hybrid MMCs using the stir casting method. This work attempts to create and describe Al6061-based hybrid MMCs reinforced with TiC/SiC/ MoS2. Tensile strength, porosity, hardness, and density (both real and theoretical) were assessed for the hybrid composites. Optical microscopy was also used for microstructural investigation.

2 Materials and methods

2.1 Materials

Materials employed in this work to construct hybrid MMCs included Al6061 alloy (matrix) and TiC, SiC, and MoS2 (reinforcement material). A great option for a reinforcing material, SiC has high levels of hardness, toughness, toughness, chemical and thermal stability, and resistance to oxidation and erosion.29, remarkable hardness, higher melting point, low density, excellent mechanical properties, and chemical resistance characterize TiC ceramic. High melting point, outstanding hardness, remarkable chemical and thermal stability, and strength and resilience characterize MoS2.30. Table 1 displays the chemical composition of the Al6061 alloy as measured by optical emission spectroscopy. The primary alloying elements are silicon and magnesium, which increase the alloy's casting and wetting when mixed with reinforcement. The base aluminum alloy 6061 was created by melting scrap Al6061 material.

2.2 Fabrication

The stir casting method was used to fabricate hybrid aluminum Al6061 MMCs. First, scrap aluminum alloy was placed in the crucible and melted at 750 degrees Celsius. To lessen clustering, the reinforcements were applied in two stages (50:50) and cooled to 250°C. To guarantee that the reinforcements were distributed uniformly throughout the matrix material, mechanical stirring was done during the reinforcement addition process. The stirrer shaft was driven at 600 rpm by a hand drill that was portable. ZrO2 was applied to the blade, and it was preheated to 200°C. A vortex was created and the reinforcements were evenly mixed throughout the matrix material using two stirrers with blades. As a degassing agent, 2% NaCl powder was utilized. After mechanical stirring, this degassing agent was applied to the molten Al alloy and reinforcement powder combination to reduce the gas concentration. The matrix and reinforcement mixture was poured into the 25 mm diameter by 150 mm length mold cavity after mechanical stirring and degassing, as indicated in Figure 1. The mixture of melting aluminium Al6061 alloy along with reinforcements was poured into the casting container (Figure 2) after it had been preheated to 200°C. The MMC composite samples have been removed out of the sand mould by departing from the mould apart following casting.

2.3. Theoretical density

Equation 1 illustrates the application of the rule of mixtures to determine the theoretical density (ρth) of hybrid aluminum Al6061 MMCs. Where ρth is the composite specimen's theoretical density. The Archimedes approach was used to calculate the composite's real density.(31) The cylindrical specimen was subjected to weight measurements both in distilled water (m_{water}) and the air (m_{air}). Equation 2 was used to determine the actual/experimental density.

$$\frac{1}{\rho^{th}} = \frac{wt\%Al6061}{\rho_{Al6061}} + \frac{wt\%B_4C}{\rho_{B_4C}} + \frac{wt\%SiC}{\rho_{SiC}} + \frac{wt\%MoS_2}{\rho_{MoS_2}} \quad Eqn.\ 1$$

$$\rho_a = \frac{m_{air}}{m_{air} - m_{water}} \rho_{water} \quad Eqn.\ 2$$

2.4 Vickers hardness test.

A 98 N load was used to conduct the Vickers hardness test. The average value was given after three measurements were made in various locations. Standard test cylindrical specimens for the tensile test, according to ATSM E832, were cut from the test castings. They had a diameter of 8 mm, a fillet radius of 6 mm, and a length of 100 mm. The assessments were carried out using a computerized.

2.5 Wear Test

Using a computerized TR-20-MICRO type micro pin-on-disc tribometer, the dry sliding wear tests were carried out in accordance with ASTM G99 guidelines. 10, 30, and 50N were the three loads that were used. The test parameters were constant wear track diameter (80 mm), disc rotation (200 rpm), sliding distance (251.2 m) and test time (5 min). The created composite specimen was processed for tribological analysis in accordance with ASTM G-99 guidelines. An electronic weighing device with an accuracy of ± 0.1 mg was employed to calculate the wear loss.

2.6 Microstructural Test

A hybrid aluminum Al6061 alloy MMC's metallurgical characteristics were investigated using an optical microscope. The samples were ground using sandpaper with grain sizes of 400, 600, 800, 1000, and 1200 after being cut to 20 mm in diameter and 10 mm in height. Additionally, it was etched using Keller's chemical etchants (190 ml distilled water, 2 ml hydrofluoric acid, 3 ml hydrochloric acid, and 5 ml nitric acid,) and smoothed through a two-wheel the system RB 204 Metpol-II simulation polishing equipment with diamond suspension. An optical microscope from the Huvitz HR-300 series type was used to perform the macrostructure test. Using a scanning electron microscope (SEM), the wear morphology of Sample 1, Sample 2, and Sample 4 at a 50 N load was examined following the wear test.

3. Results and Discussion

3.1 Microstructural Study

The optical micrograph of hybrid aluminum Al6061 alloy MMCs (Sample 2–Sample 6) and Al6061 (Sample 1) at varying reinforcement levels is displayed in Figure 3. The macrostructure demonstrated that the reinforcing particles were evenly dispersed throughout the aluminum Al6061 alloy base material. The efficiency of the vortex movement and the use of a suitable composite production process were credited for this. Owing to the identical quantity of reinforcements in each composite, the reinforcements were observed to cluster in all samples (Sample 2–Sample 6). Size, quantity, and stirring duration during the stir casting process all affect how much clustering occurs. Eleven percent (TiC and SiC) and four percent (MoS_2 of the reinforcement in samples Sample 2–Sample 6 totals sixteen percent. Owing to the same quantity of reinforcement, the clustering in the entire sample is nearly identical.

3.2. Physical behavior of Hybrid MMC Composites

Figure 4 shows how dense the hybrid composite is. Sample 3 had the hybrid composite's maximum theoretical density, as observed. For the produced hybrid aluminum Al6061 alloy MMCs, the theoretical density was Sample 3 > Sample 4 > Sample 6 > Sample 5 > Sample 2 > Sample 1 in that sequence. The reduction in low-density TiC reinforcing content and the rise in high-density SiC particle content were cited as the causes of this. The manufactured hybrid aluminium Al6061 alloy MMCs had an actual density in the following order: Sample 3 > Sample 4 > Sample 5 > Sample 6 > Sample 2 > Sample 1.

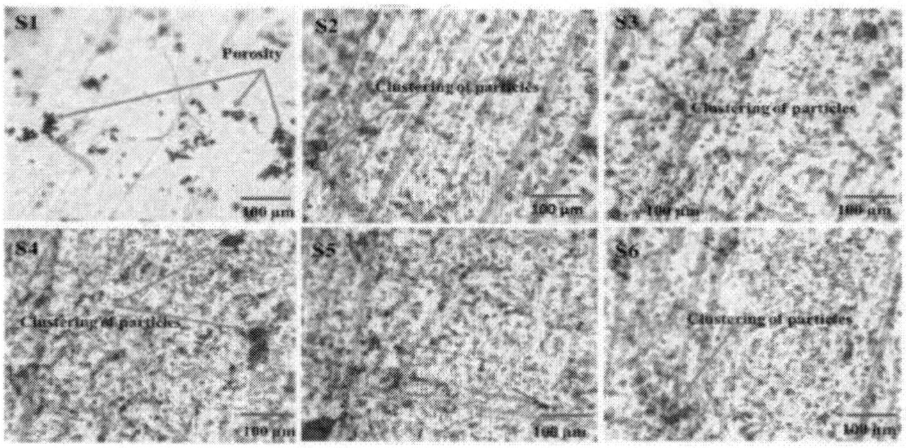

Fig. 1. Optical micrographs of the hybrid MoS_2 Composites.

Within the range of 1.4% to 3.4%, the produced hybrid composites had porosity (Figure 4). Porosity values in this range are considered appropriate for cast composites, according to the research.34 Owing to an increase in voids in the area surrounding the SiC reinforcement, the porosity level increased along with the SiC reinforcement. Given its magnitude in relation to other reinforcements, the explanation for the increased vacancy in the region of SiC is its size. As for Sample 3 hybrid MMC (12% SiC), the maximum porosity was noted. The lowest impact on porosity was observed with reinforcement TiC. Thus, stir casting is a productive method technique for creating composites with hybrid aluminum alloys. When binary and ternary reinforcements are added, the composites' low porosity and few casting faults are produced.

3.3 Mechanical behavior of hybrid MoS_2 Composites

Table 4 displays the UTS and percentage elongation of the manufactured hybrid aluminum Al6061 alloy MMCs. Al6061 that was not strengthened had the lowest UTS (76.66 MPa). The hybrid aluminum alloy Al6061/12% TiC/4% MoS_2 has the greatest UTS value (94.01 MPa) .The robust interface bonding between the matrix phase and reinforcements was ascribed to this. UTS is raised by the interface bonding, which shifts the load from the matrix to the reinforcements. Other researchers saw similar things.34, 35 the addition of 4% MoS_2 and 12% TiC increased UTS by 22.63%. In comparison to the non-reinforced Al6061 alloy, the tensile strength of the hybrid composites Sample 2, Sample 5, Sample 6, Samle4, and Sample 3 rose by 22.63%, 19.73%, 15.21%, 13.28%, and 3.74%, respectively. When reinforcements were added, the ductility of hybrid MMCs reduced. Figure 6 shows that

Al6061 reinforced with 9% TiC/3% SiC/4% MoS_2 had the lowest percentage elongation. Figure 7 displays the hardness value of the MMCs and cast alloy. The hardness of the aluminum Al6061 alloy matrix rose with the addition of reinforcements. Because of the incorporation of hard TiC particles and their superior wettability, the hardness of hybrid MMCs improved as the percentage of TiC rose. The non-reinforced Al alloy hybrid composite Sample 2, Sample 3, Sample 4, Sample 5, and Sample 6 showed an improvement in hardness of 28.31%, 16.9%, 16.94%, 20.06%, and 20.20%. From hardest to softest, the hybrid MMCs' hardness was ranked Sample 2 > Sample 5 > Sample 6 > Sample 4 > Sample 3 > Sample 1. Compared to aluminum 6061 alloy, sample 2 had a higher weight percentage of TiC and a comparable density. As a result, during stirring, it was blended more uniformly than SiC and MoS_2, which have larger densities. According to Halil et al. (26), TiC had an impact on the production of TiC. The base alloy and reinforcements exhibited increased interfacial bonding as a result of the production of TiC. By putting tension on the matrix, TiC particles in reinforced composite materials raised hardness and UTS.

Table: 1. UTS and Elongation Results of Hybrid MMC Aluminum Al6061 Alloy

Designation	Al6061/TiC/SiC/MoS_2	UTS (MPa)	YS (MPa)	Elongation (%)
Sample 1	Al6061	76.66 ± 2.5	46.45	11.90±0.35
Sample 2	Al6061/12% TiC/SiC/ 4%MoS_2	94.01 ± 4.2	56.50	10.32±0.42
Sample 3	Al6061/12% SiC/ 4%MoS_2	80.27 ± 3.4	48.24	11.43±0.57
Sample 4	Al6061/3% TiC/9%SiC/ 4%MoS_2	88.08 ± 5.3	52.93	10.83±0.44
Sample 5	Al6061/9% TiC/3%SiC/ 4%MoS_2	91.63 ± 6.8	55.62	10.08±0.31
Sample 6	Al6061/6% TiC/6%SiC/ 4%MoS_2	88.98 ± 4.5	53.47	10.82±0.54

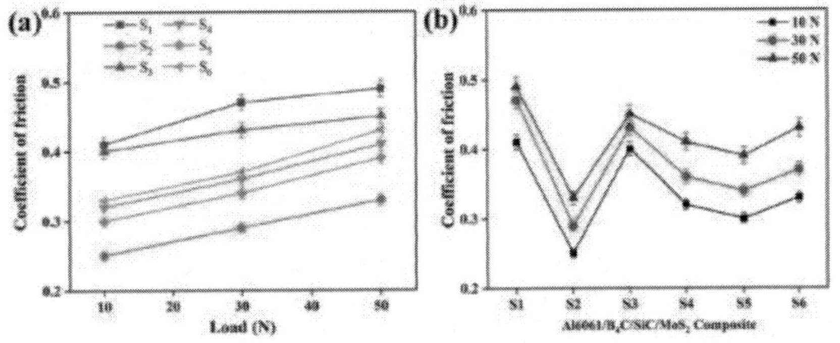

Fig 2. Dry wear loss with (a) applied load (b) weight composition.

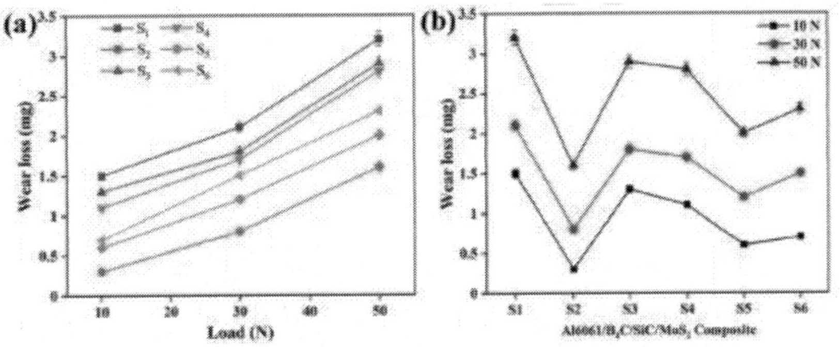

Fig 3. COF of (a) Applied load and (b) Hybrid Al6061 MMCs.

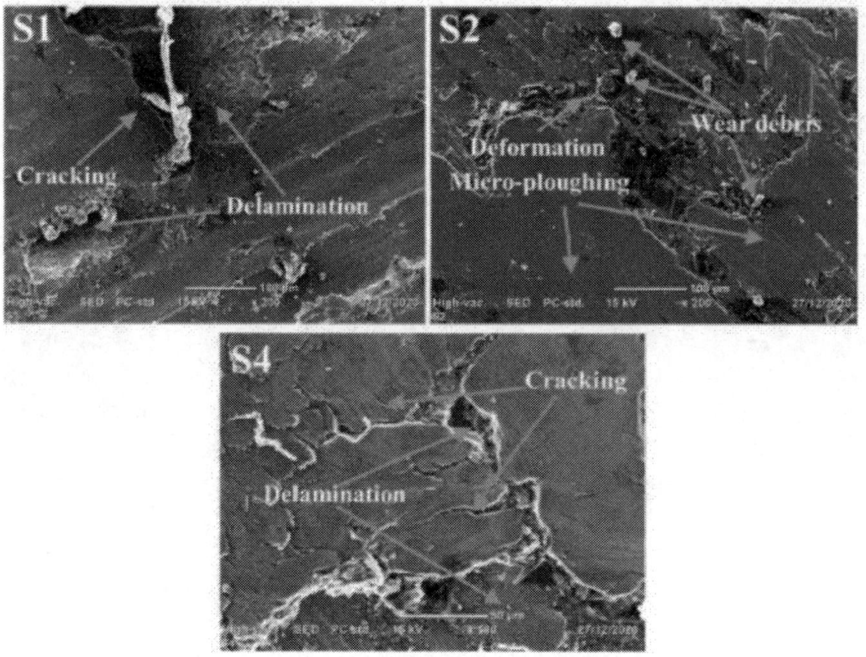

Fig.4. SEM image of dry wear morphology of sample 1, 2 and 4 at 50 N applied load.

3.4 Friction and dry Wear Test

The wear loss (g), the coefficient of friction, and the friction as well as dry wear outcomes of tests on the pin-on-the-disc device were calculated using the manufactured hybrid metal matrix composites (The coefficient of friction raised as the force being applied increased from 10 to 50 N. The rise was brought about by the frictional force rising as the average load increased and the contact area between the specimen's outer layer and the disc's inner surface increased. Within comparison with hybrid MMCs, the non-reinforced Al6061 alloy had a higher COF. The presence of firm lubricant MoS_2, and these created thin coatings or lubricating films throughout the outermost layer throughout dry sliding, was identified as the reason for the reduction in coefficient of friction. Through preventing subsurface contact between sliding surfaces, this thin film that has developed reduces coefficient of friction. due to TiC generated a layer of TiC at the contact zone, Sample 2 had the lowest COF (0.25 at 10 N load).The TiC layer formed when TiC particles had been separated to the composite and interacted with the surroundings as a result of rising temperatures.
The weight composition of the reinforcements and the variance of wear loss towards average applied load are shown in Figure 4. As a result of the larger contact area, dry wear loss increased gradually as the load increased from 10 to 50 N. Deformation, fatigue, along with a rise in the temperature followed, which resulted in the deformation, propagation, as well as formation cracks as well as the pull-out of wear debris and reinforcement particles. The resistant to dry wear of the material is negatively impacted by the increase in wear debris.The lubricating the multiple layers tear down and composite surfaces get ploughed due to the accumulation of dry wear debris. The non-reinforced sample decreased 3.2 mg at 50 N applied load, while the double-reinforced the sample (Al6061/12% TiC/4% MoS_2) decreased 1.6 mg at the lowest weight. Therefore, while reinforced, the dry wear resistance of Al6061 alloy increased by 68.42% at 50 N applied load. Improved samples performed the best with respect to of composite dry wear, with the most damaging specimens being Sample 2 > Sample 5 > Sample 6 > Sample 4 > Sample 3 > Sample 1.

3.5 Wear Mechanism

SEM was used for examining the wear morphology of Sample 1, sample 2, and sample 4 at 50 N applied load (Figure 10). MoS_2 is a solid lubricant. This small tribo-layer developed in the outermost layer to be the result of having the solid lubricant MoS_2, and these reduced the Coefficient of friction throughout sliding around. As a result, there was less chance of an adhesive wear mechanism and tribological the interaction on the hybrid metal matrix composites. Abrasive wear along with exterior worn out wear are potential wear mechanisms because abrasion wear typically occurs in tribological structures in which the tribo-elements show widely disparate the hardness parameters.
Wear mechanisms such as delamination wear; pull-out, deformation, ploughing, cracking, and wear debris were observed. The hybrid Aluminium metal metal matrix composite's fundamental wear mechanisms had been use delamination as well as wear debris. This tribo-layer splits into smaller pieces or disintegrates through additional wear debris and appears ploughing, and these are facilitated by the wear debris created when two in opposition areas rub against one another. Hence, wear caused by abrasive substances increases by this mechanism. The proposed fabricated hybrid Al6061 Metal matrix composite's primary wear mechanisms have been found to include abrasive and the surface fatigue wear. Less wear debris, ploughing, along with delamination was seen in the direction of sliding when reinforcements were present. Hard wear-resistant particles (SiC) and solid lubricant reinforcements (MOS_2) were a result of this issue.

4. Conclusion

In the present study, stir casting technique was used to create hybrid Al6061 MMCs with different percentages of weight of MoS2, SiC, and TiC. The Al6061 hybrid MMCs' their microstructure, as well as the mechanical and physical properties were evaluated. The main findings from the current testing and evaluation are as follows: Because of the high-density silicon carbide reinforcements in Sample 3 hybrid metal matrix composite (12% SiC), the hybrid MMC composite density was most prevalent, and because of the low in density TiC reinforcements for structures in Sample 2 hybrid MMC composite (12% TiC), the hybrid composite the density proved lowest level. In the present study, stir casting technique was used to create hybrid Al6061 MMCs with different percentages of weight of SiC, MoS_2, and TiC. This produced hybrid MMCs was a density of the order Sample 3 > Sample 4 > Sample 5 > Sample 6 > Sample 2 > Sample 1. The three have been found to have the greatest porosity. Because that was more void within the region around the SiC reinforcement in the Sample 3 hybrid composite (12% SiC) and Sample 4 hybrid composite (9% SiC), the UTS value was an indication of the weight percentage of SiC. The produced hybrid composites had porosities ranging from 1.5% to 3.5%, with 5 being the lowest and the highest possible values. For the Sample 2 hybrid MMC composite, the lowest the coefficient of friction is noted. From best to the very least the composite specimen's wear performance was as follows: Sample 2 > Sample 5 > Sample 6 > Sample 4 > Sample 3 > Sample1. Micro ploughing, Cracking, delamination, along with wear debris on the composite the sample are revealed by the scanning electron microscope (SEM) examination of the dry wear track. Because of the solid lubricant MoS_2, the double- and triple-reinforced samples had less wear reduction in comparison to the non-reinforced samples.

References

1. Garg, P.; Jamwal, A.; Kumar, D.; Sadasivuni, K.K.; Hussain, C.M.; Gupta, P. Advance research progresses in aluminium matrix composites: Manufacturing & applications. J. Mater. Res. Technol, 8, 4924–4939. (2019)
2. Samal, P.; Vundavilli, P.R.; Meher, A.; Mahapatra, M.M. Recent progress in aluminum metal matrix composites: A review on processing, mechanical and wear properties. J. Manuf. Processes, 59, 131–152. (2020)
3. Mahesh, K.V.; Venkatesh, C.V. A comprehensive review on material selection, processing, characterization and applications of aluminium metal matrix composites. Mater. Res. Express, 6, 072001. (2019)
4. Pai, B.C.; Ramani, G.; Pillai, R.M.; Satyanarayana, K.G. Role of magnesium in cast aluminium alloy matrix composites. J. Mater. Sci. 1995, 30, 1903–1911.
5. Subba, R.E.; Ramanaiah, N. Experimental Study on Mechanical Properties of Aluminium Metal Matrix Composites (AA 6061 Reinforced with MoS_2). Int. J. Mater. Sci. 6, 91–99. (2016)
6. Kumar, S.S.; Pandian, R.S.; Pitchipoo, P.; Rajakarunakaran, S.; Rajesh, S. Investigation of Al-Mg based composite incorporated with MoS_2 through powder metallurgy. Proc. Inst. Mech. Eng. Part E J. Process Mech. Eng., 235, 986–996. (2021)
7. J.F.Zheng , D. Pieniak and J.L.Mo,studied "Study on rotational fretting wear of bonded MoS_2 solid lubricant coating prepared on medium carbon steel", archives of civil and mechanical engineering, Vol.226, pp.1-8. (2011)
8. Ramadoss, N., Pazhanivel, K., Ganeshkumar, A. et al. Effect of SiC and MoS_2 Co-reinforcement on Enhancing the Tribological and Anti-Corrosive Performance of Aluminum Matrix (Al6063-T6) Nanocomposites. Silicon 14, 5471–5479 (2022). https://doi.org/10.1007/s12633-021-01334-7.
9. Gonfa BK, Sinha D, Vates UK, Badruddin IA, Hussien M, Kamangar S, Singh GK, Ahmed GMS, Kanu NJ, Hossain N. Investigation of Mechanical and Tribological Behaviors of Aluminum Based Hybrid Metal Matrix Composite and Multi-Objective Optimization. Materials (Basel). 2022 Aug 16;15(16):5607. doi: 10.3390/ma15165607. PMID: 36013744; PMCID: PMC9413696.
10. Ragupathy, K., Velmurugan, C., Ebenezer Jacob Dhas, D.S. et al. Prediction of Dry Sliding Wear Response of AlMg1SiCu/Silicon Carbide/Molybdenum Disulphide Hybrid Composites Using Adaptive Neuro-Fuzzy Inference System (ANFIS) and Response Surface Methodology (RSM). Arab J Sci Eng 46, 12045–12063 (2021).
11. Velmurugan, C.; Muthukumaran, V.; Ragupathy, K., et al.: Modelling Volume loss of heat treated Al6061 composites using ANN. Proced. Mater. Sci 5, 31–40 (2014). https://doi.org/10.1016/j.mspro.2014.07.239.
12. Gajalakshmi, K.; Senthilkumar, N.; Prabu, B.: Multi-response optimization of dry sliding wear parameters of Al6026 using hybrid gray relational analysis coupled with response surface method. Meas. Cont. 52(5–6), 540–553 (2019). https://doi.org/10.1177/0020294019842603
13. Hamidreza, G.; Mohd, H.I.; Norhayati, A.; Navid, M.: Microstructure development mechanical and tribological properties of a semisolid A356/xSiCp composite. J. Appl. Res. Technol. 15, 533–544 (2017). https://doi.org/10.1016/j.jart.2017.06.002.
14. Manjunath-Patel, G.C.; Shettigar, A.K.; Mahesh-Parappagoudar, B.: A systematic approach to model and optimize wear behaviour of castings produced by squeeze casting process. J. Manuf. Proc 32, 199–212 (2018). https://doi.org/10.1016/j.jmapro.2018.02.004.
15. Vijayakumar, S.; Karunamoorthy, L.: Modelling wear behaviour of Al–SiC metal matrix composites: soft computing technique. Tribol. Mater. Surf. Interfaces 6, 25–30 (2013). https://doi.org/10.1179/1751584X12Y.0000000002

16. Kumar, S.; Balasubramanian, V.: Effect of reinforcement size and volume fraction on the abrasive wear behaviour of AA7075 Al/SiCpP/M composites—a statistical analysis. Tribol. Int. 43(1–2), 414–422 (2010). https://doi.org/10.1016/j.triboint.2009.07.003
17. Baradeshwaran, A.; Vettivel, S.C.; Elayapermal, A., et al.: Experimental investigation on mechanical behavior, modelling and optimization of wear parameters of B4C and graphite reinforced aluminum hybrid composites. Mater. Design 63, 620–632 (2014). https://doi.org/10.1016/j.matdes.2014.06.054.
18. Vinoth, K.S.; Subramanian, R.; Dharmalingam, S., et al.: Mechanical and tribological characteristics of stir-cast Al–Si10Mg and self-lubricating Al–Si10Mg/MoS_2 composites. Mater. Tech 46(5), 497–501 (2012).
19. Singh, N.K., Sethuraman, B. Development and Characterization of Aluminium AA7075 Hybrid Composite Foams (AHCFs) Using SiC and TiB2 Reinforcement. Inter Metalcast 18, 212–227 (2024). https://doi.org/10.1007/s40962-023-01009-6
20. Singh, N.K., Sethuraman, B. Development and Characterization of Aluminium AA7075 Hybrid Composite Foams (AHCFs) Using SiC and TiB2 Reinforcement. Inter Metalcast 18, 212–227 (2024). https://doi.org/10.1007/s40962-023-01009-6

MATEC Web of Conferences **393**, 01008 (2024) https://doi.org/10.1051/matecconf/202439301008
STAAAR-2023

Hydrothermal synthesis and characterization of selenium-doped MoS_2 for enhanced opto-electronic properties

Soumya Rai[1], *Shreya*[1*], *Peeyush* Phogat[1], *Ranjana* Jha[1] and *Sukhvir* Singh[1]

[1]Research Lab for Energy Systems, Department of Physics, Netaji Subhas University of Technology, New Delhi, India

Abstract. Transition metal dichalcogenides show layered crystal structure with Van der Waals forces spanning the adjacent layers. Robust covalent bonds within each 2D sheet contribute to their outstanding electrical and optical properties. The incorporation of a chalcogen atom as a composite leads to enhanced optical and electrical characteristics of the TMDs, which could make them potential candidates for opto-electronic applications. In the present study, we have described the synthesis and characterizations of Se doped MoS_2 nanolayers using a facile one-step hydrothermal method. XRD analysis revealed the diffraction peaks corresponding to MoS_2 as well as Se which suggested the successful incorporation of selenium into MoS_2 nanolayers. Optical analysis revealed an increased band gap of 1.01 eV after the addition of selenium, determined through UV-visible spectroscopy with a refractive index of 3.36. The morphological analysis by FESEM showed the formation of nanolayers and the EDX pattern showed the presence of Mo, S and Se. The enhanced bandgap of the as-synthesized material and the increased absorption in the UV region ensures its usage for the fabrication of efficient broadband photodetectors. Our future research focuses on the potential application of Se doped MoS_2 nanolayers in the field of efficient photodetectors.

1 Introduction

The pursuit of efficient and adaptable photodetectors has driven the investigation of novel materials and synthesis methods. Due to their extraordinary mechanical, electrical, and optical capabilities, 2D materials such as graphene, TMDs, etc. have received a wide interest in the last decade [1]. A significant number of research on the application of 2D materials has been published in optoelectronics, energy storage, sensors, electronics, and other fields [2]. Semiconductor nanocrystals have been employed to create photoelectric devices recently

* Corresponding author : shreyasharma.aug15@gmail.com

due to their high aspect ratio and tunable band gap [3]. One of the suitable choice among these is molybdenum disulfide (MoS_2), a TMD, due to its extraordinary electrical and optical properties [4]. Previous studies have investigated the large-scale growth of MoS_2 by chemical vapour deposition to fabricate energy storage devices such as supercapacitors [5]. Since CVD method can be complex and requires precise control, more facile and cost-effective methods such as the hydrothermal method is employed to synthesize 2D MoS_2-based thin layers with band gap varying from 1.5 eV to 0.8 eV [6]. Due to more metallic nature of selenium, high sensitivity towards light and chemical stability, its incorporation into the frameworks of MoS_2 has been explored previously to improve electrical conductivity [7].

In the present paper, an effort has been made to synthesise selenium doped MoS_2 nanomaterial by a facile one step hydrothermal method to increase absorption in the ultraviolet region and enhanced electrical conductivity for achieving high efficiency and responsivity of photodetectors.

X-ray diffraction (XRD) technique was employed to determine the crystal structure and lattice parameters of the as-synthesized sample. Debye scherrer method, williamson hall method and the size strain method were employed to determine the crystallite size and strain in the sample respectively. The optical properties and band gap of the material were determined using UV-Vis spectroscopy. The morphology of the sample was examined using a field emission scanning electron microscope (FESEM). Elemental composition of the as-synthesized sample was examined using the energy dispersive x- ray (EDX) technique as an attachment of FESEM.

2 Experimental Section

2.1 Chemicals and Materials Used

Ammonium molybdate tetrahydrate [$(NH_4)_6Mo_7O_{24}.4H_2O$] (>99%), thiourea (CH_4N_2S) (>99%) and Polyvinylpyrrolidone (PVP) (>99%) were purchased from M/s Sigma-Aldrich Pvt. Ltd. Selenium powder (>99%) was procurred from M/s Loba Chemie Pvt. Ltd. Distilled water (>5MΩ) was used for synthesis and cleaning purposes. Absolute ethanol used for cleaning and washing was purchased from Changshu Hongsheng Fine Chemical Co. Ltd.

2.2 Synthesis Process

Se doped MoS_2 was synthesized by one step hydrothermal method. Initially, 0.0021 mol ammonium molybdate tetrahydrate was added to 70 ml distilled water and the solution was stirred for 10 minutes. Thereafter, 0.0301 mol of thiourea was added to the solution and further stirred for few minutes. Finally, 0.021 moles of selenium powder and 1.08 grams of PVP were introduced into the solution, and the entire mixture was stirred for a duration of 30 minutes. Afterwards, the solution was then heated to 220°C for 24 hours in an autoclave. It was later allowed to cool down by natural process. The sample was later repeatedly cleaned using ethanol and distilled water by centrifugation. It was then dried in a vacuum oven at 60°C. Finally, the sample was ground to produce the fine powder.

3 Results and Discussion

3.1 Characterizations

The XRD technique was used to learn about the crystal composition of the sample (Rigaku make Smartlab 3kW model). UV-visible spectroscopy was carried out by Shimadzu UV-2600i UV-Vis spectrometer to study the optical properties of the as synthesized sample. The morphology of the powdered sample was assessed by employing FESEM (7610F plus/JEOL).

3.2 X-Ray Diffraction

Fig. 1 a) depicts the XRD pattern for the as synthesized powdered sample in the range of 10° to 90°. The distinct peaks for both MoS_2 and Selenium could be identified clearly which shows the presence of both the phases. The crystalline nature of the sample was determined by the sharp peaks as shown. A good match of the sample was observed with the JCPDS card *(98-004-0018)* for selenium and *(01-075-1539)* for MoS_2. The space group for selenium is $P3_121$ with group number 152 which belongs to the hexagonal crystal system. Here, *P* refers to the primitive lattice, 3_1 denotes a threefold rotational symmetry axis, 2 represents a twofold rotational symmetry axis and 1 signifies the absence of any symmetry operations beyond the ones mentioned above. For MoS_2, the space group is $P6_3/m2/m2/c$ (shortened to $P6_3/mmc$) with group number 194 which depicts a hexagonal crystal lattice. Here, *P* refers to the primitive lattice, 6_3 denotes a sixfold rotational symmetry axis and *mmc* indicates mirror planes, or reflection symmetry, in three mutually perpendicular directions.
The sample exhibited hexagonal crystal system for which the relation of lattice parameters a and c are given as [6],

$$\frac{1}{d^2} = \frac{4}{3}\left(\frac{h^2+hk+k^2}{a^2}\right) + \frac{l^2}{c^2} \qquad (1)$$

where d denotes the interplanar spacing, h, k, l are the miller indices, a and c are the lattice parameters of the crystal. The lattice parameters for MoS_2 and Se have been calculated separately and depicted in table 1 below.

Table 1. Comparison of the estimated Lattice parameter values for the sample with the standard JCPDS data.

Lattice Parameters	*a=b* (Å)		*c* (Å)	
	Calculated	Standard	Calculated	Standard
MoS_2	3.11	3.14	13.22	12.53
Se	4.37	4.36	4.93	4.95

Crystallite size has a significant impact on the properties of nanomaterials including optics and electronics. Determination of crystallite size is crucial for harnessing their potential in various applications. Debye Scherrer [8] is one of those methods for finding the crystallite size and it is given by,

$$D = \frac{K\lambda}{\beta cos\theta} \qquad (2)$$

Here, D associates to the crystal grain size (crystallite size), K is a constant of 0.9 known as Scherrer constant, λ refers to the wavelength of x-ray source given by 0.154nm, β denotes

full width at half of maxima (FWHM) and θ is half the position XRD pattern. The value of crystallite size for the sample is found to be 15.8 nm.
One of the major drawbacks of the debye scherrer method is its assumption of spherical crystallites and absence of strain. But the value of crystallite size is profoundly affected by the strain present in the sample. Therefore, we employed another method to find the crystallite size which is the Williamson Hall plot method [9] [10] and is given by,

$$\beta cos\theta = \varepsilon(4sin\theta) + \frac{K\lambda}{D} \quad (3)$$

Here, ε is the micro strain present in the material, D associates to the crystal grain size, K denotes a constant of value 0.94 known as Scherrer constant, λ refers to the wavelength of x-ray source given by 0.154nm, β is full width at half of maxima (FWHM) and θ is half the position XRD pattern. Fig. 1 b) shows the plot between $4sin\theta$ (x axis) and $\beta cos\theta$ (y axis). The y-intercept of this plot gives the crystallite size which came out to be 3.68 nm whereas its slope gives the strain present in the sample which was calculated as -0.0155.
Williamson hall method assumes strain to be uniform in all directions which is not the case in our sample. To determine more precise parameters, Size Strain plot [11] is used to determine the crystallite size and the strain present in the sample which is given by,

$$(d_{hkl}\beta_{hkl}cos\theta)^2 = \frac{K\lambda}{D} * \left(d_{hkl}{}^2\beta_{hkl}cos\theta\right) + \frac{\varepsilon^2}{4} \quad (4)$$

where ε denotes the strain present in sample, D associates to the crystallite size (nm), K is constant known as Scherrer constant (0.94), λ refers to the wavelength of the X rays source (0.154 nm), β is Full width at half maxima (FWHM), d_{hkl} is the interplanar spacing and θ is the peak position (radians). Fig. 1 c) shows the plot between $d_{hkl}{}^2\beta_{hkl}cos\theta$ (x axis) and $(d_{hkl}\beta_{hkl}cos\theta)^2$ (y axis). The slope of this plot provides the crystallite size which came out to be 2.05 nm whereas its y-intercept gives the strain present in the sample which was calculated as -0.011.
The negative value of the strain reflects the compression or contraction present in the sample. External factors such as applied stress, temperature effects, smaller atoms or defects in the lattice are often responsible for this compression. This compression results in decrease in the interplanar spacing within the crystal lattice. The compressed lattice structure contributes to the broadening of diffraction peaks when the x-rays interact with the crystal lattice.

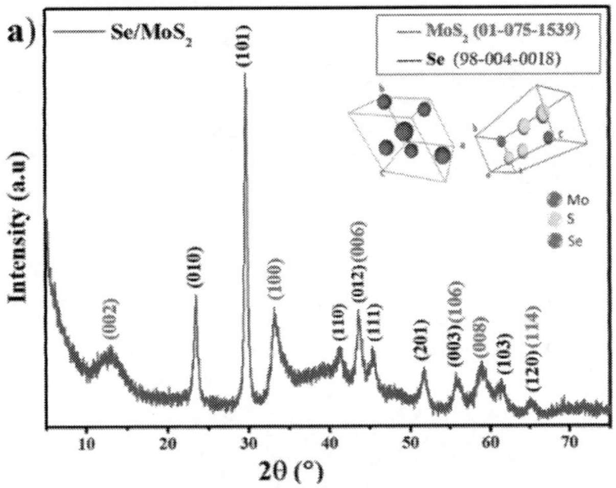

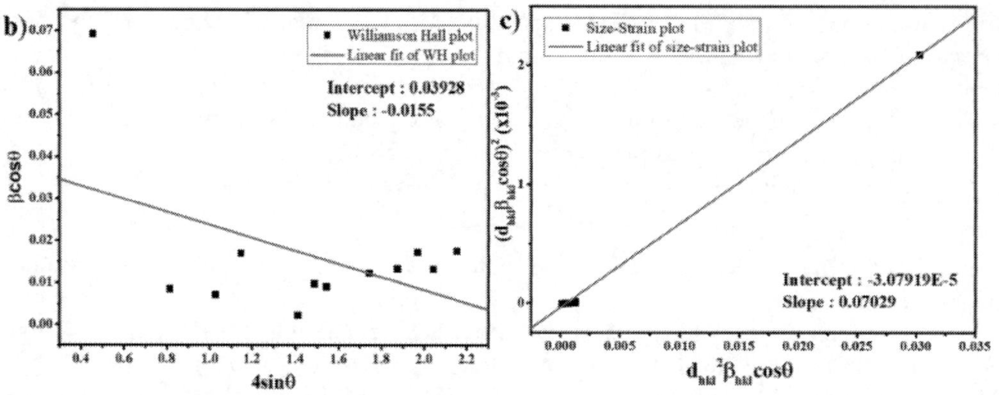

Fig. 1. a) XRD Pattern **b)** W-H Plot **c)** S-S Plot

Table 2. Crystallite size and strain calculated by using different methods

Parameters	Crystallite size (nm)	Strain
Debye Scherrer	15.8	----
W-H Plot	3.68	-0.0155
S-S Plot	2.05	-0.011

The Dislocation density [12] is the dislocations' number in a unit volume of a crystalline sample and is given by,

$$\delta = \frac{1}{D^2} \quad (5)$$

where D denotes the crystallite size (nm) and δ refers to the dislocation density which came out to be 0.004 for the as synthesized sample.

Stacking faults [12] are crystallographic flaws where the proper stacking plane order is disrupted which is given by,

$$\sigma = \frac{2\pi^2}{45(3tan\theta)^{1/2}}\beta \quad (6)$$

where σ is Stacking Fault, θ is peak position, β is Full Width at Half Maxima (FWHM). The stacking faults for the sample was calculated and found to be 0.00892.

Micro-strain [13] is defined as the ratio of the amount of deformation a material undergoes in the direction of the applied force and the material's initial dimensions which is given by,

$$\mathcal{E} = \frac{\beta}{4tan\theta} \quad (7)$$

where $\mathcal{E}$ is the micro-strain present in sample, θ is the peak position, β is the Full Width at Half Maxima (FWHM). The micro-strain was calculated to be 0.00156.

3.3 UV-Vis Spectroscopy

The optical characteristics of the as synthesized material were investigated in the 190 nm - 850 nm range of wavelength. The absorbance spectra as depicted in fig. 2 a) indicated that the as-synthesized sample show absorbance in both ultraviolet and visible region which indicated its suitability as a photodetector. The maximum absorbance peaks were identified at 360, 543, 563, 588, 650 and 777 nm for the sample. The study of band gap of a material is crucial for assessing its potential as a photodetector. Tauc's equation [14] [15] was employed to determine the band gap of the sample and is given by,

$$(\alpha h\nu)^n = A(h\nu - E_g) \quad (8)$$

where α refers to the absorption coefficient, $h\nu$ denotes the energy of photon, A depicts the absorbance constant, E_g refers to the material's band gap in eV and n is a constant indicating the kind of optical transition taking place in the material whose value is ½ for indirect band gap and 2 for direct band gap.

Fig. 2. b) shows the tauc plot of the sample i.e., the graph between $h\nu$ and $(\alpha h\nu)^2$. The band gap was calculated from this graph which came out to be 1.01 eV for the as-synthesized material. Incorporation of selenium into MoS_2 led to an increase in the bandgap which reflected absorbance in the ultraviolet region, suitable for an efficient photodetector.

3.3.1 Refractive Index, HOMO and LUMO Level

Determining the refractive index of a material is essential for its usage as an efficient photodetector since it optical performance. Matching the refractive index to substrates or surrounding media minimizes reflections, enhancing light absorption. The formula for refractive index [16][17] is given by,

$$\frac{n^2-1}{n^2+2} = 1 - \sqrt{\frac{E_g}{20}} \quad (9)$$

where n refers to the refractive index of the material and E_g denotes its calculated band gap. By putting the band gap as 1.01 eV, the refractive index of the sample came out to be 3.36.

By knowing a material's HOMO-LUMO energy levels, it is possible to design and modify photodetectors to suit specific wavelength ranges and applications.

The following formulae was used to HOMO and LUMO level [9],

$$E_{VB} = \chi - E_e + 0.5\, E_g \quad (10)$$

$$E_{CB} = E_{VB} + E_g \quad (11)$$

where χ depicts the absolute electronegativity, E_e refers to the free electron energy i.e. 4.5 eV, E_{CB} and E_{VB} are the energy of HOMO and LUMO levels respectively, and E_g is the calculated band gap. The value of absolute electronegativity for MoS_2 is 5.32 eV [18] and that for Se is 5.89 eV [19]. Table 3 summarizes the values of band gap, refractive index and HOMO - LUMO levels for the sample.

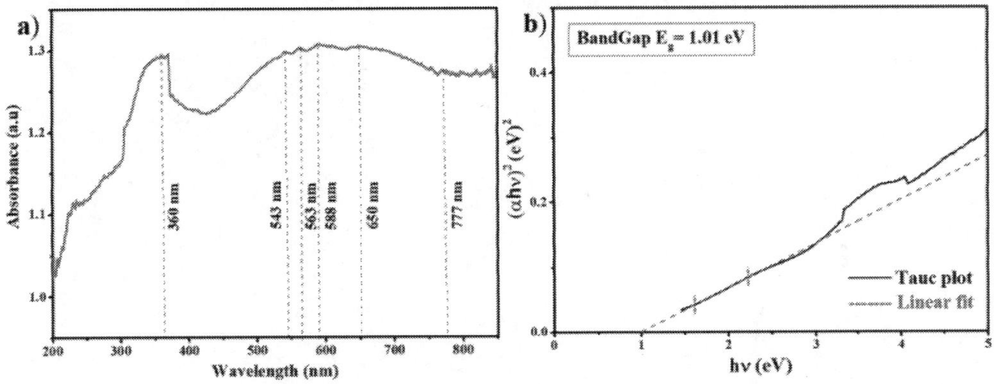

Fig. 2. a) Absorption Spectra **b)** Tauc Plot

Table 3. Band gap, refractive index and value of HOMO and LUMO levels for the sample

Sample	Band gap (eV)	Refractive Index	HOMO level (eV)		LUMO level (eV)	
			MoS_2	Se	MoS_2	Se
Se-MoS_2	1.01	3.36	2.33	2.90	1.32	1.89

3.4 FESEM

The morphological analysis for the as-synthesized sample was done using FESEM as shown in fig. 3 (a&b). The images obtained at different magnifications clearly show the formation of nanolayers. Due to their high surface energy, the nanolayers are aggregated into spherical shapes to lower the surface energy and form microspheres [fig.3 a)]. Microspheres can scatter light in multiple directions increasing the possibility of light absorption and enhancing the efficiency of light capture by a photodetector. Fig.3b) shows the formation of nanolayers of at least 100 nm in size. The agglomeration of the particles is reflected in the fig. 3a). Small fragments of selenium nanoparticles adhered to the nanolayers could be distinguished. Fig. 3c) shows the EDX pattern for the as-synthesized sample. The intensity peaks for Mo, S and Se are clearly visible which indicates the presence of these elements in the sample.

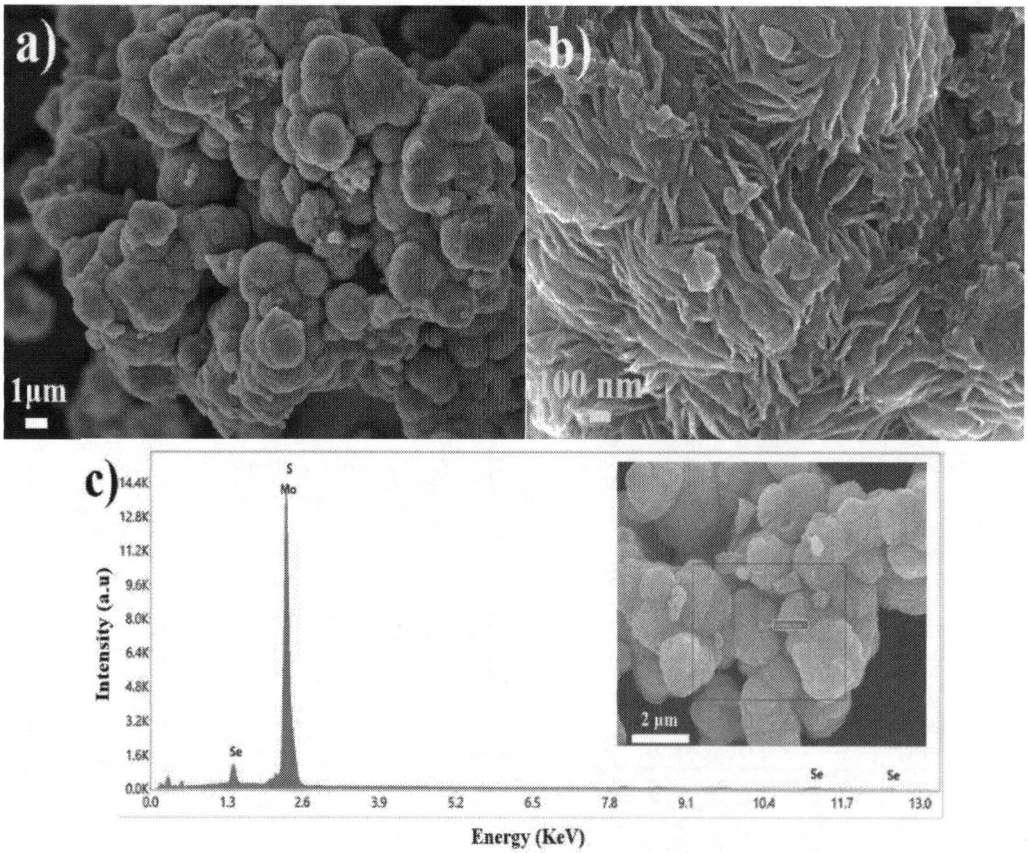

Fig. 3. (a&b) FESEM images **c)** EDX spectra

4 Conclusion

Se doped MoS_2 was successfully synthesized via a facile hydrothermal route as confirmed by various characterization techniques. Structural analysis revealed the crystalline nature of the material and verified the existence of Selenium and MoS_2 phases. A thorough morphological study of the sample showed nanolayer-like morphology. The production of microspheres was also seen as a result of agglomeration. The material showed a band gap of 1.01 eV through optical analysis. While MoS_2 shows absorption solely in visible region, the doping of selenium into it increased its absorption in the ultraviolet region as well. This wide absorption and layered morphology are suitable for broadband photodetector applications.

The authors greatly acknowledge crucial resources provided by Prof. Anand Srivastava, Vice Chancellor of Netaji Subhas University of Technology (former Netaji Subhas Institute of Technology, University of Delhi).
No conflicts of interest exist among the authors.

References

1. Y. Xu, J. Yuan, L. Fei, X. Wang, Q. Bao, Y. Wang, K. Zhang, Y. Zhang, Selenium-doped black phosphorus for high-responsivity 2d photodetectors. Small. **12**, no. 36, pp. 5000–5007 (2016) https://doi.org/10.1002/smll.201600692
2. G. Yang, Y. Gu, P. Yan, J. Wang, J. Xue, X. Zhang, N. Lu, G. Chen, Chemical vapor deposition growth of vertical mos2 nanosheets on p-gan nanorods for photodetector application. ACS App. Mat. Int. **11**, no. 8, pp. 8453–8460 (2019) https://doi.org/10.1021/acsami.8b22344
3. S. P. Chang, K. J. Chen, Zinc oxide nanoparticle photodetector. J Nano. **2012**, 602398 (2012) https://doi.org/10.1155/2012/602398
4. Y. Xie, B. Zhang, S. Wang, D. Wang, A. Wang, Z. Wang, H. Yu, H. Zhang, Y. Chen, M. Zhao, B. Huang, L. Mei, J. Wang, Ultrabroadband MoS2 photodetector with spectral response from 445 to 2717 nm. Adv. Mat. **29**, no. 17, p. 1605972 (2017) https://doi.org/10.1002/adma.201605972
5. N. Choudhary, M.D. Patel, J. Park, B. Sirota, & W. Choi, Synthesis of large scale MoS2 for electronics and energy applications. J Mat. Res. **31,** no. 7, pp. 824–831 (2016) https://doi.org/10.1557/jmr.2016.100
6. Shreya, P. Phogat, R. Jha, S. Singh, Elevated refractive index of MoS2 amorphous nanoparticles with a reduced band gap applicable for optoelectronics. pp. 431–439 (2023) https://doi.org/10.1007/978-981-99-2349-6_39
7. X. Ren, Q. Ma, H. Fan, L. Pang, Y. Zhang, Y. Yao, X. Ren, and S. (F.) Liu, A Se-doped MoS2 nanosheet for improved hydrogen evolution reaction. Chem. Comm. **51,** no. 88, pp. 15997–16000 (2015) https://doi.org/10.1039/C5CC06847A
8. P. Phogat, Shreya, R. Jha, S. Singh, Optical and microstructural study of wide band gap ZnO@ZnS core–shell nanorods to be used as solar cell applications. pp. 419–429 (2023) https://doi.org/10.1007/978-981-99-2349-6_38

9. Shreya, P. Phogat, R. Jha, S. Singh, Microwave-synthesized γ-WO3 nanorods exhibiting high current density and diffusion characteristics. Trans. Met. Chem. **48,** no. 3, pp. 167–183 (2023)
https://doi.org/10.1007/s11243-023-00533-y

10. D. Kumari, Shreya, P. Phogat, Dipti, S. Singh, R. Jha, Enhanced electrochemical behavior of C@CdS core-shell heterostructures. Mat. Sci. and Eng. B. **301,** p. 117212 (2024)
https://doi.org/10.1016/j.mseb.2024.117212

11. P. Phogat, Shreya, R. Jha, S. Singh, Diffusion controlled features of microwave assisted ZnS/ZnO nanocomposite with reduced band gap. ECS J. of S.S.S. and Tech. **12,** no. 3, p. 034004 (2023)
10.1149/2162-8777/acc426

12. T. Kumar, Shreya, P. Phogat, V. Sahgal, R. Jha, Surfactant-mediated modulation of morphology and charge transfer dynamics in tungsten oxide nanoparticles. Phys. Scr. **98,** no. 8 (2023)
http://dx.doi.org/10.1088/1402-4896/ace566

13. S. Sharma, P. Phogat, R. Jha, S. Singh, Electrochemical and optical properties of microwave assisted mos2 nanospheres for solar cell application. Int. J. of S.G. and C.E. pp. 66–72 (2023)
http://dx.doi.org/10.12720/sgce.12.3.66-72

14. Dipti, P. Phogat, Shreya, D. Kumari, S. Singh, Fabrication of tunable band gap carbon-based zinc nanocomposites for enhanced capacitive behaviour. Phys. Scr. **98,** no. 9 (2023)
10.1088/1402-4896/ACF07B

15. J. Dahiya, P. Phogat, A. Hooda, S. Khasa, Investigations of praseodymium doped LiF-ZnO-Bi2O3-B2O3 glass matrix for photonic applications. AIP Conf Proc. **2995,** no.1 (2024)
https://doi.org/10.1063/5.0178197

16. P. Phogat, S. Shreya, R. Jha, S. Singh, Impedance study of zinc sulphide quantum dots via one step green synthesis. Mat. Sci. For. **1099,** pp. 119–125 (2023)
http://dx.doi.org/10.4028/p-G1CCxq

17. P. Phogat, S., R. Jha, S. Singh, Electrochemical analysis of thermally treated two-dimensional zinc sulphide hexagonal nano-sheets with reduced band gap. Phys. Scr. **98,** no.12 (2023)
http://dx.doi.org/10.1088/1402-4896/ad0d93

18. Y. Xu, M. A. A. Schoonen, The absolute energy positions of conduction and valence bands of selected semiconducting minerals. Am. Min. **85,** no. 3–4, pp. 543–556 (2000)
https://doi.org/10.2138/am-2000-0416

19. R. G. Pearson, Absolute electronegativity and hardness: application to inorganic chemistry. Inorg. Chem. **27,** no. 4, pp. 734–740 (1988)
http://dx.doi.org/10.1021/ic00277a030

Integrated Renewable Energy-Based Wastewater Management System: A Sustainable Approach

Mahesh Kumar K[1], *Sri Akaash* S.R[1*], *Sree Adithya* S.R[1], *and Sri Rakshaga* S.R[2]

[1]Department of Mechanical and Automation Engineering, Agni College of Technology, Chennai, India
[2]Department of Electronics and Communication Engineering, Agni College of Technology, Chennai, India

Abstract. The escalating global demand for water and the increasing need for sustainable wastewater treatment systems have spurred innovative approaches. This paper proposes an integrated renewable energy-based wastewater management system that harnesses solar and wind energy to power the treatment process. Renewable energy sources are aimed at reducing wastewater treatment's environmental impact while promoting energy efficiency. This study outlines the methodology, presents results, and discusses the implications of the proposed system.

1 Introduction

This study presents an original approach to addressing the increasing pressure on water resources and the need for environmentally friendly wastewater treatment solutions. This study also introduces a novel wastewater treatment system, and it is important to note that the concepts and methodologies outlined herein are the result of original research and innovation. Conventional wastewater treatment processes often rely on non-renewable energy sources, contributing to the carbon footprint and escalating operational costs. In response to this challenge, our research endeavours to integrate renewable energy sources, specifically solar and wind power, into the wastewater treatment process. And Studies on [1] & [3] have highlighted the major role played by solar energy usage.

The primary objective of this study is to design and implement a wastewater management system that not only treats sewage efficiently but also minimizes its environmental impact through the utilization of renewable energy. By employing solar panels and windmills, we aim to power key components of the treatment process, including the raw effluent pump for transporting wastewater from the collection tank to the anaerobic digester and the rotational mechanism of the Rotating Biological Contactor (RBC). The surplus energy generated during periods of high solar and wind activity is intelligently diverted to a battery storage system, providing a reliable power source during low energy production periods, such as night-time.

* Corresponding author: adithyaakaash8@gmail.com

2 Method of sewage water treatment

In this study, an innovative wastewater treatment process is proposed, starting with an anaerobic digestion phase. The biological treatment occurs within the anaerobic digester using collected sewage wastewater. Microorganisms break down organic matter in the absence of oxygen, producing some amount of methane, H2S, SOX, and NOX during the treatment process.

Following anaerobic digestion, the partially treated effluent is directed to the Rotating Biological Contactor (RBC). The RBC consists of rotating discs that facilitate the growth of aerobic microorganisms, further breaking down remaining organic compounds. Unlike traditional RBC systems, our innovative approach utilizes wind energy to drive the rotation of the discs directly. A series of bevel gears and mechanical linkages ensure a seamless and efficient transfer of mechanical energy from windmills to the RBC, eliminating the need for electrical conversion. Also, using current from the solar system to rotate the shaft of RBCs using the motor connected at other end of the shaft which is away from the windmill side which is used to rotate the shaft if there is no energy produced by the windmills.

The sludge generated during the wastewater treatment process is collected daily from the bottom of the anaerobic digester, RBCs, and settling tank. This sludge undergoes further treatment in a sludge drying bed, where moisture is removed, resulting in a nutrient-rich dried sludge. This dried sludge serves as valuable manure, closing the loop of resource recovery within the system.

After settling in the tank, the treated water, although significantly improved, may still contain suspended solids. To remove these suspended solids, the water undergoes a final treatment stage using semi-rapid sand filters. These filters effectively remove residual suspended solids, ensuring the quality of the treated water.

The entire system is monitored and controlled using microcontrollers. These devices regulate the speed of motors, manage energy distribution, and monitor the level in the collection tank. The microcontrollers play a crucial role in optimizing the efficiency of the wastewater treatment process, ensuring that energy is utilized judiciously and the system operates at peak performance.

3 Methodology

3.1 Explanation

Our proposed system begins with the collection of sewage wastewater in a collection tank, where the initial separation of solids occurs. The wastewater then enters the anaerobic digester for biological treatment, which breakdown the organic compound. This partially treated effluent is directed to the Rotating Biological Contactor (RBC), which, unlike traditional systems, is powered by wind energy using a system of Bevel gears and solar energy using a DC motor. Simultaneously, solar panels generate electrical energy, which is directly used to power the raw effluent pump for transporting wastewater to the anaerobic digester. Excess electrical energy from the solar panels is intelligently directed to a battery storage system, ensuring a consistent power supply during periods of low solar activity, such as night-time. Microcontrollers play a pivotal role in regulating motor speeds, controlling energy distribution, and monitoring the level in the collection tank. This intelligent control enhances the overall efficiency of the system by adapting to varying energy inputs and optimizing resource utilization.

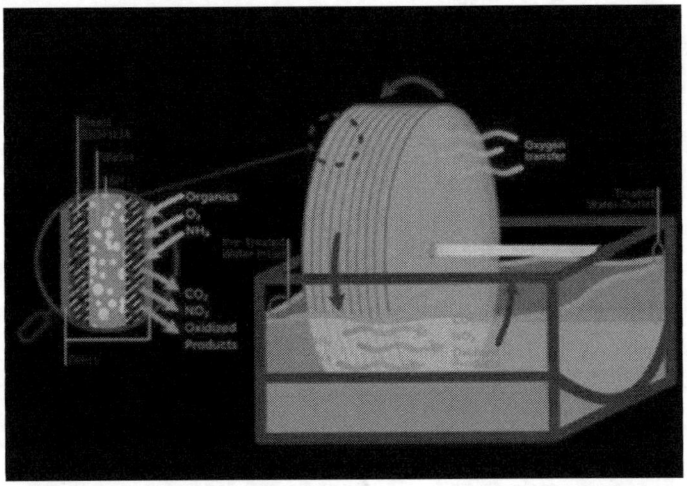

Fig. 1. Rotatory Biological Contractor

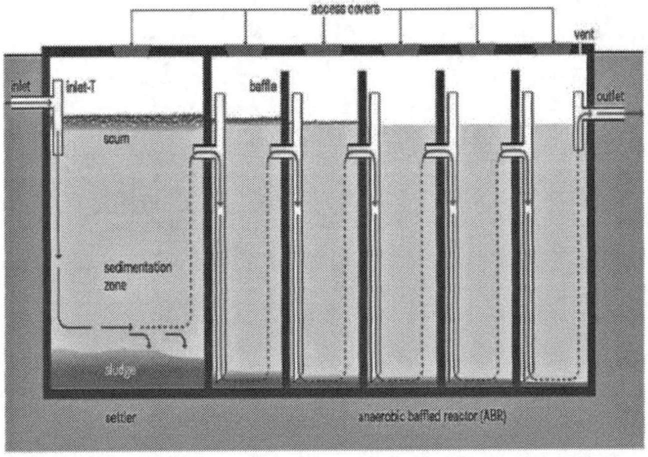

Fig. 2. Anaerobic Digestor

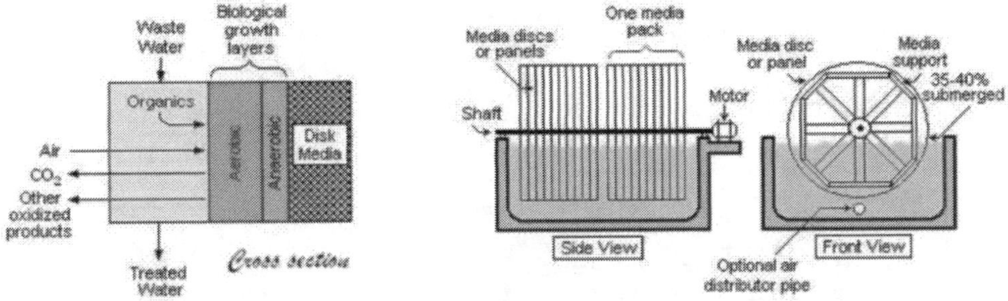

Fig.3. Schematic View of RBCs Colour illustrations

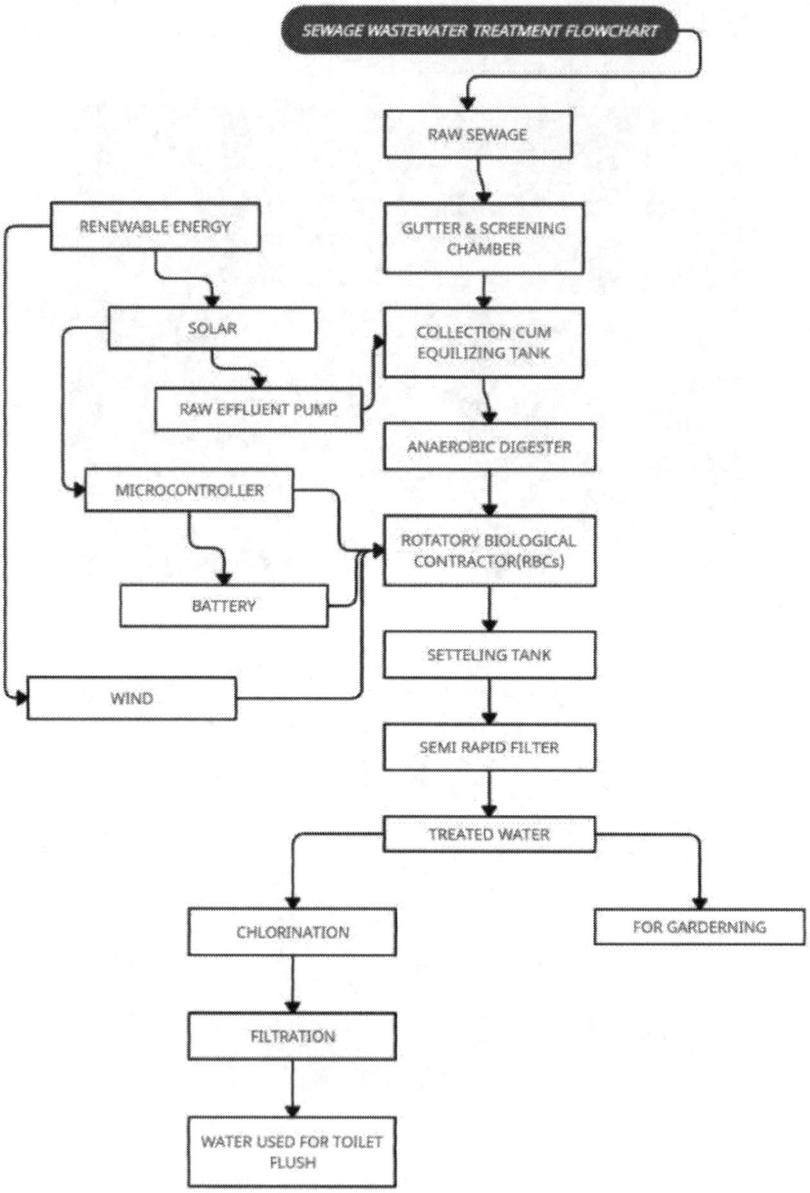

Fig. 4. Flowchart of Sewage Water Treatment

4 Discussion

The integrated renewable energy-based wastewater management system stands as a testament to its environmental sustainability and treatment efficiency. A distinctive feature of this approach lies in the direct utilization of solar and wind energy within the Rotating Biological Contactor (RBC), circumventing the need for electrical conversion and marking itself as an original and highly efficient solution. In the realm of energy generation and storage, the system achieves optimal electrical energy generation through solar panels, coupled with intelligent storage in a battery system to ensure a continuous power supply. Notably, an original innovation within this framework involves the utilization of stored

energy specifically for lighting during night-time, addressing the challenges posed by intermittent renewable energy sources.
Microcontrollers assume a pivotal role in the seamless operation of the wastewater treatment process, facilitating adaptive energy utilization. The incorporation of microcontrollers for real-time monitoring and motor speed control constitutes an original system control feature, contributing significantly to the overall energy efficiency of the system. In the domain of sludge management and resource recovery, the system employs a sludge drying bed that effectively transforms sludge into nutrient-rich dried sludge suitable for agricultural use. This integration of a sludge drying bed as a resource recovery loop stands out as an original and sustainable solution.
The final treatment stage of the system involves semi-rapid sand filters, which exhibit proficiency in removing suspended solids, ultimately yielding high-quality treated water. An original contribution within this phase lies in the high efficiency of the semi-rapid sand filters, ensuring the suitability of the treated water for various non-potable purposes. In essence, each facet of this integrated wastewater management system showcases not only efficiency but also originality in its approach, innovation, system control, and resource recovery, underscoring its comprehensive and sustainable impact.

5 Result

The wastewater management system is comprehensively assessed through various phases, starting with the Anaerobic Digestion Phase. This phase involves the analysis of initial sewage characteristics, including COD, BOD, and suspended solids, prior to anaerobic digestion. Additionally, the phase monitors the production of methane, H2S, SOX, and NOX during anaerobic digestion while identifying and quantifying residual organic compounds in the effluent.
Moving to the Rotating Biological Contactor (RBC) phase, the study delves into the impact of wind energy on RBC disc rotation, evaluating the efficiency and reliability of solar-powered shaft rotation, and assessing the mechanical energy transfer efficiency through mechanical linkages and bevel gears.
The Sludge Management phase involves the daily monitoring of sludge collection in terms of quantity and characteristics. Furthermore, the efficiency of the sludge drying bed in producing nutrient-rich dried sludge is thoroughly assessed.
The Final Treatment Stage focuses on the efficiency of semi-rapid sand filters in removing residual suspended solids and evaluates the quality of treated water after the final treatment stage.
Energy Distribution and Storage play a critical role in the system. The solar panel energy generation, including production and variability, is documented. Wind energy variability and its impact on system performance are studied, alongside an evaluation of the effectiveness of the battery storage system.
The Microcontroller System, integral to adaptive energy utilization, is investigated for its role in motor speed regulation and its impact on energy efficiency. Real-time monitoring of energy distribution and consumption provides insights into energy utilization optimization. Altogether, this multifaceted assessment underscores the system's thorough examination, ensuring a holistic understanding of its performance across various crucial parameters.

6 Conclusion

In conclusion, our integrated renewable energy-based wastewater management system presents a holistic and sustainable approach to sewage treatment. The incorporation of direct wind and solar energy utilization in the RBC, solar power integration, and intelligent control through microcontrollers collectively contribute to an energy-efficient and environmentally friendly system.

The successful implementation of this system not only reduces the reliance on conventional energy sources but also addresses the challenges associated with weather dependent energy production. The closed-loop resource recovery, where sludge is transformed into valuable manure, further underscores the sustainable nature of the proposed wastewater treatment process.

While the system has demonstrated promising results, ongoing monitoring and optimization are essential to address any potential challenges and further enhance its performance. Future research endeavours could explore scalability, adaptability to different environmental conditions, and potential applications in various geographical regions.

In summary, our proposed integrated renewable energy-based wastewater management system marks a significant step towards achieving sustainable and efficient wastewater treatment practices, contributing to the broader goal of environmental conservation and resource optimization.

References

1. M.S. Mousavi, A. Ahmadi, A. Entezari, Forecast of using renewable energies in the water and wastewater industry of Iran. New Energy Exploitation and Application **1**(2), 1-9 (2022).
2. S. Zahmatkesh, K.T. Amesho, M. Sillanpaa, C. Wang, Integration of renewable energy in wastewater treatment during COVID-19 pandemic: challenges, opportunities, and progressive research trends. Cleaner Chemical Engineering **3**, 100036 (2022).
3. B. Shyam, P. Kanakasabapathy, Renewable energy utilization in India—policies, opportunities and challenges. In: 2017 International Conference on Technological Advancements in Power and Energy (TAP Energy), pp. 1-6 (2017).
4. B. J. Gungor, I. Dincer, A renewable energy based waste-to-energy system with hydrogen options. International Journal of Hydrogen Energy **47**(45), 19526-37 (2022).
5. C. Liu, J. Zhuo, Y. He, S. Li, Q. Yao, Synergetic process between wind power and sewage sludge pyrolysis in a novel renewable-energy microgrid system. Clean Energy **7**(3), 493-508 (2023).
6. G. Daronco, A. Beluco, A wind PV hybrid system for power supply of a sewage treatment plant in a small town in southern Brazil. Sewage Treatment Plants, 341 (2020).
7. Y. Wang, J. Li, S. Wang, J. Yang, C. Qi, H. Guo, X. Liu, H. Zhang, Operational optimization of wastewater reuse integrated energy system. Energy **200**, 117309 (2020).
8. F. Haces-Fernandez, GoWInD: Wind Energy Spatiotemporal Assessment and Characterization of End-of-Life Activities. Energies **13**(22), 6015 (2020).
9. A.M. Soliman, A. Al-Falahi, S. Mohamed AEldean, M. Elmnifi, A new system design of using solar dish-hydro combined with reverse osmosis for sewage water treatment: Case study Al-Marj, Libya. Desalination Water Treat **193**, 189-211 (2020).

10. M.D. Nguyen, M. Thomas, A. Surapaneni, E.M. Moon, N.A. Milne, Beneficial reuse of water treatment sludge in the context of circular economy. Environmental Technology & Innovation **28**, 102651 (2022).
11. Y. Zhao, A. Nzihou, B. Ren, N. Lyczko, C. Shen, C. Kang, B. Ji, Waterworks sludge: An underrated material for beneficial reuse in water and environmental engineering. Waste and Biomass Valorization **12**, 4239-51 (2021).
12. K.H. Bachanek, B. Tundys, T. Wiśniewski, E. Puzio, A. Maroušková, Intelligent street lighting in a smart city concepts—A direction to energy saving in cities: An overview and case study. Energies **14**(11), 3018 (2021).
13. A. Ayat, S. Arris, A. Abbaz, M. Bencheikh-Lehocine, A.H. Meniai, Application of response surface methodology for modeling and optimization of a bio coagulation process (sewage wastewater treatment plant). Environmental Management **67**(3), 489-97 (2021).

Advancement and characterization of Nitride Coating Materials on Aluminum Alloy-based Bipolar plates in PEMFC's Applications

S. J. Narasimharaju[1*], *Poorna Chandra Rao B*[1], *and K. Annamalai*[1]

[1]Department of Automobile Engineering, Madras Institute of Technology Campus, Anna University, Chennai, Tamil Nadu, India.

Abstract. This review article delves into the recent strides made in the realm of Proton Exchange Membrane Fuel Cell through the exploration of nitride coating materials on Aluminium (Al) alloy-based Bipolar plates (BPs). The study focuses on the advancements achieved in enhancing the performance and durability of these crucial components. Through a comprehensive characterization process, the nitride coatings' impact is assessed, emphasizing critical parameters such as corrosion current density, potentiostat results, and Internal Contact Resistance (ICR). The review highlights the notable advantages brought about by nitride coatings, particularly their corrosion resistance and durability, in mitigating challenges posed by the demanding operational conditions of fuel cells. Additionally, it sheds light on existing challenges, such as coating porosity and adhesion issues, underscoring the importance of ongoing research efforts to optimize these materials. As the exploration of nitride-coated Al alloy BPs progresses, this review anticipates contributing valuable insights to the broader goal of advancing sustainable and efficient energy solutions in PEMFC applications.

Keywords: Nitride Coating, Al Alloy, Bipolar plates, PEMFCs.

1. Introduction

Fuel cells are innovative and sustainable energy conversion devices that have attracted noteworthy interest in recent times due to their promising resolution for tackling the global obstacles of energy efficiency, environmental sustainability, and the transition away from fossil fuels. These electrochemical devices generate electricity by facilitating a controlled reaction between hydrogen and oxygen, producing water and releasing energy in the process. What sets fuel cells apart from traditional combustion-based power generation is their remarkable efficiency, low environmental impact, and versatility, making them a compelling candidate for a diverse array of applications, spanning from transportation to stationary power generation purposes. In Fig. 1. Shows Basic Fuel Cell Operation This introduction will delve into the fundamental principles, types, and potential advantages of fuel cells, shedding light on their pivotal role in shaping a more sustainable and cleaner energy future [1], [2].

*Corresponding Author: narasimharajusj@mitindia.edu

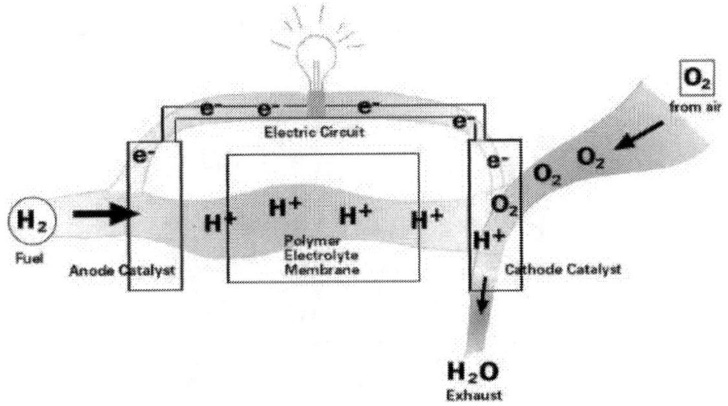

Fig. 1. Basic Fuel Cell operation [1].

The PEMFC represents a cutting-edge electrochemistry device that produces electrical power via the chemical reaction of hydrogen and oxygen, with water as the primary by-product. It is a key technology in the realm of clean energy and is renowned for its high energy efficiency and low environmental impact. **Fig. 2.** shows typical working principles of PEMFC exhibit a broad spectrum of potential utilizations, encompassing powering vehicles, stationary energy output generation, and portable electronics, making them a significant player in the transition to more sustainable energy sources. At the heart of a PEMFC constitutes the core component known as the proton exchange membrane, which is commonly composed comprising a solid polymer electrolyte that enables the selective passage of positively charged protons (H+ ions) simultaneously impeding the flow of electrons. Upon introduction of hydrogen gas to the anode, a process unfolds where the gas undergoes a separation into protons and electrons a catalytic process. The protons then travel across the membrane to the cathode side, where they unite with oxygen and the electrons, which have followed an external circuit, resulting in the generation of water and the release of electrical energy in the process. PEMFCs offer several advantages, such as exhibiting high-power density, rapid start-up capabilities, and the ability to operate efficiently at low temperatures. These attributes make them particularly suitable for applications where rapid response and compact, lightweight design are crucial, like in automotive fuel cells and portable electronic devices. One of the primary challenges in the development of PEMFCs lies in finding durable and efficient catalysts, typically platinum or platinum-group metals, to facilitate the electrochemical reactions. Additionally, maintaining the integrity of the proton exchange membrane, optimizing the water management system, and managing impurities are all essential for the enduring operational efficiency and reliability of the PEMFC. Overall, PEMFCs show great promise as versatile technology holding the potential to diminish greenhouse gas emissions and lessen reliance on fossil fuels, while simultaneously providing efficient and clean power solutions for a wide range of applications. Continuous research and development endeavours persist in enhancing the performance, cost-effectiveness, and scalability of PEMFCs, contributing to their growing significance in the clean energy landscape [3], [4].

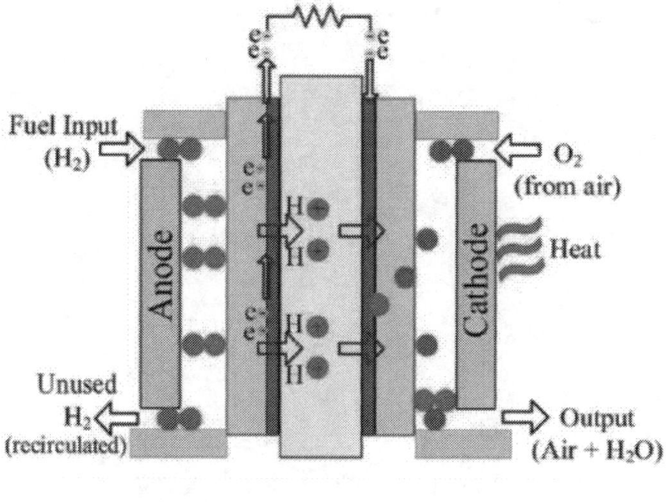

Fig. 2. Typical working principles of PEMFC [5].

1.1 Components of fuel cells

Fuel cells consist of several essential components working together to facilitate the transformation of chemical energy into electrical power, and heat. **Fig. 3.** provides a simplified schematic representation of these individual fuel cell components. These components include an anode, where hydrogen fuel is introduced and catalytically split into protons and electrons; a cathode, which facilitates the reaction of oxygen with electrons and protons to form water; an electrolyte, typically a membrane that separates the anode and cathode while allowing the passage of protons; catalysts to expedite the electrochemical reactions at the electrodes; BPs to interconnect and collect current from individual cells within a stack; the fuel cell stack itself, composed of multiple interconnected cells to increase voltage and power output; a balance of plant that includes auxiliary components like pumps, fans, and humidifiers to maintain optimal conditions; a gas supply system for hydrogen and oxygen; a cooling system to dissipate excess heat generated during operation; hydrogen storage in applicable cases; and a control and monitoring system to regulate and oversee the fuel cell's performance. These components collectively enable fuel cells to generate electricity efficiently and sustainably [6].

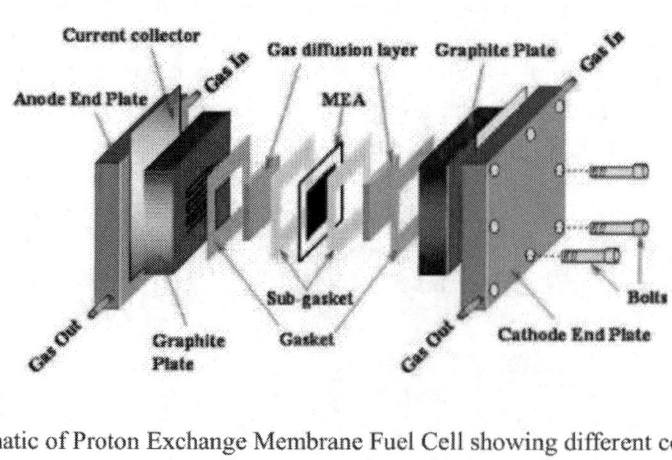

Fig. 3. Schematic of Proton Exchange Membrane Fuel Cell showing different components [7].

1.2 Bipolar plates

In **Fig. 4.** Shows BPs constitute crucial elements within PEMFCs, fulfilling a significant role in vital function within the efficient generation of electricity through the electrochemical conversion of hydrogen and oxygen. These plates are essential for both the structural integrity and the electrical and thermal management of the fuel cell stack. BPs serve as the physical separators between individual cells, providing a means to create a series of interconnected cells in the stack. One of the primary functions of BPs is to distribute a supply of hydrogen and oxygen gases to the anode and cathode sides of each cell while also allowing the removal of water and heat generated during the electrochemical reactions. Additionally, they play a critical role in establishing an electrical connection between adjacent cells, forming a continuous electrical circuit. This enables the flow of electrons produced during the chemical reactions, ultimately leading to the generation of electrical power. The design and material composition of BPs are essential considerations in PEMFC development. Ideally, these plates must possess several key attributes, including high electrical conductivity to minimize power losses, corrosion resistance to withstand the harsh chemical environment within the fuel cell, and durability to ensure a long operational lifespan. Light weight is also desirable to diminish the overall weight associated with the fuel cell configuration, particularly in applications like vehicles. Furthermore, effective thermal management properties are crucial to dissipate excess heat generated during operation. Various materials have been explored for BP construction, such as graphite, metal alloys, and advanced composites. Each material choice comes with its own set of advantages and trade-offs, with ongoing research aimed at finding the optimal balance of properties while keeping manufacturing costs in check. BPs serve as essential elements within PEMFCs, serving an essential multifaceted and playing a crucial part in ensuring the efficient, reliable, and cost-effective operation of these clean energy devices. Ongoing research and innovation in BP design and materials are essential for advancing the performance and affordability of PEMFCs, contributing to the broader adoption of this technology in various applications, including transportation and stationary power generation [8], [9].

Fig. 4. BP in PEMFC Stack [10].

2. Aluminum Alloy

Al alloys have limited applications within the PEMFC fuel cells because of their lower electrical conductivity compared to materials commonly used for BPs, such as graphite, coated metals, or composites. While Al and its alloys offer advantages like lightweight properties and corrosion resistance, their relatively lower electrical conductivity makes them less suitable for BP applications in PEM fuel cells. Researchers primarily focus on materials with higher electrical conductivity, such as graphite, stainless steel, or titanium, to ensure efficient current collection and overall performance of fuel cells. Consequently, Al alloys are not the preferred choice for BPs within PEMFC Al metal blends are versatile metallic materials composed primarily of Al, combined with various alloying elements to enhance specific properties. Engineered for a diverse array of uses, these alloys exhibit a favorable equilibrium in characteristics, including low density, excellent thermal conductivity, and corrosion resistance. The alloying process involves the addition of elements such as copper, zinc, magnesium, manganese, and silicon, each contributing distinct attributes to the final material. Classified into wrought and cast alloys, they are further distinguished by series designations using a four-digit numbering system. Widely employed in industries like aerospace, automotive, construction, packaging, and electronics, Al alloys play a pivotal role in manufacturing aircraft components, automotive parts, beverage cans, building facades, and electronic housings. Notable for their lightweight nature, corrosion resistance, and machinability, Al alloys are a preferred choice in diverse applications, although considerations of strength and susceptibility to certain forms of corrosion are essential in material selection [11], [12].

2.1 Electrical conductivity of Al alloy-based Bi-polar plate

The electrical conductivity within Al alloy-based BPs electrical conductivity is a crucial factor in the performance of fuel cell systems, particularly in PEMFCs and Alkaline Fuel Cells. These plates fulfill a significant pivotal function in enabling the flow of the electrical current between the electrodes of the fuel cell. Al alloys are favored for BPs due to their advantageous properties, including high conductivity, lightweight nature, corrosion resistance, and cost-effectiveness. The alloy composition significantly influences the electrical conductivity of these plates. Commonly used Al alloys for BPs include the 1xxx series (pure Al), 3xxx series (Al-manganese), and 6xxx series (Al-magnesium-silicon) alloys. The electrical conductivity within the Al alloy-based BPs electrical conductivity is influenced by factors such as alloy composition, microstructure, processing methods, and surface treatments. For instance, alloying elements like copper and silicon can enhance the conductivity, but the overall impact depends on the alloy composition and the presence of other elements. Microstructure fulfills a significant critical function in determining the electrical conductivity of Al alloys. Heat treatment processes and alloying can be tailored to optimize the microstructure for improved conductivity. Additionally, surface treatments, such as coatings or coatings with conductive materials like graphite, can further enhance the electrical performance of the BPs [13], [14].

BPs are integral components in fuel cell technology, serving as conductive interfaces between individual cells within a fuel cell stack. Composed of materials like Al alloys, these plates play a crucial role in enabling the movement of electrons between both the anode and cathode, essential for the electrochemical reactions powering the fuel cell. The choice of materials, such as Al alloys, is strategic, considering their commendable electrical conductivity, lightweight nature, and corrosion resistance. BPs contribute significantly to the overall enhancement of the effectiveness of a fuel cell system by providing a conductive pathway for electrons, minimizing electrical resistance, and ensuring uniform distribution of reactants. Ongoing research in this field focuses on optimizing the design and composition improving the electrical conductivity of BPs while addressing challenges like mechanical durability and corrosion susceptibility, aiming for continuous improvements in fuel cell performance [15], [16].

Table. 1. DOE technical specification targets for PEMFC BPs.

Characteristic	**Units**	**2025 DOE Target** [17]
Electrical conductivity	Scm^{1}	> 100
Areal specific resistance	$\Omega\ cm^2$	<0.01
H2 permeability	$cm^3sec^{1}cm^{2}$	2 × 10 6
Thermal conductivity	$Wm^{1}K^{1}$	-
Corrosion, anode	$\mu A\ cm^{2}$	< 1
Corrosion, cathode	$\mu A\ cm^{2}$	< 1 & no active peak
Life expectancy	Hours	8000
Plate weight	kg/kW	0.18
price	$\$kW^{1}$	2
Flexural strength	MPa	> 40

Hence, it is crucial to create materials suitable for bipolar plates that allow for easy and cost-effective manufacturing, while simultaneously fulfilling specific criteria related to electrical, thermal, mechanical, and chemical properties. This is essential for the successful commercialization of PEMFCs. Complying with the guidelines established by the US Department of Energy and other industry stakeholders, **Table. 1.** delineates various technical requirements relevant to the progress of bipolar plates in PEMFCs.

3. Nitride

Nitride coating is a surface treatment process that involves the deposition of nitrides onto various materials to enhance their properties. Nitrides are compounds formed by nitrogen with other elements, and the coating is typically applied to metals, ceramics, and certain polymers. This process is commonly employed in sectors including aerospace, automotive, and manufacturing to improve the performance and longevity of various elements. The nitride coating is known for its ability to provide enhanced rigidity, durability, and protection against corrosion to its substrate material. One associated with the most widely used nitride coatings is Titanium Nitride (TiN), although other variants like Chromium Nitride (CrN) and Al Nitride (AlN) are also utilized based on specific application requirements. The coating process can be achieved through various methods, including techniques such as Physical Vapour Deposition (PVD) and Chemical Vapour Deposition (CVD). During PVD, its nitride coating is formed by condensing vaporized nitride compounds onto the substrate surface, while in CVD, the coating is generated through a chemical reaction between gaseous precursors and the substrate. Nitride-coated components find applications in cutting tools, industrial machinery, and automotive parts, where increased hardness and the ability to withstand wear and corrosion are crucial. The enhanced performance and durability provided by nitride coatings contribute to the overall efficiency and lifespan of various engineered products [18], [19].

3.1 Types of Nitrides

Nitrides encompass a diverse group of compounds with unique properties, making them valuable for various industrial applications. Here are some notable types of nitrides, TiN: Renowned for its gold-coloured appearance, TiN is widely employed due to its exceptional hardness, wear resistance, and corrosion resistance. Common applications include cutting tools and metalworking components [20]. CrN: Recognized for its high hardness, low friction, and superior wear resistance, CrN coatings are often utilized in cutting tools, moulds, and components within industries like automotive and aerospace [21]. AlN: Valued for its thermal conductivity, AlN coatings are suitable for applications requiring effective heat dissipation. Electronic components like heat sinks benefit from AlN coatings [22]. Zirconium Nitride (ZrN): Offering good wear resistance and high-temperature stability, ZrN coatings find application in scenarios involving elevated temperatures and abrasive wear. Cutting tools for machining hardened steels are a common use case [23]. Boron Nitride: While not traditionally used as a coating, BN, a ceramic material, possesses excellent lubricating properties. It is occasionally employed as a dry lubricant in situations where conventional lubricants may be impractical [24]. Silicon Nitride (Si3N4): Though not frequently applied as a coating, Si3N4 is a ceramic material with remarkable mechanical and thermal properties. It is widely utilized in tools used for cutting purposes, bearings, and other elements owing to its elevated hardness and resistance to thermal shock [25]. The application of these nitride coatings is facilitated through methods like PVD or CVD, allowing for precise control over coating thickness and uniformity. The selection of a specific nitride coating depends on the distinct requirements of the application, encompassing factors including properties like hardness, wear resistance, corrosion resistance, and thermal stability.

3.2 Nitride-based coating of Al substrate

Nitride-based coatings on Al substrates offer a transformative solution to enhance the surface properties of Al components, addressing inherent limitations such as low hardness and wear resistance. Through specialized coating processes, various nitride compounds, including materials such as TiN AlN, and CrN, are applied to the Al surface, imparting improved mechanical characteristics and extending the material's operational lifespan. The primary objective of nitride-based coatings is to fortify Al substrates against wear, corrosion, and other forms of deterioration. Nitrogen-rich environments, achieved through methods like PVD or CVD techniques, facilitate the formation of robust nitride layers on the Al surface. PVD methods involve vaporizing nitride compounds in a vacuum, followed by their deposition onto the substrate, while CVD relies on chemical reactions between gaseous precursors and the Al surface. The resulting nitride-coated Al exhibits heightened hardness, wear resistance, and corrosion resistance, making it suitable for a wide range of applications. Common uses include cutting tools, industrial machinery components, and automotive parts where durability and performance under challenging conditions are paramount [26], [27].

3.3 Types of strategies used for nitride coating

PVD: Widely employed, PVD is a technique for nitride coating. In this process, nitride compounds are vaporized within a vacuum, subsequently condensing on the surface of the substrate. PVD methods include cathodic arc deposition, sputtering, and evaporation [28].

Materials like TiN, CrN, and ZrN are often applied using PVD. CVD is another prevalent method for nitride coating. It involves the chemical reaction of gaseous precursors to form nitride compounds on the substrate's surface. This process allows for exact command of coating thickness and depth uniformity. CVD is frequently employed for coatings like TiN and AlN [29]. Plasma Nitriding: Plasma nitriding is a thermochemical process that introduces nitrogen into the surface of a metal by using plasma. The substrate is exposed to a nitrogen-rich environment at elevated temperatures, promoting the formation of nitrides on the material's surface. This method is often used for ferrous materials [30]. Ion Nitriding: Ion nitriding involves bombarding the substrate with nitrogen ions in a vacuum chamber at elevated temperatures. This process enhances the hardness and wear resistance of the material by forming nitride layers. Ion nitriding is commonly applied to steel components [31]. Salt Bath Nitriding: This is a diffusion-based nitriding method where the substrate is immersed in a bath of molten salts containing nitrogen. The nitrogen diffuses into the material's surface, forming nitrides. This method is suitable for certain types of steels [32]. Gas Nitriding: In gas nitriding, the substrate is exposed to ammonia gas in a controlled atmosphere at elevated temperatures. The nitrogen from the ammonia diffuses into the material, forming nitrides. This method is commonly used for alloy steels [33]. Laser Nitriding: Laser nitriding is a selective surface hardening technique where a laser beam is used to locally heat the material, followed by nitrogen diffusion. This precise method is suitable for localized surface treatment [34]. PVD methods such as cathodic arc deposition and sputtering are commonly utilized. PVD allows for precise control over coating thickness and uniformity, enabling the deposition of nitride compounds like TiN or CrN onto the BPs [35]. CVD involves the chemical reaction of gaseous precursors to form nitride compounds on the substrate's surface. CVD techniques, including low-pressure CVD or plasma-enhanced CVD, facilitate the formation of conformal and adherent nitride coatings, enhancing properties like hardness and wear resistance [36]. Plasma Nitriding is a thermochemical process where nitrogen-rich plasma is used to introduce nitrogen into the surface of the metal at elevated temperatures. This treatment enhances surface hardness, wear resistance, and corrosion resistance of the metallic BPs, making them more resilient in challenging fuel cell environments [37].

3.3.1 Physical Vapour Deposition

Physical Vapour Deposition (PVD) methods for nitride layering on metallic BPs involve the deposition of nitride compounds onto the substrate through physical processes in a vacuum environment. In Cathodic Arc Deposition, a high-current, low-voltage electric arc is generated between a cathode target (made of the nitride material) and the substrate. The nitride material is vaporized from the cathode target due to the arc discharge. These vaporized atoms then condense on the BP, forming a nitride coating [38]. Sputter Deposition, Sputter deposition involves bombarding a target (made of the nitride material) with energetic ions, resulting in the expulsion of atoms from the target. The ejected atoms then deposit on the metallic BP, forming a nitride coating. Reactive sputtering, where nitrogen is introduced during the process, is often used for nitride coatings [39]. Evaporation, in this method, the nitride material undergoes high-temperature heating within a vacuum, inducing it to evaporate. The evaporated material then condenses onto the substrate surface, forming a nitride coating. This method is less common for nitride coatings compared to other PVD methods [40]. Plasma Enhanced Physical Vapour Deposition combines the principles of PVD with plasma enhancement. A plasma is created in the deposition chamber, enhancing the energy and reactivity of the vaporized nitride material. This results in a more uniform and adherent nitride coating on the BP [41]. Filtered Cathodic Vacuum Arc is an advanced form of cathodic arc deposition that includes additional filtering mechanisms. The filtering helps to control the composition of the deposited film, ensuring a high-quality nitride coating on the BP with improved properties [42].

3.3.2 Chemical Vapour Deposition

Chemical Vapour Deposition (CVD) methods for nitride coating on metallic BPs are integral to enhancing surface characteristics, such as resistance to corrosion and wear durability. Reduced pressure Chemical Vapour Deposition operates under reduced pressure conditions, allowing precise control over the deposition process. Gaseous precursors containing nitrogen and the desired metal are introduced into a vacuum chamber at low pressure. A chemical reaction occurs on the surface of the metal BP, resulting in the formation of a nitride coating [43]. Atmospheric Pressure Chemical Vapour Deposition constitutes a simpler and more cost-effective CVD method operating at atmospheric pressure. Gaseous precursors are introduced at atmospheric pressure, initiating a chemical reaction on the exterior of the metallic BP. This process leads to the deposition of a nitride coating [44]. Plasma-Enhanced Chemical Vapour Deposition combines traditional CVD with plasma activation to enhance the deposition process. A plasma is generated within the deposition chamber, increasing the reactivity of the gaseous precursors. This heightened reactivity promotes a controlled and adherent nitride coating on the metallic BP [45]. Metalorganic Chemical Vapour Deposition involves the use of metalorganic precursors containing the desired metal and nitrogen sources. Metalorganic precursors are introduced to the deposition chamber, in which they undergo chemical transformations on the surface of the metallic BP. This results in the formation of a nitride coating [46]. Hot-Wall CVD, in hot-wall CVD, the substrate is externally heated, enhancing the chemical reaction between precursors. Gaseous precursors containing nitrogen and the desired metal are introduced into the heated chamber, leading to a chemical reaction on the surface of the metal BP. This results in the deposition of a nitride coating [47].

3.3.3 Plasma nitriding

Plasma nitriding is a thermochemical surface treatment method used to introduce nitrogen into the surface of metallic components, forming nitrides. While plasma nitriding is not a traditional deposition method like PVD or CVD, it is an important process for enhancing the surface properties of metal BPs. Plasma Nitriding Process, Plasma nitriding involves the use of a nitrogen-rich plasma at elevated temperatures to introduce nitrogen into the surface of the metal. The process is typically applied to ferrous materials, such as stainless steels, and may not be suitable for all metallic substrates. The treatment depth and properties of the nitride layer depend on factors such as treatment time, temperature, and gas composition [48], [49].

4. Base Material on Al alloy BPs.

Table. 2. Chemical composition (wt.%) of Al alloy metal materials for BPs.

Series / Grade	Mg	Cr	Mn	Si	Cu	Zn	Fe	Ti	Ref
Al 356	0.20-0.45		0.35	6.5-7.5	0.25	0.35	0.60	0.25	[50]

Al 5052 (magnesium)	2.60	0.25	0.10	0.23	0.12	0.15	0.28	-	[51]
Al 6061 (Magnesium & silicon)	2.44	0.18	0.02	0.13	0.02	0.04	0.35	-	[52]
Al 7075 (zinc)	2.376	0.238	0.273	0.22	1.605	5.572	0.275	0.020	[53]

Table. 2. provides a chemical composition of Al alloy materials utilized for BPs in fuel cells that can be customized to meet specific performance criteria. Typically, Al serves as the primary constituent, constituting a substantial portion of the alloy. To enhance conductivity, copper is commonly added, though its concentration is carefully controlled to prevent adverse impacts on corrosion resistance. Silicon is introduced to improve strength and casting fluidity, while manganese contributes to both strength and corrosion resistance. The inclusion of magnesium enhances overall alloy strength, especially in heat-treatable variations. Zinc and titanium may also be incorporated to further bolster strength and hardness, with the precise alloy composition dependent on the desired balance of properties. Beyond these primary elements, trace amounts of chromium, nickel, and other elements may be present based on specific alloy formulations.

5. Nitride Coating on Al alloy BPs.

Table. 3. Al alloy data reported in the literature.

Sl. No.	Substrate	Surface modification	Electrolyte	Corrosion Current Density (mA/cm^2)	Potentiostatic result (mA/cm^2)	ICR ($m\Omega\ cm^2$)	Ref
1	Al 356	TiN/CrN (PVD)	0.5 M H_2SO_4 H2 bubbled	-	4.6 at -0.1 V (vs. SCE)	-	[54]
			0.5 M H_2SO_4 air bubbled	-	29.03 at 0.6 V (vs. SCE)	-	
2	Al 5052	TiN (PVD)	0.001 M H_2SO_4 + 0.1 ppm NaF at 70 °C	34.4 (H_2)	-	20.08 at 150 N/cm^2	[55]
		CrN (PVD)	0.001 M H_2SO_4 + 0.1 ppm NaF at 70 °C	36.8 (H_2)	-	7.76 at 150 N/cm^2	
		C/TiN (PVD)	0.001 M H_2SO_4 + 0.1 ppm NaF at 70 °C	0.4 (H_2) 36 (air)	-	-	

		C/CrN (PVD)	0.001 M H_2SO_4 + 0.1 ppm NaF at 70 °C	0.5 (H_2) 40.7 (air)	10 at -0.47 V (vs. MSE)	4.08 at 150 N/cm^2	
3	Al 6061	TiN/Ti	0.1 M H2SO4/2 ppm HF	0.6	-	0.022	[56]
3	Al 7075	TiN/CrN (PVD)	0.5 M H_2SO_4 H_2 bubbled	-	60.29 at -0.1 V (vs. SCE)	-	[54]
			0.5 M H_2SO_4 air bubbled	-	89.68 at 0.6 V (vs. SCE)	-	

Table. 3. provides a concise summary of data on nitride-based coatings on Al alloy BPs within PEMFCs. applications. The focus is on crucial aspects such as the Al substrate, surface modification through nitride coatings, the impact on the electrolyte, corrosion current density, potentiostat results, and ICR. The "Al Substrate" column specifies the type of Al alloy used as the substrate for BPs. "Surface Modification" outlines the specific nitride-based coating applied to enhance the plates' properties. The "Electrolyte" column details the type of electrolyte used in the PEMFC setup. "Corrosion Current Density" provides insights into the corrosion rate, a critical parameter for evaluating the effectiveness of the nitride coating. The "Potentiostat Results" section encapsulates the outcomes of potentiostatic tests, shedding light on the coating's performance under controlled electrochemical conditions. Finally, "ICR" highlights the Internal Contact Resistance, a key metric reflecting the coating's impact on the electrical conductivity of the BPs. This table serves as a comprehensive reference for researchers and practitioners seeking a consolidated overview of critical data points in the context of nitride-based applications of coatings on Al alloy BPs for PEMFC applications [57].

6. Conclusion

In conclusion, the advancement and characterization of nitride coating materials on Al alloy-based BPs in PEMFCs represent a significant stride in enhancing the performance and longevity of these critical components. The research and development in this domain, incorporating nitride coatings on Al substrates, have yielded promising outcomes. The nitride coatings, known for their corrosion resistance and durability, play a pivotal role in mitigating the challenges posed by aggressive fuel cell environments. The comprehensive characterization of these coatings involves evaluating parameters such as corrosion current density, potentiostat results, and ICR, providing valuable insights into their effectiveness. However, challenges such as coating porosity and adhesion issues have been identified, underscoring the need for continued research to optimize these materials further. The ongoing efforts in advancing nitride-coated Al alloy BPs showcase the commitment to pushing within the realm of fuel cell technology, with a focus on improving efficiency., prolonged lifespan, and increased reliability in PEMFC applications. As these advancements progress, they have the potential to make a substantial contribution to the broader goal of developing environmentally friendly and efficient power solutions.

References

[1] C. Rayment and S. Sherwin, "Introduction to Fuel Cell Technology," 2003.

[2] B. Cook, "Introduction to fuel."

[3] A. Kazim, "Introduction of PEM fuel-cell vehicles in the transportation sector of the United Arab Emirates." [Online]. Available: www.elsevier.com/locate/apenergy

[4] J. Wu *et al.*, "A review of PEM fuel cell durability: Degradation mechanisms and mitigation strategies," *Journal of Power Sources*, vol. 184, no. 1. pp. 104–119, Sep. 15, 2008. doi: 10.1016/j.jpowsour.2008.06.006.

[5] A. Fathy and A. Alanazi, "An Efficient White Shark Optimizer for Enhancing the Performance of Proton Exchange Membrane Fuel Cells," *Sustainability (Switzerland)*, vol. 15, no. 15, Aug. 2023, doi: 10.3390/su151511741.

[6] S. M. Haile, "Fuel cell materials and components," *Acta Mater*, vol. 51, no. 19, pp. 5981–6000, Nov. 2003, doi: 10.1016/j.actamat.2003.08.004.

[7] S. Basu, "Proton exchange membrane fuel cell technology: India's perspective," *Proceedings of the Indian National Science Academy*, vol. 81, no. 4. Indian National Science Academy, pp. 865–890, Sep. 01, 2015. doi: 10.16943/ptinsa/2015/v81i4/48301.

[8] E. Middelman, W. Kout, B. Vogelaar, J. Lenssen, and E. De Waal, "Bipolar plates for PEM fuel cells," in *Journal of Power Sources*, May 2003, pp. 44–46. doi: 10.1016/S0378-7753(03)00070-3.

[9] X. Li and I. Sabir, "Review of bipolar plates in PEM fuel cells: Flow-field designs," *Int J Hydrogen Energy*, vol. 30, no. 4, pp. 359–371, 2005, doi: 10.1016/j.ijhydene.2004.09.019.

[10] Y. A. Dobrovol'skii *et al.*, "Materials for bipolar plates for proton-conducting membrane fuel cells," *Russ J Gen Chem*, vol. 77, no. 4, pp. 752–765, Apr. 2007, doi: 10.1134/S1070363207040366.

[11] A. E. Fetohi, R. M. Abdel Hameed, K. M. El-Khatib, and E. R. Souaya, "Study of different aluminum alloy substrates coated with Ni-Co-P as metallic bipolar plates for PEM fuel cell applications," *Int J Hydrogen Energy*, vol. 37, no. 14, pp. 10807–10817, Jul. 2012, doi: 10.1016/j.ijhydene.2012.04.066.

[12] W. F. Miao and D. E. Laughlin, "PRECIPITATION HARDENING IN ALUMINUM ALLOY 6022," 1999.

[13] D. Alexander *et al.*, "In-situ Diagnostics of Composite Filament Material Suitable for Bi-Polar Plate Using Additive Manufacturing," *ECS Trans*, vol. 104, no. 8, pp. 283–292, Oct. 2021, doi: 10.1149/10408.0283ecst.

[14] A. Sanjid, M. R. Anisur, and R. K. Singh Raman, "Durable degradation resistance of graphene coated nickel and Monel-400 as bi-polar plates for proton exchange

membrane fuel cell," *Carbon N Y*, vol. 151, pp. 68–75, Oct. 2019, doi: 10.1016/j.carbon.2019.05.071.

[15] T. Wilberforce *et al.*, "A comprehensive study of the effect of bipolar plate (BP) geometry design on the performance of proton exchange membrane (PEM) fuel cells," *Renewable and Sustainable Energy Reviews*, vol. 111, pp. 236–260, Sep. 2019, doi: 10.1016/j.rser.2019.04.081.

[16] S. Porstmann, T. Wannemacher, and W. G. Drossel, "A comprehensive comparison of state-of-the-art manufacturing methods for fuel cell bipolar plates including anticipated future industry trends," *Journal of Manufacturing Processes*, vol. 60. Elsevier Ltd, pp. 366–383, Dec. 01, 2020. doi: 10.1016/j.jmapro.2020.10.041.

[17] Y. Song *et al.*, "Review on current research of materials, fabrication and application for bipolar plate in proton exchange membrane fuel cell," *Int J Hydrogen Energy*, vol. 45, no. 54, pp. 29832–29847, Nov. 2020, doi: 10.1016/j.ijhydene.2019.07.231.

[18] I. Bertoti, "Characterization of nitride coatings by XPS," 2002. [Online]. Available: www.chemres.huyAKKL

[19] A. Liu, J. Deng, H. Cui, Y. Chen, and J. Zhao, "Friction and wear properties of TiN, TiAlN, AlTiN and CrAlN PVD nitride coatings," *Int J Refract Metals Hard Mater*, vol. 31, pp. 82–88, Mar. 2012, doi: 10.1016/j.ijrmhm.2011.09.010.

[20] P. Patsalas, N. Kalfagiannis, and S. Kassavetis, "Optical properties and plasmonic performance of titanium nitride," *Materials*, vol. 8, no. 6, pp. 3128–3154, 2015, doi: 10.3390/ma8063128.

[21] A. Lippitz and T. Hübert, "XPS investigations of chromium nitride thin films," *Surf Coat Technol*, vol. 200, no. 1-4 SPEC. ISS., pp. 250–253, Oct. 2005, doi: 10.1016/j.surfcoat.2005.02.091.

[22] R. M. R. Pinto, V. Gund, R. A. Dias, K. K. Nagaraja, and K. B. Vinayakumar, "CMOS-Integrated Aluminum Nitride MEMS: A Review," *Journal of Microelectromechanical Systems*, vol. 31, no. 4. Institute of Electrical and Electronics Engineers Inc., pp. 500–523, Aug. 01, 2022. doi: 10.1109/JMEMS.2022.3172766.

[23] S. Grigoriev *et al.*, "Influence of tribological properties of Zr-ZrN-(Zr,Cr,Al)N and Zr-ZrN-(Zr,Mo,Al)N multilayer nanostructured coatings on the cutting properties of coated tools during dry turning of Inconel 718 alloy," *Wear*, vol. 512–513, Jan. 2023, doi: 10.1016/j.wear.2022.204521.

[24] Z. Qin, Y. Zeng, Q. Hua, Q. Xu, X. Shen, and Y. Min, "Synergistic effect of hydroxylated boron nitride and silane on corrosion resistance of aluminum alloy 5052," *J Taiwan Inst Chem Eng*, vol. 100, pp. 285–294, Jul. 2019, doi: 10.1016/j.jtice.2019.04.029.

[25] M. Okayasu, M. Hitomi, and H. Yamazaki, "Mechanical and fatigue strengths of silicon nitride ceramics in liquid aluminum alloys," *J Eur Ceram Soc*, vol. 29, no. 11, pp. 2369–2378, Aug. 2009, doi: 10.1016/j.jeurceramsoc.2009.01.013.

[26] C. Wang, H. Huang, H. Wu, J. Hong, L. Zhang, and J. Yan, "Ultra-low wear of titanium alloy surface under lubricated conditions achieved by laser texturing and simultaneous nitriding," *Surf Coat Technol*, vol. 474, Dec. 2023, doi: 10.1016/j.surfcoat.2023.130083.

[27] J. Li, Y. Huang, Y. Zhou, and F. Zhu, "Role of boron nitride nanosheet coatings on aluminum substrates during the nanoindentation from the atomic perspective," *Appl Surf Sci*, vol. 608, Jan. 2023, doi: 10.1016/j.apsusc.2022.155126.

[28] S. Y. Lee and S. H. Kim, "Metal nitride coatings by physical vapor deposition (PVD) for a wear resistant aluminum extrusion die," *J Nanosci Nanotechnol*, vol. 14, no. 12, pp. 8993–8998, Dec. 2014, doi: 10.1166/jnn.2014.10058.

[29] Y. G. Roman and A. P. M. Adriaansen, "241 248 PREPARATION AND CHARACTERIZATION 241 ALUMINIUM NITRIDE FILMS MADE BY LOW PRESSURE CHEMICAL VAPOUR DEPOSITION: PREPARATION AND PROPERTIES," 1989.

[30] F. Abdi, H. Aghajani, A. Taghizadeh Tabrizi, L. Nasimi, and F. Fazli Shokouhi, "Study on the effect of the crack closing of AlCoCrFeMnNi high entropy alloy electro-spark deposited coating by plasma nitriding on the corrosion resistance," *J Alloys Compd*, vol. 966, Dec. 2023, doi: 10.1016/j.jallcom.2023.171629.

[31] M. Okumiya, Y. Tsunekawa, H. Sugiyama, Y. Tanaka, N. Takano, and M. Tomimoto, "Surface modification of aluminum using ion nitriding and barrel nitriding," *Surf Coat Technol*, vol. 200, no. 1-4 SPEC. ISS., pp. 35–39, Oct. 2005, doi: 10.1016/j.surfcoat.2005.02.110.

[32] E. Richter *et al.*, "Nitriding of stainless steel and aluminium alloys by plasma immersion ion implantation," 1281.

[33] G. Aktaş Çelik, Ş. H. Atapek, Ş. Polat, A. Obrosov, and S. Weiß, "Nitriding Effect on the Tribological Performance of CrN-, AlTiN-, and CrN/AlTiN-Coated DIN 1.2367 Hot Work Tool Steel," *Materials*, vol. 16, no. 7, Apr. 2023, doi: 10.3390/ma16072804.

[34] M. Shahien, M. Yamada, T. Yasui, and M. Fukumoto, "Cubic aluminum nitride coating through atmospheric reactive plasma nitriding," in *Journal of Thermal Spray Technology*, Mar. 2010, pp. 635–641. doi: 10.1007/s11666-010-9469-0.

[35] J. Barranco *et al.*, "Cr and Zr/Cr nitride CAE-PVD coated aluminum bipolar plates for polymer electrolyte membrane fuel cells," *Int J Hydrogen Energy*, vol. 35, no. 20, pp. 11489–11498, Oct. 2010, doi: 10.1016/j.ijhydene.2010.05.050.

[36] J. Bi *et al.*, "Development and evaluation of nitride coated titanium bipolar plates for PEM fuel cells," *Int J Hydrogen Energy*, vol. 46, no. 1, pp. 1144–1154, Jan. 2021, doi: 10.1016/j.ijhydene.2020.09.217.

[37] M. Xu, S. Kang, J. Lu, X. Yan, T. Chen, and Z. Wang, "Properties of a plasma-nitrided coating and a crnx coating on the stainless steel bipolar plate of PEMFC," *Coatings*, vol. 10, no. 2, Feb. 2020, doi: 10.3390/coatings10020183.

[38] S. Anders *et al.*, "Formation of metal oxides by cathodic arc deposition," 1995.

[39] S. D. Ekpe and S. K. Dew, "Investigation of thermal flux to the substrate during sputter deposition of aluminum," *Journal of Vacuum Science & Technology A: Vacuum, Surfaces, and Films*, vol. 20, no. 6, pp. 1877–1885, Nov. 2002, doi: 10.1116/1.1507342.

[40] A. Jankowski and J. Hayes, "The evaporative deposition of aluminum coatings and shapes with grain size control," in *Thin Solid Films*, Jan. 2004, pp. 568–574. doi: 10.1016/j.tsf.2003.07.018.

[41] F. Böke, I. Giner, A. Keller, G. Grundmeier, and H. Fischer, "Plasma-Enhanced Chemical Vapor Deposition (PE-CVD) yields better Hydrolytical Stability of Biocompatible SiOx Thin Films on Implant Alumina Ceramics compared to Rapid Thermal Evaporation Physical Vapor Deposition (PVD)," *ACS Appl Mater Interfaces*, vol. 8, no. 28, pp. 17805–17816, Jul. 2016, doi: 10.1021/acsami.6b04421.

[42] B. K. Tay, Z. W. Zhao, and D. H. C. Chua, "Review of metal oxide films deposited by filtered cathodic vacuum arc technique," *Materials Science and Engineering R: Reports*, vol. 52, no. 1–3. pp. 1–48, May 30, 2006. doi: 10.1016/j.mser.2006.04.003.

[43] W.-H. Kim, W. J. Maeng, M.-K. Kim, and H. Kim, "Low Pressure Chemical Vapor Deposition of Aluminum-Doped Zinc Oxide for Transparent Conducting Electrodes," *J Electrochem Soc*, vol. 158, no. 8, p. D495, 2011, doi: 10.1149/1.3599055.

[44] R. G. Gordon, D. M. Hoffman, and U. Riaz, "Atmospheric pressure chemical vapor deposition of aluminum nitride thin films at 200-250 °C 3Zr(NEt 2) 4 + 4NH 3-»> Zr 3 N 4 + 12HNMe 2 Al 2 (NMe 2)6 + 2NH 3-* 2A1N + 6HNMe 2 (1) (2) Herein we report the CVD of A1N films at 200-250 °C using Al 2 (NMe 2)," 1991. [Online]. Available: http://journals.cambridge.org

[45] G. Dingemans, M. C. M. Van De Sanden, and W. M. M. Kessels, "Plasma-enhanced chemical vapor deposition of aluminum oxide using ultrashort precursor injection pulses," *Plasma Processes and Polymers*, vol. 9, no. 8, pp. 761–771, Aug. 2012, doi: 10.1002/ppap.201100196.

[46] S. Leone, I. Streicher, M. Prescher, P. Straňák, and L. Kirste, "Metal-Organic Chemical Vapor Deposition of Aluminum Yttrium Nitride," *Physica Status Solidi - Rapid Research Letters*, Oct. 2023, doi: 10.1002/pssr.202300091.

[47] G. Wagner, W. Leitenberger, K. Irmscher, F. Schmid, M. Laube, and G. Pensl, "Aluminum incorporation into 4H-SiC layers during epitaxial growth in a hot-wall CVD system," *Materials Science Forum*, vol. 389–393, no. 1, pp. 207–210, 2002, doi: 10.4028/www.scientific.net/msf.389-393.207.

[48] H.-Y. Chen, H.-R. Stock, and P. Mayr, "Plasma-assisted nitriding of aluminium," 1994.

[49] P. Visuttipitukul, T. Aizawa, and H. Kuwahara, "Advanced Plasma Nitriding for Aluminum and Aluminum Alloys."

[50] M. C. Mehta, D. Mandal, and S. K. Chaudhury, "Microstructural Changes and Quality Improvement of Al7Si0.2Mg (356) Alloy by Die Vibration," *International Journal of Metalcasting*, vol. 14, no. 4, pp. 987–998, Oct. 2020, doi: 10.1007/s40962-020-00408-3.

[51] E. B. Moustafa and A. O. Mosleh, "Effect of (Ti–B) modifier elements and FSP on 5052 aluminum alloy," *J Alloys Compd*, vol. 823, May 2020, doi: 10.1016/j.jallcom.2020.153745.

[52] W. Xiao and Y. Wang, "Corrosion resistance of aluminum fluoride modified 6061 aluminum alloy," *Mater Lett*, vol. 298, Sep. 2021, doi: 10.1016/j.matlet.2021.129932.

[53] B. Zhou, B. Liu, S. Zhang, R. Lin, Y. Jiang, and X. Lan, "Microstructure evolution of recycled 7075 aluminum alloy and its mechanical and corrosion properties," *J Alloys Compd*, vol. 879, Oct. 2021, doi: 10.1016/j.jallcom.2021.160407.

[54] A. Bolouri and C. G. Kang, "Study on dimensional and corrosion properties of thixoformed A356 and AA7075 aluminum bipolar plates for proton exchange membrane fuel cells," *Renew Energy*, vol. 71, pp. 616–628, 2014, doi: 10.1016/j.renene.2014.06.021.

[55] Z. Li, K. Feng, Z. Wang, X. Cai, C. Yao, and Y. Wu, "Investigation of single-layer and multilayer coatings for aluminum bipolar plate in polymer electrolyte membrane fuel cell," in *International Journal of Hydrogen Energy*, Elsevier Ltd, May 2014, pp. 8421–8430. doi: 10.1016/j.ijhydene.2014.03.136.

[56] W. M. Yan, J. C. Lin, C. Y. Chen, and M. Amani, "Performance evaluation of TiN/Ti coatings on the aluminum alloy bipolar plates for PEM fuel cells," *Renew Energy*, vol. 216, Nov. 2023, doi: 10.1016/j.renene.2023.119042.

[57] P. Yi *et al.*, "Effect of plasma electrolytic nitriding on the corrosion behavior and interfacial contact resistance of titanium in the cathode environment of proton-exchange membrane fuel cells," *J Power Sources*, vol. 418, pp. 42–49, Apr. 2019, doi: 10.1016/j.jpowsour.2019.02.043.

Microwave Assisted Zinc Sulphide Quantum Dots for Energy Device Applications

Aman Rai[1], Peeyush Phogat[1], Shreya[1], Ranjana Jha[1], and Sukhvir Singh [1]*

[1]Research Lab for Energy Systems, Department of Physics, Netaji Subhas University of Technology, Delhi, India

Abstract.

Zinc sulfide (ZnS) quantum dots (QDs) have received a lot of attention because of their potential usage in solar cell applications. The present work illustrates a straightforward green synthesis route for ZnS QDs, which is both cost-effective and environmentally friendly. The phase identification of the synthesized material was performed through the X-ray diffraction technique which revealed the presence of single-phase zinc sulfide. The crystallite size was calculated through various techniques such as Debye Scherrer's, Williamson Hall and Size-Strain plot revealing quantum confinement effects. High-resolution transmission electron microscope (HR-TEM) revealed the presence of quantum dots within the quantum size range and demonstrated excellent quantum yield. However, the decrease in the particle size has increased the band gap for ZnS QDs to 3.4 eV with a refractive index of 2.29. The polycrystalline character of the as-synthesized ZnS was shown by the selected area electron diffraction (SAED) pattern of the corresponding TEM images. Morphological analysis, carried out via field emission scanning electron microscopy (FESEM), unveiled the existence of zinc sulfide quantum dots agglomerates. Elemental composition analysis was performed using energy dispersive spectroscopy (EDS) as an attachment to FESEM, which confirmed the existence only of zinc and sulphur.

Keywords: Zinc Sulfide, reduced band gap, Microwave, Solar cell

1 Introduction

Nanoparticles due to their unique characteristics and versatility, have gained significant importance in different fields. Among these, Quantum dots (QDs) have shown outstanding efficacy in photonic devices, attracting significant interest in a variety of applications. QDs offer several benefits over standard organic fluorophores,
Such as having high fluorescence intensity, broad absorption coefficients, narrow emission spectra, size-dependent emission, and strong photostability[1]. Over the years, efforts have been made for the

* Corresponding author: peeyush.phogat@gmail.com

synthesis of high fluorescent QDs. The quantum confinement effect further amplifies these characteristics, resulting in narrow emission spectra and a wide range of excitation wavelengths[2]. Such characteristics attracted significant interest in technical applications such as biological fluorescence and imaging[3]. These nanoparticles have also been used in wastewater treatment as well as water-splitting photocatalytic degradation and disinfection methods.

ZnS is a wide band gap II–VI semiconductor nanomaterial. Over the years many synthesis routes have been taken to prepare zinc sulfide QDs which include the hydrothermal method, co-precipitation method, microwave-assisted method, and many more the quantum confinement effect limits the spatial mobility of the exciton within a particle[4]–[6]. This confinement causes the formation of discrete energy levels, also known as energy sub-bands, with greater energy band gaps. As a result, when compared to the analogous bulk materials, these altered energy characteristics have apparent impacts on the electrical and optical properties. The structure and morphology of ZnS QDs are certainly the most important in achieving enhanced photoelectrical characteristics. Over several years, the synthesis of ZnS quantum dots has been the focus of research, with attempts made to improve the material's optical and structural features[7]–[10].

The present study accentuates the synthesis of zinc sulfide QDs by microwave-assisted method. The novelty of the presented work lies in microwave synthesis, where particle size reduction has been seen. Furthermore, the given work focuses on low-cost green synthesis. The X-ray diffraction pattern (XRD) was used to study structural information. UV-Vis spectroscopy was used to analyse the sample's optical and electrical properties. A transmission electron microscope (HR-TEM) and a field emission scanning electron microscope (FE-SEM) were used for microstructural and morphological study.

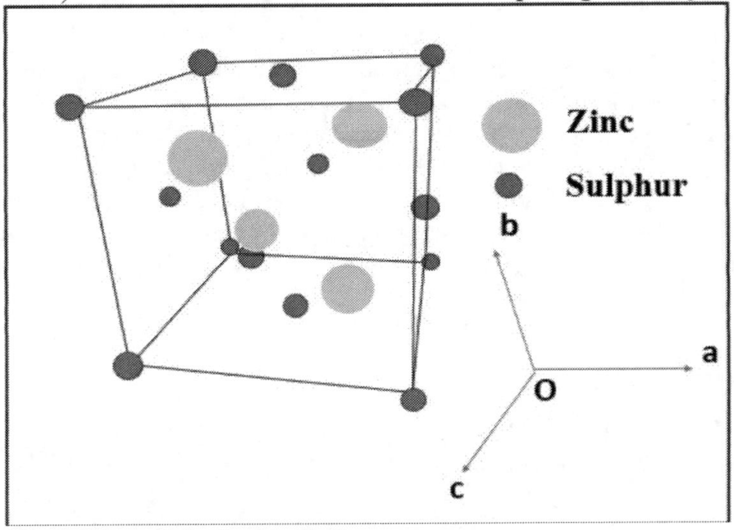

Fig.1 3D model of cubic ZnS.

2 Experimental Information

2.1 Chemical Used

Distilled water (DI) of near 5 mega-Ohm resistance and absolute alcohol was used for cleaning and washing purposes. All the solutions were prepared in DI water if not specified otherwise. All other chemicals used in synthesis are listed in Table 1.

Table 1 Chemicals Used

Chemicals Name	Company
Zinc oxide [ZnO]	M/s THOMAS BAKER
Sodium Sulfide [Na_2S]	M/s THOMAS BAKER

2.2 Synthesis

A 0.15 M colloidal solution of zinc oxide was synthesized by dispersing zinc oxide nanoparticles of average particle size of 30nm into 80 ml DI. The solution was then given microwave heating for 4 hours at 60° C along with ultraviolet (UV) radiation and ultrasonic vibrations. In the meantime, 0.17 M of sodium sulphide was prepared in 70 ml DI and was stirred on a magnetic stirrer for 30 minutes. After 30 minutes sodium sulfide solution was added to the zinc oxide solution through a burette as slowly as possible. After four hours of agitation, the mixture was allowed to cool naturally in a room temperature of room temperature. Following that, the sample was centrifuged and washed with a water and ethanol to remove any probable impurities. Following that, the sample was heated in a vacuum oven to dry it, resulting in the production of a powdered sample. However, there are drawbacks associated with microwave synthesis. One such limitation is the nature of microwave heating, which can result in temperature fluctuations within the reaction container. Consequently, the size distribution of ZnS quantum dots (QDs) may be wider compared to heating techniques. Additionally, the quick heating process, in microwaves can cause nucleation and growth rates posing challenges in achieving precise control over the final size of the QDs. On the hand slower heating enables more regulated growth and enhances uniformity in size.

3 Result and Discussion

3.1 Characterization Details

The fine powdered of synthesized sample was analyzed using an X-Ray Diffractometer, make: Rigaku, model: smart lab 3KW X-Ray Diffractometer. Copper (Cu) Kα-radiation (λ=1.5406A°) was used to record the x-ray diffraction pattern. FESEM (model:7610FPlus, make:

JEOL) with EDS as an attachment was used for the morphological and elemental compositional analysis. SHIMADZU UV-2600i Uv-Vis spectrometer was used to record the optical response of the as-synthesized material. The TEM images were taken by TECNAI(HR-TEM) operated at 200 KV.

3.2 X-ray Diffraction

The X-ray diffraction pattern for synthesized zinc sulfide was recorded over the range of 5^0 to 90^0 for $2\theta(^0)$ as shown in Fig-2. The obtained XRD spectra was then analysed and peaks for zinc sulphide were cross-referred with ***JCPDS file 98-007-7082*** showing a good match with single phase ZnS with cubic crystal system. The peaks were identified and marked with Miller indices (hkl) of (111), (002), (022), (113), (222), (004), (133), (024), and (244) shown in the figure. The as-synthesised ZnS X-ray pattern indicates a single-phase with space group F-43 m and space group number 216 which depicts that the synthesized zinc sulfide has FCC structure and symmetry along the axis rotation mirror plane and is centrosymmetric. Sharp and distinct peaks of zinc sulphide depicted a high crystalline nature.

The sample shows a cubic crystal structure for which the lattice parameter is given by equation-1 as given below ,

$$\frac{1}{d^2} = \frac{h^2}{a^2} + \frac{k^2}{b^2} + \frac{l^2}{c^2} \qquad \text{.....................(1)}$$

Here interplanar spacing is denoted by d, while the Miller indices are denoted by h,k,l. and for cubic structures, it is a well-known fact that a=b=c and α=β=γ. Table 2 presents the comparison of calculated lattice parameters and standard values from the JCPDS file.

Table2. Lattice Parameters

Parameters	**As synthesized ZnS**	**ZnS in JCPDS file: *00-001-1047***
a (Å)	5.397	5.4090
b(Å)	5.397	5.4090
C(Å)	5.397	5.4090

3.3 Lattice Parameters, Crystallite Size and Strain

Crystallite size of a material is a significant parameter in material science which provides information about material characterization, Grain size distribution, mechanical properties, and optical properties. The crystallite size was determined using the widely known Debye-Scherrer formula given in equation 2 as[11],

$$D = \frac{K\lambda}{\beta Cos\theta} \qquad \text{......... (2)}$$

In this equation, D denotes the crystal grain size (crystallite size), K is a constant of 0.9 referred to as the Scherrer constant λ is the wavelength of the x-ray source given by 0.154 nm, β is full with at half of the maxima (FWHM), and θ is half of the peak positions in the XRD pattern. The value of crystallite size applying Scherer's formula was found to be 14 nm, which was near the quantum confinement region, indicating that there were quantum electronic states and that they were constrained. Debye-Scherrer's analysis was predicated on the notion that the material is devoid of any strain and crystallite is spherical. However, that is not the case here.

The observed deviation in lattice parameters from their standard values indicates the existence of microstructural strain within the materials (as shown in Table 1). The strain was identified using a well-known method called the Williamson Hall plot. The broadening of XRD is attributed to strain and crystallite size, according to the Williamson-Hall plot (W-H plot). The W-H equation has the following formula[12] ,

$$\beta_T Cos\theta = (\varepsilon \times 4Sin\theta) + \frac{K\lambda}{D} \qquad ....(3)$$

Where is ϵ strain, D refers to crystal grain size (crystallite size), K is a constant of 0.9 known as the Scherrer constant, and λ is the wavelength of the x-ray source, which is supplied by 0.154 nm. β is FWHM, or full width at half maximum, and θ is the half position of the XRD pattern also included in equation 3. Visualizing the graph of the 4Sinθ against Cosθ relation. The strain calculated through the W-H plot provides a negative strain or compression, with a crystallite size of 10 nm.

As the Williamson-Hall plot assumes stain in a material isisotropic.In reality,materials often have anisotropic strain which can make results less accurate therefore Size Strain(S-S) approach was also taken for the calculation of crystallite size and strain by using eq as[13],

$$(d\beta Cos\Theta)^2 = \frac{K\lambda}{D} + (d^2\beta Cos\Theta) + \frac{\varepsilon^2}{4} \qquad .... (4)$$

Here, where ε is strain, D is related to crystal grain size (crystallite size), K is a constant of 0.9 known as the Scherrer constant, and λ is the wavelength of the x-ray source, which is given by 0.154 nm. β is FWHM, or full width at half maximum, and θ is half position of the XRD pattern

calculated values were found to be 10.2 nm and 0.00147 respectively, comparing the value of slope depicted in figure-3(a&b) reveals the micro strain to be 0.0027 by Williamson hall plot and 0.00147 by Size Strain plot. Due to the negative strain present, the material displays compression. Table 3 presents all the values calculated from the Debye-Scherrer equation-H plot and Size-Strain plot.

Table3. Crystallite Size and strain

Parameters	Debye Scherrer	Williamson Hall plot	Size Strain Plot
Size	14 nm	10 nm	10 nm
Strain	-	0.0027	0.00147

Dislocation is described as a flaw and imperfection in the crystal structure caused by atom displacement from their ideal lattice positions. Dislocation density is the number of dislocations per unit volume and was derived using the equation.

$$\delta = \frac{1}{D^2} \qquad ....(5)$$

Where D is crystallite size and δ is dislocation density which was found to be 0.00368201.

A stacking fault is a form of crystallographic defect that develops when the usual stacking sequence is disrupted, which can occur during atom movement. Which is provided by,

$$\sigma = \frac{2\pi^2}{45(tan\theta)^{1/2}} \times \beta \qquad ....(6)$$

Where σ sacking fault,θ is peak position, β is full width half maxima. The stacking fault is found to be 0.224.

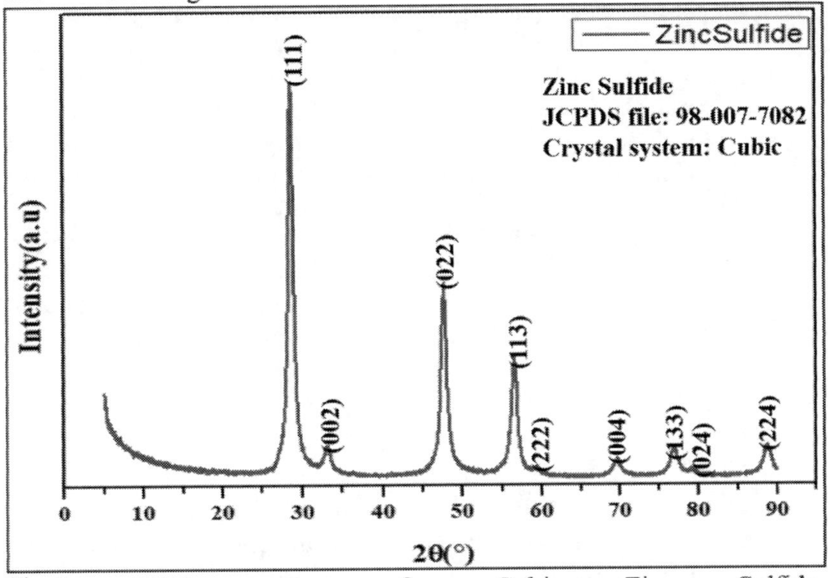

Fig.2 XRD pattern for Cubic Zinc Sulfide

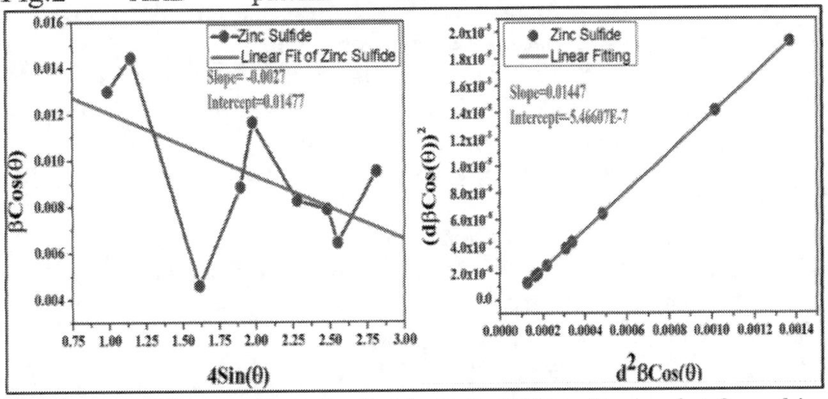

Fig.3.**(a)** Williamson Hall for Cubic ZnS. **(b)**Size Strain plot for cubic ZnS

3.4 UV-Spectroscopy

The UV-Vis spectra of as synthesized zinc sulfide was recorded in the wavelength range of 190 nm to 850 nm. The as-synthesized zinc sulfide showed absorption in the UV region only, which indicates that zinc sulfide has future applications which can impact the efficiency and performance of solar cells. UV absorption implies an increased capacity of photovoltaic materials to capture a broader range of sunlight and achieve more efficient energy conversion. In uv spectrum [Fig.3] highest absorbance peak is identified at 324 nm .

Size dependent band gap in QDs is a result of quantum effects. Smaller particles have larger band gap and vice versa. this unique behaviour of QDs allows precise control over band gap .Tauc plot shown in Fig.4 was obtained from further examination of absorption data and using beer lambert's principle which is given as[14],

$$\alpha h\upsilon = A[h\upsilon - E_g]^{\frac{1}{n}} \qquad \text{......... (7)}$$

Where A = Absorbance(a.u); α= Molecular extinction coefficient ; h =Plank's constant ; υ=Frequency ; E_g= direct band gap ; n=1/2(direct band gap),2 (indirect band gap). For Zns (direct band gap) n=1/2.Extrapolation of plotted graph was done by linear fitting the data points to get the band gap value of 3.4 eV. The 3.4 eV band gap of ZnS allows it to efficiently absorb UV radiation in the shorter wavelength range, making it a valuable choice for applications that require sensitivity to UV light. Standard bandgap of bulk zinc sulfide is 3.68 eV. The refractive index plays an important role in the solar cell application of material. The refractive index can affect how efficiently light interacts with the layer of solar cell. When the refractive indices of the material and the medium are matched, more light can penetrate the medium and reach the absorbing layer, increasing the efficiency of solar cells. To determine the refractive index of the synthesized samples equation (5) has been taken into consideration[15],

$$\frac{n^2-1}{n^2+2} = 1 - \sqrt{\frac{E_g}{20}} \qquad \text{.........(8)}$$

Where n represents the refractive index, while Eg signifies the direct band gap of zinc sulfide of 3.4 eV. The calculated refractive index from the 3.4 eV band gap came to be 2.29 which is relatively high indicating that the Synthesized sample is a strong optical medium and can significantly alter the path of light. High refractive indices are often associated with materials that are transparent and have optical properties useful for various optical applications.

HOMO represents the molecular orbit with the highest energy that contains electrons. When at ground state HOMO is the highest energy that an electron can stay at .LUMO can be described as the lowest energy in a molecule that does not contain any electrons. Calculation of HOMO(Valance band) and LUMO(Conduction band) band gap was done by equation below.

$$E_{VB} = \chi + E_e + 0.5 * E_g \qquad \text{.............(9)}$$

$$E_{CB} = E_{VB} + E_g \qquad \text{................(10)}$$

Where χ is called absolute electronegativity which has a value of 5.26 eV for Zinc sulfide E_e is the free electron energy on the hydrogen scale, E_{VB} and E_{CB} are the energy of the valance band and conduction band respectively and E_g is band gap of Synthesized sample. Calculated energies of valance and conduction band came out to be 2.46 eV and 5.86 eV respectively. Band gap of 3.4 eV depicts that such material falls under the category of wide range bandgap material. The wide band gap allows this material to significantly absorb and emit high- energy radiation in ultra violet (UV) region. wide band gap of materials are beneficial in photovoltaic and solar cells where they capture high energy photons efficiently.

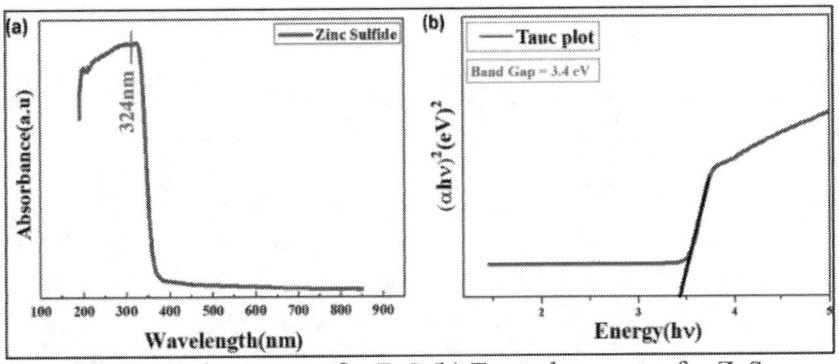

Fig.4.(a) Absorption spectra for ZnS,(b) Tauc plot spectra for ZnS.

3.5 FESEM

The morphological analysis of synthesized zinc sulfide was conducted using a High-Resolution Field Emission Scanning Electron Microscope(FE-SEM) equipped with Energy Dispersive Spectroscopy (EDX) (model-7610F plus, make-JEOL). As shown in Figure 4(a) and EDX spectrum in Fig.4(b) SEM images of Zns at high resolution and magnification depict that particles are in quantum confinement range which are aggregated due to high surface energy. While the materials were being synthesized, an EDX pattern was captured to confirm the elements present, and it showed the presence of high-intensity peaks due to Zn and S.

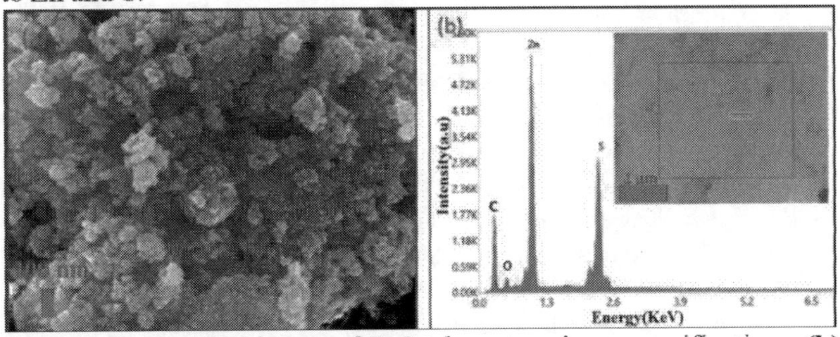

Fig.5**(a)** shows SEM image of ZnS taken at various magnifications. **(b)** EDX spectra of selected area for ZnS

3.6 HR-TEM

In this study, transmission electron microscopy (TEM) analysis was employed to visually assess the size, morphology, and dispersion characteristics of the as-prepared nanoparticles. As depicted in Figure 6(a), the TEM images reveal nanoparticles characterized by a nano-scaled size, spherical morphology, and a uniform spatial distribution. The average particle diameter is estimated to be approximately 15.8 nanometers. which is in agreement with the XRD results. The results are noteworthy, however, because a percentage of total particles had a diameter larger than 15 nm. Furthermore, large particles that appear to be the result of aggregation of Quantum Dots(QDs) are noticeable in Fig.6.These particles may only appear at rather high QD concentrations. Figure 6(b) demonstrates the corresponding area of the material's selected area electron diffraction (SEAD) pattern and depicts the peak with hkl values of (111),(002), and (022)[16].

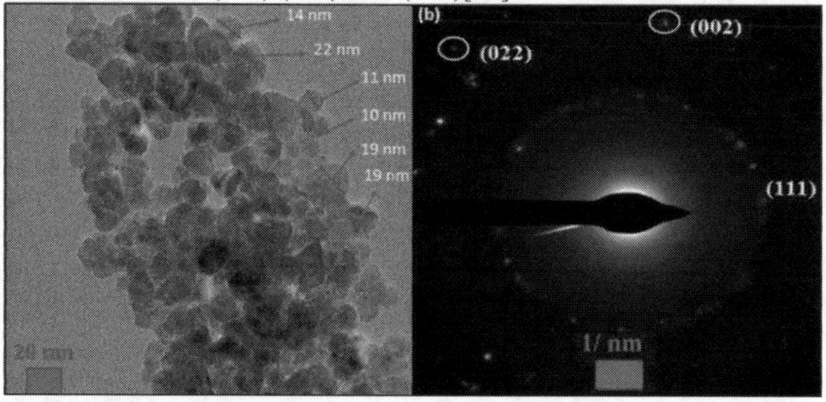

Fig.6**(a).** HR-TEM images of synthesized ZnS. (b) SEAD pattern of ZnS.

4 Conclusion

The current work is concentrated on the synthesis of ZnS with a reduced band gap. The ZnS was produced by conventional microwave synthesis. Structural study shows the presence of both zinc and sulfur. Scherer's formula determined the crystallite size to be 14 nm, which is within to the quantum confinement area. The W-H plot was used to calculate a negative strain associated with a crystallite size of 10 nm. In addition, dislocation density and stacking fault calculations were carried out. Band gap and refractive index ZnS were found to be 3.4eV and 2.29 respectively. SEM images show ZnS particles aggregated due to increased surface energy in the quantum confinement range. The EDX pattern indicates the elements by exhibiting high-intensity peaks caused by Zn and S. The transmission electron microscopy (TEM) analysis revealed nanoparticles with nano-scaled size, spherical morphology, and uniform distribution. The average particle diameter was 15.8 nanometers, with a percentage bigger than 15 nm. Large particles aggregating Quantum Dots occurred in great numbers.

References

[1] E. K. Goharshadi, S. H. Sajjadi, R. Mehrkhah, and P. Nancarrow, "Sonochemical synthesis and measurement of optical properties of zinc sulfide quantum dots," *Chemical Engineering Journal*, vol. 209, pp. 113–117, Oct. 2012, doi: 10.1016/J.CEJ.2012.07.131.

[2] E. K. Goharshadi, R. Mehrkhah, and P. Nancarrow, "Synthesis, characterization, and measurement of structural, optical, and phtotoluminescent properties of zinc sulfide quantum dots," *Mater Sci Semicond Process*, vol. 16, no. 2, pp. 356–362, Apr. 2013, doi: 10.1016/J.MSSP.2012.09.012.

[3] K. T. Al-Rasoul, I. M. Ibrahim, I. M. Ali, and R. M. Al-Haddad, "Synthesis, Structure And Characterization Of Zns Qds And Using It In Photocatalytic Reaction," *INTERNATIONAL JOURNAL OF SCIENTIFIC & TECHNOLOGY*

RESEARCH, vol. 3, no. 5, 2014, Accessed: Dec. 05, 2023. [Online]. Available: www.ijstr.org

[4] M. Madkour *et al.*, "Nano-heterostructured photo-stable Cd$_x$Zn$_{1\&\#x2212;x}$S heterojunction as a non-photocorrosive visible light active photocatalyst," *Optical Materials Express, Vol. 6, Issue 9, pp. 2857-2870*, vol. 6, no. 9, pp. 2857–2870, Sep. 2016, doi: 10.1364/OME.6.002857.

[5] C. Ippen *et al.*, "ZnSe/ZnS quantum dots as emitting material in blue QD-LEDs with narrow emission peak and wavelength tunability," *Org Electron*, vol. 15, no. 1, pp. 126–131, Jan. 2014, doi: 10.1016/J.ORGEL.2013.11.003.

[6] W. Zhang *et al.*, "InP/ZnS/ZnS Core/Shell Blue Quantum Dots for Efficient Light-Emitting Diodes," *Adv Funct Mater*, vol. 30, no. 49, p. 2005303, Dec. 2020, doi: 10.1002/ADFM.202005303.

[7] H. Kumar, A. Kumari, and R. R. Singh, "Tunable narrow emission in ZnS/CdS/ZnS quantum well structures prepared by aqueous route," *Opt Mater (Amst)*, vol. 69, pp. 23–29, Jul. 2017, doi: 10.1016/J.OPTMAT.2017.04.009.

[8] R. Zhang, Y. Wang, M. Jia, J. Xu, and E. Pan, "One-pot hydrothermal synthesis of ZnS quantum dots/graphene hybrids as a dual anode for sodium ion and lithium ion batteries," *Appl Surf Sci*, vol. 437, pp. 375–383, Apr. 2018, doi: 10.1016/J.APSUSC.2017.12.110.

[9] T. I. Chanu, D. Samanta, A. Tiwari, and S. Chatterjee, "Effect of reaction parameters on photoluminescence and photocatalytic activity of zinc sulfide nanosphere synthesized by hydrothermal route," *Appl Surf Sci*, vol. 391, pp. 548–556, Jan. 2017, doi: 10.1016/J.APSUSC.2016.05.045.

[10] Komal, P. Shikha, and T. S. Kang, "Facile and green one pot synthesis of zinc sulphide quantum dots employing zinc-based ionic liquids and their photocatalytic activity," *New Journal of Chemistry*, vol. 41, no. 15, pp. 7407–7416, Jul. 2017, doi: 10.1039/C7NJ01373F.

[11] T. Kumar, Shreya, P. Phogat, V. Sahgal, and R. Jha, "Surfactant-mediated modulation of morphology and charge transfer dynamics in tungsten oxide nanoparticles," *Phys Scr*, vol. 98, no. 8, p. 085936, Jul. 2023, doi: 10.1088/1402-4896/ACE566.

[12] Dipti, P. Phogat, Shreya, D. Kumari, and S. Singh, "Fabrication of tunable band gap carbon based zinc nanocomposites for enhanced capacitive behaviour," *Phys Scr*, vol. 98, no. 9, p. 095030, Aug. 2023, doi: 10.1088/1402-4896/ACF07B.

[13] P. Phogat, Shreya, R. Jha, and S. Singh, "Diffusion Controlled Features of Microwave Assisted ZnS/ZnO Nanocomposite with Reduced Band Gap," *ECS Journal of Solid State Science and Technology*, vol. 12, no. 3, p. 034004, Mar. 2023, doi: 10.1149/2162-8777/ACC426.

[14] Shreya, P. Phogat, R. Jha, and S. Singh, "Microwave-synthesized γ-WO3 nanorods exhibiting high current density and diffusion characteristics," *Transition Metal Chemistry*, vol. 48, no. 3, pp. 167–183, Jun. 2023, doi: 10.1007/S11243-023-00533-Y/FIGURES/16.

[15] S. Sharma, P. Phogat, R. Jha, and S. Singh, "Electrochemical and Optical Properties of Microwave Assisted MoS2 Nanospheres for Solar Cell Application Electrochemical and Optical Properties of Microwave Assisted MoS 2 Nanospheres for Solar Cell Application," *Article in International Journal of Smart Grid and Clean Energy*, 2023, doi: 10.12720/sgce.12.3.66-72.

[16] P. Phogat, S. Shreya, R. Jha, and S. Singh, "Impedance Study of Zinc Sulphide Quantum Dots via One Step Green Synthesis," *Materials Science Forum*, vol. 1099, pp. 119–125, Oct. 2023, doi: 10.4028/P-G1CCXQ.

Tensile and Hardness behaviour of AlMg1SiCu alloy reinforced with SiC/ MoS_2 metal matrix composite

Ragupathy *K*[1], Anand *T*[1], Vishvanathperumal *S*[2], Arun *M*[1],

[1] Agni college of Technology, Chennai, Tamilnadu, India.
[2] S.A. Engineering College, Chennai, Tamilnadu, India.

Abstract. The current work examines the AlMg1SiCu alloy material through the stir casting process, where the Molybdenum Disulphide (MoS2) using three different variations of weight percentage (3 MoS_2%, 6% MoS_2, 9% MoS_2) and 10% uniform proportion silicon carbide (SiC). The distribution of reinforcements such as Hard particle Silicon Carbide (SiC) and Soft particleMolybdenum Disulphide in the matrix was studied by looking at the microstructure of the composites before Tensile and Hardness test using Backscattered Secondary Electron image (BSE). A Brinell hardness tester was employed to determine the hardness of samples made of AlMg1SiCu alloy and hybrid metal matrix composites. Tensile test carried out using a Universal test Machine for AlMg1SiCu alloy and HMMC Samples. With an increase in MoS_2 particle percentage, hybrid composites' tensile strength improves. With a AlMg1SiCu hybrid composite containing 10% SiC and 9% MoS2, the maximum tensile strength has been achieved.

1 Introduction

Modern composites are additionally growing in the hybrid metal matrix area, where the base AlMg1SiCu metal which may be aluminum is combined with non-ceramic reinforcements such as Boron Carbide ,Aluminium oxide , Titanium carbide, and Silicon Carbide . A composite material is produced when ceramic material is added to a metal matrix. The result exhibits an appealing blend of mechanical and physical properties that wouldn't be possible with a solid alloy [1]. In the term "Hybrid matrix" implies, in order to improve the composite's qualities, solid lubricants such as the graphite(Gr) or MoS2 non-ceramic materials are also added to supplement the base metal. The wear-resistant and anti-frictional characteristics of the composite would be improved by adding soft lubricants within the metal matrix. When specific reinforcement is blended through a hybrid composite material, physical and mechanical properties improve along with the composite's tribological behaviours [2]. A variety of methods, including a liquid infiltration, squeeze casting, stir casting, the use of powder metallurgy, diffusion bonding, and laser composite surfacing, were used to create these hybrid metal matrix composites [3]. The most technically sophisticated method for casting the HMMC from all of the above is stir the casting process, and it provides the material the right characteristics. Molybdenum Disulphide addition improves component density and improves compressibility; 3% of Molybdenum Disulphide will result in optimal dry sliding wear behaviour, and the volume percentages in excess of 5% lower the composite's hardness.

* Corresponding author: ragupathy86@gmail.com

According to Maivizhi Selvi et al. [4], MoS2 improved the mechanical characteristics of the Al6061 alloy composites, AA 6061-6% aluminium(III) oxide and Al6061-6% aluminium(III) oxide -2% Molybdenum Disulphide. These characteristics included flexural strength , toughness, and hardness values. Major improvements have been

made on various mechanical characteristics, such as the hardness and tensile properties. Singh J et al.'s investigation [5] on the mechanical characteristics of stir-cast T1 temper 6063 alloy reinforced with Molybdenum Disulphide revealed that the composite's density had increased and the reinforcement particles had been evenly distributed. It has been noticed compared to the extensive literature which there is a lack of reporting on the impact of adding reinforcing materials, such as Molybdenum Disulphide, to the aluminium matrix. It is commonly known that because aluminium alloy is mainly utilised to be a structural material, its tribological properties are not a design function. Few studies have examined the impact of Silicon Carbide and Molybdenum Disulphide particles on the tribological characteristics and mechanical characteristics of aluminium matrix composites. Tensile strength and hardness of hybrid composite materials raised with increasing molybdenum disulphide material.

2.1 Materials and methods

2.1.1 Materials

AlMg1SiCu alloy was chosen as the matrix material in the present study, and Silicon carbide and molybdenum disulphide, or MoS2 particles were considered as reinforcement. Using a spectrometer, the theoretical chemical composition of the AlMg1SiCu alloy was determined

Sample Number	Fabrication of composites Sample
1	AlMg1SiCu alloy + 10 wt.% SiC+ 3% MoS_2
2	AlMg1SiCu alloy + 10 wt.% SiC +6 % MoS_2
3	AlMg1SiCu alloy + 10 wt.% SiC +9% MoS_2

to be (wt.%).97.57 aluminium,0.63 Silicon ,0.89 Magnesium,0.24 Copper, 0.16 Iron; 0.48 Mn;0.014 Ti; 0.007 Zn; and 0.003 Cr. Particles of reinforcements have an average size of 75 microns.

Table 1. Fabrication of composites Sample

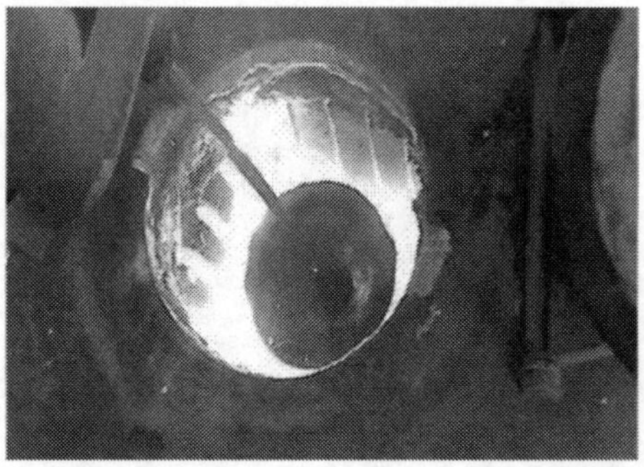

Fig. 1. Stir casting setup

2.1.2 Microstructural Analysis

The distribution of reinforcement particles in the matrix was studied by looking at the microstructure of the composites before wear test using Backscattered Secondary Electron image (BSE). The SEM images were taken for all the materials tested under specified conditions after the experiments. These SEM images were used to explain the surface morphology of the wear specimen. For the purpose of to show the boundaries that surround the grains inside the HMMC composite at the micron size level, the specimens were etchant-treated with Keller's reagent. For examining the microstructure, a JOEL SEM (Model: JSM 6390) had been utilised.

2.1.3 Tensile Test

Stir-cast aluminium AlMg1SiCu MMC has been manufactured with the incorporation of powdered SiC and MoS2 composites. Stir cast metal matrix composite castings have been produced using a variety of various combinations that had been prepared. After processing these stir-cast composites, the required ASTM E8 / ASTM E8M standard samples were created. For tensile test, 20 mm diameter and 110 mm length specimens were fabricated as show in Figure 3. Table 1 shows the composition of specimens used in tensile tests. The load-displacement data are obtained in the system attached to the tensile tester, and a plot of stress vs strain for each specimen is obtained. The tensile testing machine used for the experiments is shown in Figure 4.According to ASTM E8 standard, Figure 2 shows the standard specimen size for a test. Using a computer-controlled servo system, the tensile examination was performed at room temperature on the test specimens with an outer Bore size of 10 mm and a length of the gauge size of 36 mm.

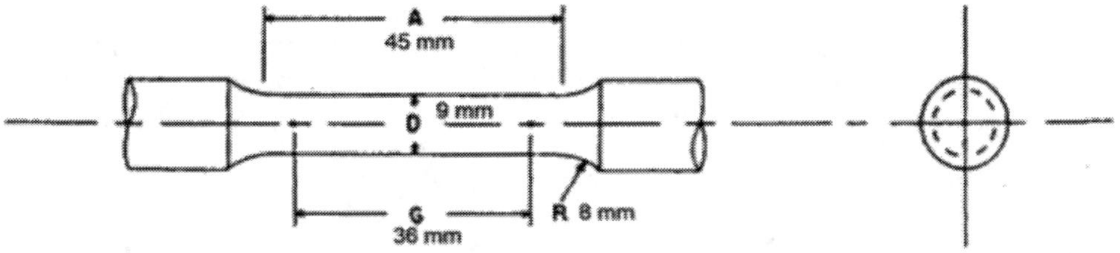

Fig. 2 The standard spécimen size for Tensile test

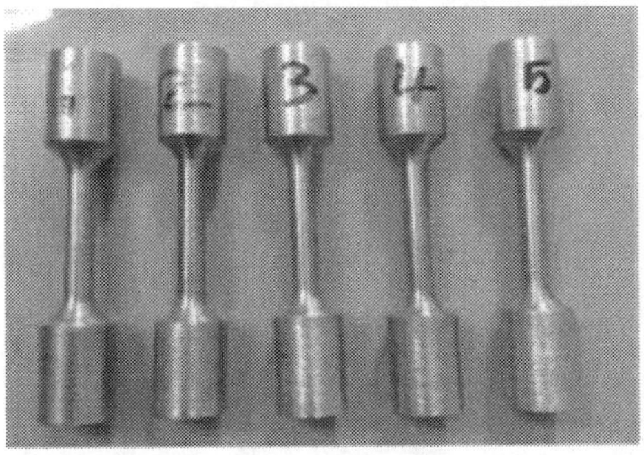

Fig. 3. Tensile test samples

Fig.4 Tensile testing machin

2.1.4 Brinell Hardness Test

The samples made of HMMC had been polished with standard grit (400–1000) rough sheets. A Brinell hardness tester has been utilised to determine the hardness of samples made of single and hybrid metal matrix composites. As indicated in Figure 5, a trial applied force of

550 kg was used with a carbide ball indenter over a duration of 15 s at 5 distinct locations on the samples' layer in accordance with ASTM E10-18.

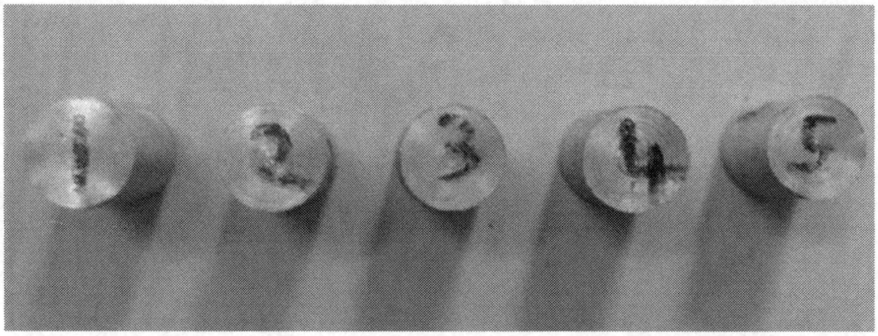

Fig. 5 Hardness test samples reinforced composites

2.1.5 Microstructural Analysis

The figures 6–8 display the line EDS, which is results for the AlMg1SiCu/silicon carbide/MoS_2 composite specimens. The element aluminium(al), magnesium(mg), manganese(mn), silicon(si), along with iron are among the functional elements of aluminium that exhibit high intensity peaks. Low peaks are observed for other impurities like nitrogen, oxygen, and carbon,. The elements molybdenum, silicon, aluminium, and sulphur are confirmed with the EDS elemental plot.

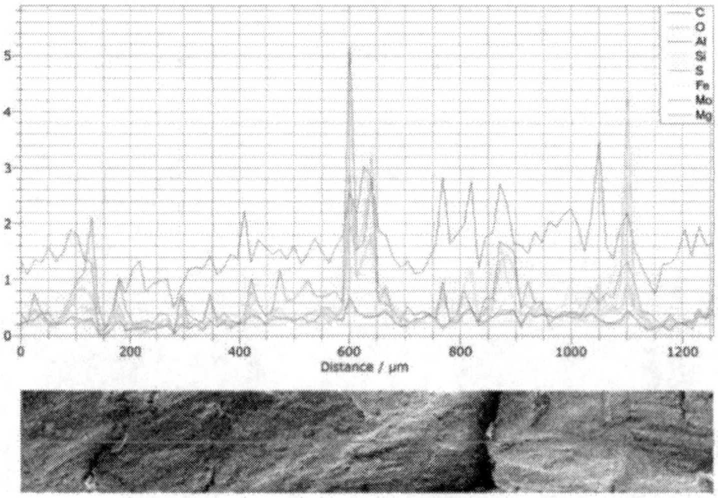

Fig. 6 Line surface EDS of 3 wt.% MoS_2 Sample

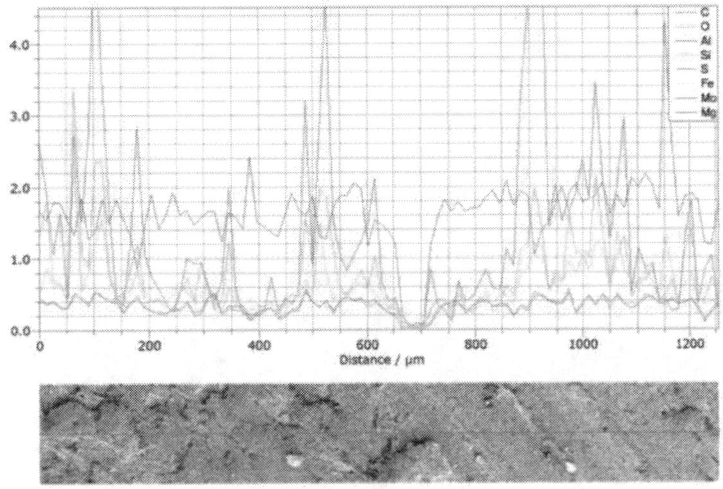

Fig.7 Line surface EDS of 6 wt.% MoS2 Sample

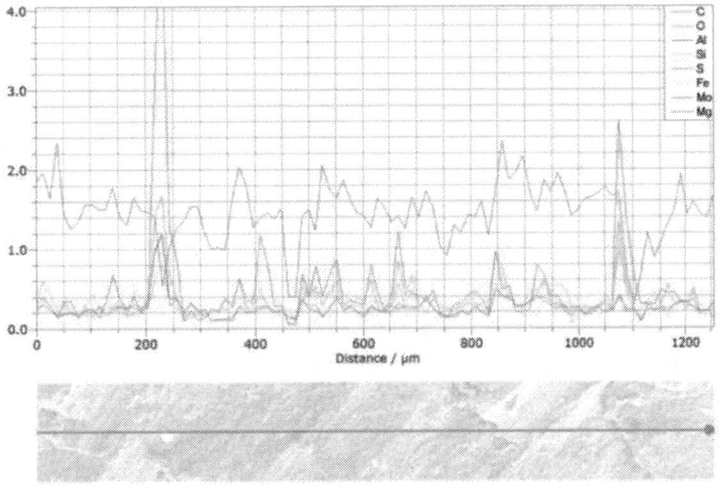

Fig. 8 Line surface EDS of 9 wt.% MoS_2 Sample

2.1.6 Hardness of as cast composites

Results of the Brinell Hardness Test are displayed in Figure 9. as-cast composites with varying reinforcement content. Due to the thermal mismatch during the solidification of the molten metal, the hardness values of AlMg1SiCu alloy composites increase with the addition of SiC particles (Habibur Rahman 2014).The increased dislocation density is due to the differing thermal expansion of the AlMg1SiCu alloy and the silicon carbide particles. As dislocation density increases, dislocation movement is blocked, leading to an increase in hardness values. The presence of hard Silicon carbide particles also resists dislocation motion, increasing the composite's hardness (Dora Siva Prasad 2014).

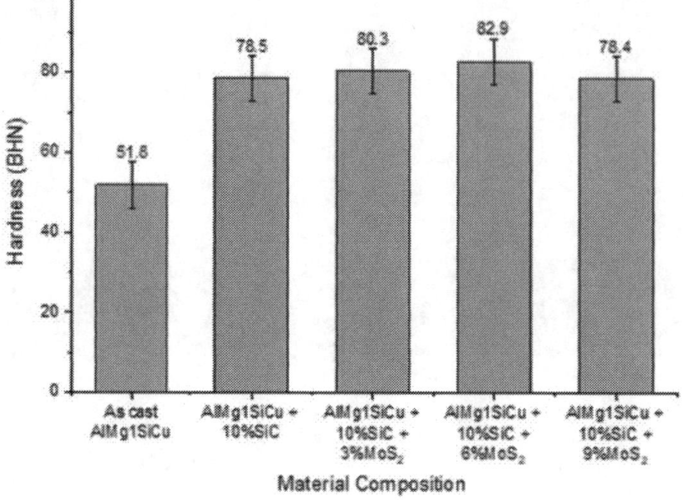

Fig. 9 Hardness Values of hybrid MMC composites

The effect of the weight percentage of reinforcements on the Hardness of the various AlMg1SiCu composite samples appears in Figure 9. The hardness of the reinforced AlMg1SiCu HMMC composites is found to be higher than that of the base AlMg1SiCu alloy, and it rises gradually when additional reinforcement particles are added, to a maximum a weight percentage of 6 wt%. of MoS_2 and additionally declines with the incorporation of Molybdenum Disulphide particles; however, the composite specimens' strain value progressively rises to be the Molybdenum Disulphide added additionally . The incorporation of reinforcement is what gives the lubricating property. According to Rebba and Ramanaiah (2014), Molybdenum Disulphide particles promote rapid grain shifting along direction of the slip planes, Permitting the specimens of HMMC composite to move and deform under the pressure of the indenter.

2.1.6 Tensile studies

Stress Vs strain curves are used to interpret the, ultimate stress, and tensile stress variation elongation of As cast AlMg1SiCu alloy, AlMg1SiCu +10% Silicon Carbide Composite, AlMg1SiCu +10%SiC +3% Molybdenum Disulphide , AlMg1SiCu 10 % Silicon Carbide +6 % Molybdenum Disulphide and AlMg1SiCu +10% Silicon Carbide + 9% Molybdenum Disulphide hybrid composites. Figure 10 shows the tensile strength of the AlMg1SiCu alloy reinforced HMMC composites . The graphical representation indicates evident that when the weight percentage of MoS_2 Reinforcement increases, correspondingly increases the tensile strength Values of AlMg1SiCu alloy Reinforced HMMC composites. contrasted with more refined particles, SEM images of the worn-out surfaces show that the bond that exists among coarse fragments in deeper thin craters possesses strengthened. As the number of sliding grooves and shallow craters increases, the hybrid composites' elastic modulus gradually decreases, increasing their tensile strength in comparison to as-cast AlMg1SiCu alloy. To identify the hybrid composite consisting of 10% SiC and 9% Molybdenum Disulphide, the optimum tensile strength was achieved.

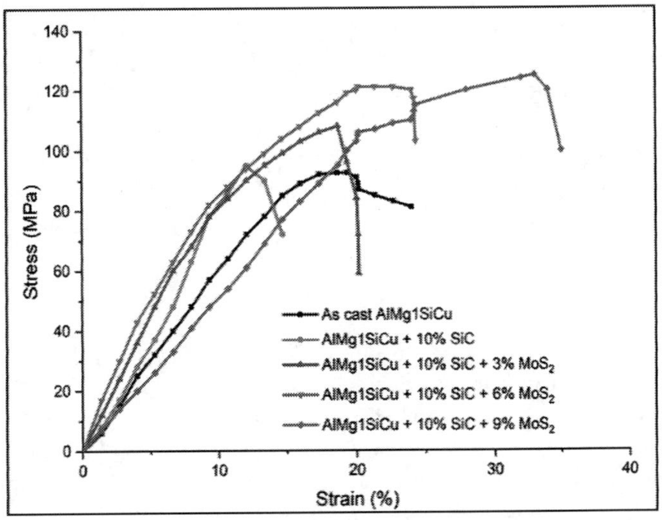

Fig. 10 Stress strain curves of hybrid composites

The theoretical Young's modulus was evaluated according to Callister (2007), which reported that the rule of mixture equation predicts that Youngs Modulus. The modulus of elasticity for AlMg1SiCu , SiC , and MoS_2 are 68, 90 and 250 GPa, respectively. Predicted Theoretical Young's modulus is calculated based on Equation 1. Predicted Youngs Modulus vaule are listed in Table 2.

Theoretical Young's modulus

$$E_C = E_{Al}V_{Al} + E_{SiC}V_{SiC} + E_{Mo}V_{Mo} \quad (1)$$

Table 2 Density and Young's modulus of hybrid composites

S.No	Composition	Density (g/cm^3)	Youngs Modulus Gpa
1	As cast AlMg1SiCu	2.700	68
2	AlMg1SiCu +10%SiC	2.931	70.2
3	AlMg1SiCu +10%SiC +3%MoS2	2.851	75.66
4	AlMg1SiCu +10%SiC +6%MoS2	2.886	81.12
5	AlMg1SiCu +10%SiC +9%MoS2	2.927	86.58

3.Conclusion

The major conclusions drawn from the present study on experimental investigations on the Tensile and Hardness behaviour of stir casted AlMg1SiCu hybrid aluminium composites are summarized.The Back scattered electron micrograph reveals a shows that the AlMg1SiCu matrix has been covered within a homogeneous blend of AlMg1SiCu alloy , Hard particles of SiC and soft particles of MoS2. silicon, molybdenum, sulphur and Aluminium, have been confirmed by the EDS line surface elemental plot. The results of the hardness test values, which HMMC AlMg1SiCu composites increased with an increase in MoS2 content up to 6 weight percent, at which point further addition slightly decreased the hardness. The lubricating property can be obtained by the incorporation of molybdenum disulphide (MoS2) particles. MoS2(molybdenum disulphide) particles promote rapid rapid grain shifting along direction of the slip planes, Permitting the specimens of HMMC composite to move and deform under the pressure of the indenter. It is clear that the presence of MoS2 reinforcement particles increases the tensile strength. 129.76 MPa maximum value of tensile strength has been reached for hybrid composites containing 10% silicon carbide and 9% molybdenum disulphide. Also, adding solid molybdenum disulphide (MoS2) lubricant increased the maximum strain.

References

1. Arun Kumar Sharma, Rakesh Bhandari, Amit Aherwar & Ruta Rimašauskiene, 'Matrix materials used in composites: A comprehensive study', Materials Today: Proceedings, vol. 21, no. 3, pp. 1559-1562(2020).
2. Chak, V, Chattopadhyay, H & Dora, TL , 'A review on fabrication methods, reinforcements and mechanical properties of aluminum matrix composites', Journal of Manufacturing Processes , vol. 56, pp. 1059–1074(2020).
3. Selvi, S. & A., Saravanakumar & Raj, G. & Keerthana, N. & Madhanagopal, G.D. & Nikhilesh, G.. Influence of MoS2 on Microstructure and Mechanical Behaviours of Al6061-Al2O3 Hybrid Composite. Applied Mechanics and Materials. Vol 62-66,pp.813-814. (2015). www.scientific.net/AMM.813-814.62.
4. Boostani AF, Tahamtan S, Jiang ZY, et al. "Enhanced tensile properties of aluminium matrix composites reinforced with graphene encapsulated SiC nanoparticles". Compos Part A: Appl Sci Manuf 2015; 68: pp 155–163(2015)..
5. Singh J and Chauhan A. "Characterization of hybrid aluminum matrix composites for advanced applications–a review". J Mater Res Technol 2016; 5: 159–169.
6. Mussatto A, Ahad IU, Mousavian RT, Delaure Y, Brabazon D. Advanced production routes for metal matrix composites. Eng Reports 2021;3:1–25. https://doi.org/10.1002/eng2.12330.
7. Devaraju A, Kumar A, Kumaraswamy A, Kotiveerachari B. Wear and mechanical properties of 6061-T6 aluminum alloy surface hybrid composites [(SiC + Gr) and (SiC + Al2O3)] fabricated by friction stir processing. J Mater Res Technol 2013;2: 362–9. https://doi.org/10.1016/j.jmrt.2013.10.004.
8. M. A. Hassan, O. T. Charles, B. Ahmed, and M. Samuel, ""influence of silicon carbide particulate loading on tensile,compressive and impact strengths of al-sicp composite for sustainable development," Chemical engineering transactions, vol. 63, pp. 619–624, 2018.
9. Ramadoss, N., Pazhanivel, K., Ganeshkumar, A. et al. Effect of SiC and MoS2 Co-reinforcement on Enhancing the Tribological and Anti-Corrosive Performance of Aluminum Matrix (Al6063-T6) Nanocomposites. Silicon 14, 5471–5479 (2022). https://doi.org/10.1007/s12633-021-01334-7.
10. Gonfa BK, Sinha D, Vates UK, Badruddin IA, Hussien M, Kamangar S, Singh GK, Ahmed GMS, Kanu NJ, Hossain N. Investigation of Mechanical and Tribological Behaviors of Aluminum Based Hybrid Metal Matrix Composite and Multi-Objective Optimization. Materials (Basel). 2022 Aug 16;15(16):5607. doi: 10.3390/ma15165607. PMID: 36013744; PMCID: PMC9413696.
11. Ragupathy, K., Velmurugan, C., Ebenezer Jacob Dhas, D.S. et al. Prediction of Dry Sliding Wear Response of AlMg1SiCu/Silicon Carbide/Molybdenum Disulphide Hybrid Composites Using Adaptive Neuro-Fuzzy Inference System (ANFIS) and Response Surface Methodology (RSM). Arab J Sci Eng 46, 12045–12063 (2021).

12. Velmurugan, C.; Muthukumaran, V.; Ragupathy, K., et al.: Modelling Volume loss of heat treated Al6061 composites using ANN. Proced. Mater. Sci 5, 31–40 (2014). https://doi.org/10.1016/j.mspro.2014.07.239.
13. Baradeshwaran, A.; Vettivel, S.C.; Elayapermal, A., et al.: Experimental investigation on mechanical behavior, modelling and optimization of wear parameters of B4C and graphite reinforced aluminium hybrid composites. Mater. Design 63, 620–632 (2014). https://doi.org/10.1016/j.matdes.2014.06.054
14. Vinoth, K.S.; Subramanian, R.; Dharmalingam, S., et al.: Mechanical and tribological characteristics of stir-cast Al–Si10Mg and self-lubricating Al–Si10Mg/MoS2 composites. Mater. Tech 46(5), 497–501 (2012)
15. M. Rouhi, M. Moazami-Goudarzi, M. Ardestani, 602 Comparison of effect of SiC and MoS2 on wear 603 behavior of Al matrix composites. Trans. Nonferrous 604 Met. Soc. China. 29, 1169–1183 (2019). https://doi. 605 org/10.1016/S1003-6326(19)65025-9
16. K. Halil, O. I˙smail, D. Sibel, C¸. Ramazan, Wear and 611 mechanical properties of Al6061/SiC/B4C hybrid 612 composites produced with powder metallurgy. 613 J. Mater. Res. Technol. 8, 5348–5361 (2019). https:// 614 doi.org/10.1016/j.jmrt.2019.09.002
17. S. Liu, Y. Wang, T. Muthuramalingam, G. Anbu627 chezhiyan, Effect of B4C and MOS2 reinforcement on 628 micro structure and wear properties of aluminum 629 hybrid composite for automotive applications Compos. Part B Eng. 176, 107329 (2019). https://doi. 630 org/10.1016/j.compositesb.2019.107329.
18. Ramadoss, N., Pazhanivel, K., Ganeshkumar, A. et al. Effect of SiC and MoS2 Co-reinforcement on Enhancing the Tribological and Anti-Corrosive Performance of Aluminum Matrix (Al6063-T6) Nanocomposites. Silicon 14, 5471–5479 (2022). https://doi.org/10.1007/s12633-021-01334-7.
19. Singh, N.K., Sethuraman, B. Development and Characterization of Aluminium AA7075 Hybrid Composite Foams (AHCFs) Using SiC and TiB2 Reinforcement. Inter Metalcast 18, 212–227 (2024). https://doi.org/10.1007/s40962-023-01009-6
20. Singh, N.K., Sethuraman, B. Development and Characterization of Aluminium AA7075 Hybrid Composite Foams (AHCFs) Using SiC and TiB2 Reinforcement. Inter Metalcast 18, 212–227 (2024). https://doi.org/10.1007/s40962-023-01009-6

Sustainable Supplier Selection in the Manufacturing Sector Using Integrated MCDM Techniques

Josy George[1] and *S* Balaguru[2*]

[1,2]VIT Bhopal University, Bhopal (M.P), India

Abstract. Selecting a supplier is a crucial task for any manufacturing company aiming to optimize its supply chain operations and gain a competitive advantage. The complexities of this decision-making process, influenced by various criteria and uncertainties, underscore the need for advanced methodologies such as Multi-Criteria Decision-Making (MCDM) techniques. These methodologies provide a structured framework for evaluating and prioritizing potential suppliers based on numerous criteria, often with conflicting considerations. This paper demonstrates the integration of the Analytical Hierarchy Process (AHP) and Simple Additive Weighting (SAW) through a practical example involving ten different alternatives and four distinct criteria. The alternatives are assessed and ranked for performance, utilizing systematic and objective approaches to address the inherent multi-criteria complexity of supplier selection. The strength of MCDM lies in its capability to handle both quantitative and qualitative factors, supported by decision support tools, making it a valuable asset for improving the supplier selection process.

Keywords: Analytical Hierarchy Process (AHP), Simple Additive Weighting (SAW), Multi-Criteria Decision-Making (MCDM) techniques, Supplier selection.

1 Introduction

In today's highly competitive market environment, the decision about supplier selection is very important for the success of production management [1]. MCDM techniques are essential tools for addressing complex decision problems in various domains. For many production units or manufacturing units, the issue of supplier selection problems is solved with the help of MCDM techniques [2]. MCDM techniques are used for solving problems with several alternatives and different criteria. The objective of the MCDM technique is to find the most suitable alternative among the given set of alternatives by comparing the available criteria [3], [4]. Decision-makers are required to make pair-wise comparisons between criteria and alternatives, indicating which one is the best and which one is the worst. This simplicity minimizes the cognitive burden on decision-makers and expedites the decision-making process.

MCDM is a widely embraced branch of decision theory. Despite the introduction of numerous MCDM techniques, each has its unique characteristics, and determining a singular method as the best remains intangible [5], [6]. Nevertheless, specific MCDM methods prove more suitable for particular problems. Conversely, the plethora of available MCDM methods transforms the selection process into an MCDM problem in itself [7], [8].

*Corresponding author: balaguru.s@vitbhopal.ac.in

1.1 Analytic Hierarchy Process (AHP)

AHP is a decision-support model created by Thomas L. Saaty at the University of Pittsburgh [9]. AHP proves effective in facilitating optimal decision-making when faced with numerous criteria [10]. The implementation of the AHP method can be as follows [11]:

1. Identify the problem, establish the solution, and construct a hierarchy of the encountered issues.
2. Establish the weight of criteria by pairwise comparisons for each criterion. This comparative process utilizes the priority scheme outlined in Table 1 [12].

Table 1. Saaty Scale

Intensity of importance	Definition
1	Equally important
3	Moderate important
5	Strongly important
7	Very strong
9	Extremely important
2, 4, 6, 8	The values between the two adjacent judgments

3. Normalize a pairwise comparison matrix is obtained by

$$\bar{a}_{jk} = \frac{a_{jk}}{\sum_{i=1}^{m} a_{jk}} \quad (1)$$

4. Calculate the weight by criteria by

$$\sum Column = k_1 + k_2 + k_3 + \cdots . + k_n \quad (2)$$

5. Calculate the eigenvalues by multiplying the values in each column of the paired matrix within the same row and then raising the result to the power of the corresponding criterion number.

$$\lambda_1 = (k_1 + k_2 + k_3 + \cdots . + k_n)^{\frac{1}{n}} \quad (3)$$

6. Determine the priority weight, maximum Eigen value, and finally consistency index

$$CI = \frac{(\lambda_{max} - n)}{n-1} \quad (4)$$

Where: CI = Consistency Index, λ_{max} = Maximum Eigen Value, n = Number of elements

7. In the next step, to validate the results of the AHP, the consistency ratio (CR) is calculated using the formula, CR = CI/RI in which the consistency index (CI) is, in turn, measured through the following formula [13]:

$$CR = \frac{CI}{IR} \quad (5)$$

Where: CR = Consistency Ratio, CI = Consistency Index, IR = Index Random

The value of RI is related to the dimension of the matrix and will be extracted from Table 2. It should be noted that a consistency ratio lower than 0.10 verifies that the results of the comparison are acceptable [14].

Table 2. Index Random

Matrix	1	2	3	4	5	6	7	8	9	10
IR	0	0	0.58	0.9	1.12	1.24	1.32	1.41	1.45	1.49

1.2 Simple Additive Weighting (SAW)

SAW is a method commonly used in Multi-Attribute Decision Making (MADM) [15]. It is also referred to as the weighted sum method [16]. The total score for an alternative is determined by summing the results by multiplying the rating and the weight assigned to each attribute [17]. The steps in utilizing the SAW method are as follows:

1. The matrix normalization is adapted based on the attribute type, generating a normalized matrix [18].

$$r_{ij} = \frac{x_{ij}}{Max(x_{ij})} \quad (6)$$

$$r_{ij} = \frac{Min(x_{ij})}{x_{ij}} \quad (7)$$

2. To calculate the weighted normalized matrix [18].

$$V_i = \Sigma_{j-1}^{n} = 1\, w_j r_{ij} \quad (8)$$

The weights of all criteria are determined using equation 2, and descending rankings according to the preference value of each alternative, representing the optimal outcome of the supplier assessment.

2 Literature Survey

MCDM become one of the most adapting and fastest-growing problem-solving techniques where decision-making becomes tuff under different conditions. Throughout the last decade, many scholars and experts in the field of industrial engineering have employed MCDM for decision-making. In the current literature, there is a widespread use of MCDM approaches to assist decision-makers in evaluating potential alternatives based on various criteria when dealing with the supplier selection problem. In some optimization problems, Taguchi methods are also used for the selection and optimization of alternatives [19]. Over the past two decades, methods centered on pairwise comparisons have gained broad acceptance among researchers for determining criteria weights and rankings.

Selecting suppliers and service providers through competitive bidding processes is a critical task for most operational organizations and manufacturers. In today's competitive markets, companies acknowledge the importance of meticulously choosing suppliers capable of meeting their requirements with the desired quality and within specified timeframes. In light of these conditions, organizations must effectively adjust to technological advancements, customer demands, and supplier management prerequisites to gain a competitive edge over other rivals [20]. Consequently, businesses evaluate the performance of their suppliers to identify the optimal supplier and enhance their supply chain efficiency, making supplier selection a pivotal element in the procurement process. Choosing a suitable supplier is considered non-trivial, leading decision-makers to rely on empirical evaluations predominantly.

3 Methodology

This section provides a numerical example featuring a manufacturing company aiming to choose a suitable supplier for raw material supply. The problem is addressed through the integration of the AHP and SAW methods. In this particular scenario, there are ten suppliers (A1 to A10) and four criteria (C1, C2, C3, and C4), with C1 representing the cost attribute and the rest considered benefit attributes. Initial information about the suppliers and criteria can be found in Table 6.

3.1 The Weighting Process with AHP Method

Determine the pairwise comparison matrix. At this stage, comparisons were made between the selections of criteria for private tutors using the Saaty Scale (1 to 9).

Table 3. Pairwise Comparison Matrix

	Price	Quality	Service	Delivery
Price	1	3	5	7
Quality	1/2	1	3	5
Service	1/5	1/3	1	3
Delivery	1/7	1/5	1/3	1

Determine the priority weight scale.

Table 4. Matrix of Criteria Scores and Priority Weights

	Price	Quality	Service	Delivery	Total	Priority
Price	0.5426	0.6618	0.5357	0.4375	2.178	0.5444
Quality	0.2713	0.2206	0.3214	0.3125	1.126	0.2815
Service	0.1085	0.0735	0.1071	0.1875	0.477	0.1192
Delivery	0.0775	0.0441	0.0357	0.0625	0.220	0.0550
Total	1.0000	1.0000	1.0000	1.0000	4.0000	1.0000

Determine the consistency ratio, after obtaining the weight of each criterion, the next step is to check the consistency ratio to verify the consistency of the pairwise comparisons. Summation matrices for each row are then determined by multiplying the priority weight values in Table 4 with the pairwise comparison matrix in Table 3. The calculation results are presented in Table 5.

Table 5. Consistency Ratio & Its Validation

	Total	Priority	Total / Priority	λ_{max}	CI	RI	CR=CI/RI

Price	2.3694	0.5444	4.3523	4.2031	0.0677	0.9	0.0752*
Quality	1.1860	0.2815	4.2138				
Service	0.4868	0.1192	4.0844				
Delivery	0.2288	0.0550	4.1619				
Total			16.8124				

*CR value ≤ 0.1 then the assessment is declared consistent

3.2 Alternative ranking using the SAW method

In this stage, the identification of the best supplier involves examining the initial decision matrix displayed in Table 6.

Table 6. Initial Decision Matrix

	Price	Quality	Service	Delivery
A1	9	9	9	8
A2	6	8	9	8
A3	8	6	8	9
A4	9	8	9	6
A5	8	9	9	8
A6	9	6	9	9
A7	6	6	9	6
A8	6	8	6	8
A9	8	9	6	9
A10	9	9	6	9

Identify criteria with attributes classified as either benefits or costs and then normalize the matrix. The results are obtained by multiplying the priority weights from the AHP method in Table 4 with the normalized matrix R using the SAW method, as shown in Table 7.

Table 7. Vector Calculation and Ranking of Alternatives

	Price	Quality	Service	Delivery	Vector	Rank
A1	0.0708	0.4448	0.2644	0.1641	0.9441	2
A2	0.1062	0.3954	0.2644	0.1641	0.9301	3
A3	0.0797	0.2965	0.2350	0.1846	0.7958	9
A4	0.0708	0.3954	0.2644	0.1230	0.8537	6
A5	0.0797	0.4448	0.2644	0.1641	0.9529	1
A6	0.0708	0.2965	0.2644	0.1846	0.8163	8
A7	0.1062	0.2965	0.2644	0.1230	0.7902	10
A8	0.1062	0.3954	0.1763	0.1641	0.8419	7

A9	0.0797	0.4448	0.1763	0.1846	0.8853	4
A10	0.0708	0.4448	0.1763	0.1846	0.8765	5

Finally, in the SAW method, the best alternative is A5, and then A1, A2, A9, and A10 will be respectively.

4 Discussion

This study utilizes the integration of AHP and SAW methods to enhance the decision-support process for selecting the best supplier. To improve efficiency and ease of use, simple software like MS Excel was employed. Evaluating suppliers based solely on criteria will be sufficient for future applications of the model. Implementing this evaluation via simple software will expedite the process.

References

[1] C. A. Weber, J. Current, A. Desai, An optimization approach to determining the number of vendors to employ, Supply Chain Management: An International Journal, **5**, 90-98, (2000).

[2] E. Timmerman, "An Approach to Vendor Performance Evaluation," IEEE Engineering Management Review, **15**, 14–20, (1987).

[3] S. Saghafian, S. R. Hejazi, Multi-criteria Group Decision Making Using A Modified Fuzzy TOPSIS Procedure, in proceeding of IEEE International Conference on Computational Intelligence for Modelling, Control and Automation and International Conference on Intelligent Agents, Web Technologies and Internet Commerce (CIMCA-IAWTIC'06), Vienna, Austria, November 28-30 (2005).

[4] P. Jain, S. Balaguru, R. Pendse, Design authentication of novel common interconvertible pallets for automobile engine: a finite element study, International Journal on Interactive Design and Manufacturing (IJIDeM), **17**, 1949–1958, (2023).

[5] E. LOKEN, Use of multicriteria decision analysis methods for energy planning problems, Renewable and Sustainable Energy Reviews, **11**, 1584–1595, (2007), http://doi.org/10.1016/j.rser.2005.11.005.

[6] E. Mulliner, N. Malys, V. Maliene, Comparative analysis of MCDM methods for the assessment of sustainable housing affordability, Omega Westport, **59**, 146–156, (2016). http://doi.org/10.1016/j.omega.2015.05.013.

[7] E. Triantaphyllou, Multi-criteria Decision Making Methods: A Comparative Study, (Springer, Boston, MA, US, 2000).

[8] E. Bernroider, V. Stix, A method using weight restrictions in data envelopment analysis for ranking and validity issues in decision making, Comput Oper Res, **34**, 2637–2647, (2007).

[9] H. Yu, Y. Ma, L. Wang, Y. Zhai, Z. Du, A method for evaluating the rescue priority level of power line post-disaster based on AHP, in Proceedings of International Conference on Mechatronics and Automation (ICMA), IEEE, Takamatsu, Japan, August 06-09 (2017). http://doi.org/10.1109/ICMA.2017.8015784

[10] W. Ran, AHP study on energy indicators system for sustainable development of Henan province, in Proceedings of 3rd International Conference on Communication Software and Networks, Xi'an, China, May 27-29 (2011).

[11] Q. Xu, M. Xiong, A method for improving consistency of judgment matrix in the AHP, in Proceedings of 2nd Asia-Pacific Conference on Intelligent Robot Systems (ACIRS), IEEE, Wuhan, China, June 16-18 (2017).

[12] R. Morgan, An investigation of constraints upon fisheries diversification using the Analytic Hierarchy Process (AHP), Mar Policy, vol. **86**, (2017). https://doi.org/10.1016/j.marpol.2017.05.037

[13] K. Aldrin Wiguna, R. Sarno, N. F. Ariyani, Optimization Solar Farm site selection using Multi-Criteria Decision Making Fuzzy AHP and PROMETHEE: case study in Bali, in Proceedings of International Conference on Information & Communication Technology and Systems (ICTS), IEEE, Surabaya, Indonesia, October 12 (2016), 237 http://doi.org/10.1109/ICTS.2016.7910305

[14] H. Taherdoost, Decision Making Using the Analytic Hierarchy Process (AHP); A Step by Step Approach, International Journal of Economics and Management Systems, **2**, 244–246 (2017).

[15] A. Pranolo, S. M. Widyastuti, Simple additive weighting method on intelligent agent for urban forest health monitoring, in Proceedings of International Conference on Computer, Control, Informatics and Its Applications (IC3INA), IEEE, Bandung, Indonesia, October 21-23 (2014), 132

[16] J. Kittur et al., Evaluating optimal generation using different multi-criteria decision-making methods, in Proceedings of International Conference on Circuits, Power and Computing Technologies [ICCPCT-2015], IEEE, Nagercoil, India, March 19-20 (2015), 1 http://doi.org/10.1109/ICCPCT.2015.7159365.

[17] H. Shakouri G., M. Nabaee, S. Aliakbarisani, A quantitative discussion on the assessment of power supply technologies: DEA (data envelopment analysis) and SAW (simple additive weighting) as complementary methods for the 'Grammar' Energy, **64**, 640-647 (2015). http://doi.org/10.1016/j.energy.2013.10.022

[18] R. E. Setyani, R. Saputra, Flood-prone Areas Mapping at Semarang City by Using Simple Additive Weighting Method, Procedia Soc Behav Sci, **227**, 378–386, (2016). https://doi.org/10.1016/j.sbspro.2016.06.089

[19] P. R. Lokhande, S. S. Salunkhe, S. Balaguru, Numerical simulation and experimentation of endodontic file using Taguchi DoE, International Journal for Simulation and Multidisciplinary Design Optimization, **12**, 32, (2021). https://doi.org/10.1051/smdo/2021032

[20] H. E. EVCİOĞLU, Supplier selection in supply chain network using MCDM methods, Sigma Journal of Engineering and Natural Sciences, **41,** 1-16, (2023).

Assessment of Wear rate and Coefficient of Friction of Al6262 WC/MoS2 under wet sliding condition

Kadapa Hemadri[1] , *Ajith* Arul Daniel S[2] , Vijayananth S[2] , Kumar R[3]

[1]Research Scholar, Department of Mechanical Engineering, Vels Institute of Science, Technology & Advanced Studies, Chennai 600117, Tamil Nadu India.
[2]Department of Mechanical Engineering, Vels Institute of Science, Technology & Advanced Studies, Chennai 600117, Tamil Nadu India.
[3]Department of Mechanical Engineering, Eritrea Institute of Technology, Eritrea

Abstract. The present paper aims at the results of an experimental examination of hybrid metal matrix composites using stir-casting technique.Aluminum alloy (Al 6062) samples reinforced with Tungsten Carbide (WC) and Molydedneum diSulphide are given. 2wt.% , 4wt.%,6wt.% of MoS2 and 3 wt.%, 6 wt.%, 9 wt.% tungsten carbide were mixed together with the aluminum alloy to create the hybrid composite. The tribological aspects of aluminium composites were studied using Pin on disc Method under wet sliding condition.Testing the hybrid composites' hardness revealed that raising the weight percentage of tungsten carbide might raise the hybrid composites' hardness. To examine the wear behavior of hybrid composites, the Taguchi method was utilized in the design of experiments (DOE) approach. The factors that were found to affect wear rate included load, sliding speed, and reinforcing percentage of WC and MoS2 are the most influencing parameters.

Keywords: Tribological Assessment, Wear Rate Measurement, Coefficient of Friction Analysis, Al6262 Alloy, WC/MoS2 Composite

1. Introduction

The wear and frictional behaviour of metals has been described in a number of research works. But there's still a need to comprehend the metal behaviour in many tribological contexts [1-3]. Studies on the tribological characteristics of several metal types have been published recently, including as brass, aluminum, and steel grades. Research on tribological characteristics of WC and Al alloy composites are in demand as machine weight reduction becomes more and more necessary in automobile sector [4]. Japan is one of the first countries in the automotive industry to adopt fiber-reinforced composites for cylinder liners, piston rings, and pistons because of their lower weight and resilience to wear [5]. However, because of the abrasive action of the fragmented fibers, fiber-reinforced composites frequently have a negative effect on the mating material. In these circumstances, materials with great strength and low weight have taken on a unique significance [6]. Nearly all composites are lightweight, yet depending on the kind of reinforcements used, strength may be compromised. Occasionally, providing service in environments with greater temperatures and significant part friction is also necessary [7].However, it is discovered that the majority of published works concentrate on the usage of metals, with experiments carried out to meet particular operational parameters, such as applied load, environmental temperature, sliding speed, and sliding distance [8].Because operational parameters have a major impact on how metals behave whether in dry or wet lubricant contact, understanding the wear and frictional behavior of the metals is therefore difficult and incomparable [9].By combining Al2O3/graphite composite in the material structure, wear resistance of Al6061 alloy is improved. According to the research, dual ceramic composites outperform than pure Al6061 in terms of wear enhancement [10]. The density, hardness, and wear resistance qualities of Al6061 alloy increase as the weight percentage of SiC increases when 2–6% SiC is added [11]. The better wear characteristics of the hybrid composite Al7075/Al2O3, which contains 2-8% Al2O3 and 5% graphite. wear resistance qualities. Additionally, when the quantity of alloying elements grows, the tribological characteristics such as compressive, flexural, and tensile strength in relation to the base alloy[12].Wear happens when material is moved as debris or when hard particles slip off the surface .

- Corresponding author: ajithdanny1989@gmail.com

The best wear parameters for MMCs can be found using the Taguchi approach, according to published research [13]. A wear test on aluminum alloys that had 20 wt% Al2O3 particles added. They consequently concluded that the wear resistance was influenced by the following parameters: wear path, wear rate, wear load, and reinforcement particle ratio [14].

2.Experimental Setup

Al 6262 has been chosen as a base material due to its excellent weldability, strong corrosion resistance, and superior material properties. MoS2 is employed as reinforcement due to its natural capacity for self-lubrication. The tungsten carbide is chosen as a primary reinforcement for its superior strength. The reinforcements are heated to 500°C for 25 minutes prior to casting. The reinforcement was warmed before the Al6262 was melted at 550°C to 650°C in a graphite furnace. After the reinforcement, magnesium is added at a rate of 2% to the aluminum matrix, and the mixture is agitated for 10 minutes at 800 rpm to avoid explosions [15]. The experiments were carried out using a computerized Pin-on-Disc wear test machine shoen in Fig.1.The 4T 20W40 Green Engine Oil lubricant was employed between the rotating disc and pin during the testing, which were conducted under wet sliding conditions. Under varied operating conditions of sliding velocity, sliding distance, and applying load, the frictional force and coefficient of friction of the various sliding surfaces were measured. The Wirecut machined sample was shown in fig. 2.The chemical combination of the base material is shown in Table 1.

Table 1. Typical chemical composition for aluminum alloy 6262

Si	Fe	Cu	Mn	Mg	Zn	Ti	Cr	Al
1.12	0.25	0.08	0.75	1.05	0.15	0.10	0.23	Bal

Fig. 2. Wet Sliding Pin On Disc Setup

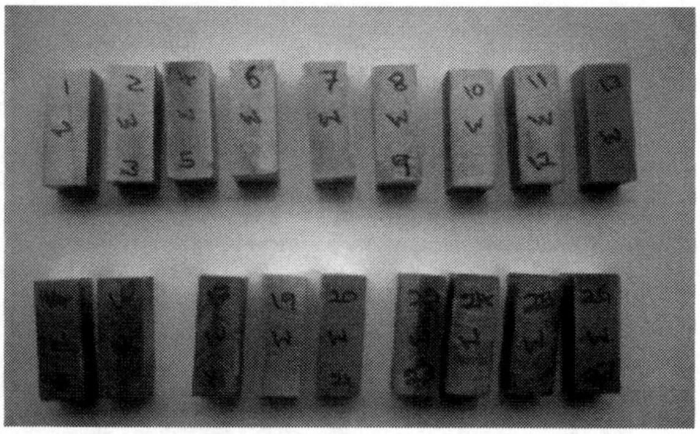

Fig. 2a. Wet wear test samples

3. Design of Experiments

The experiment design employed in this investigation, which has three levels and five factors, is displayed in Table 3. The weight % of tungsten carbide, weight percent of molybdneum disulphide, load, sliding velocity, and sliding distance are the input parameters. Table 2 displays the composites' hardness value. Hardness values rise with an increase in the weight % of the composites and an intermediate level of MoS2 percentage [16-19]. Table 3 displays the selection of the L27 orthogonal array for the studies based on these criteria. The hardness of the composites is displayed in Table 2.The level and factors of input parameters are shown in table 3.

Table 2. Hardness of the composites

WC	3	3	3	6	6	6	9	9	9
MoS2	2	4	6	2	4	6	2	4	6
BHN	76	80.2	75.6	81.5	84.3	80.2	87.6	92.2	86.7

Table 3. L27 Levels and Factors

Factors	**Unit**	**Level 1**	**Level 2**	**Level 3**
WC	%	3	6	9
MoS2	%	2	4	6
Load	N	15	25	35
Sliding Speed	m/sec	1.2	1.6	2
Sliding Distance	m	500	1000	1500

4. Result and Discussion

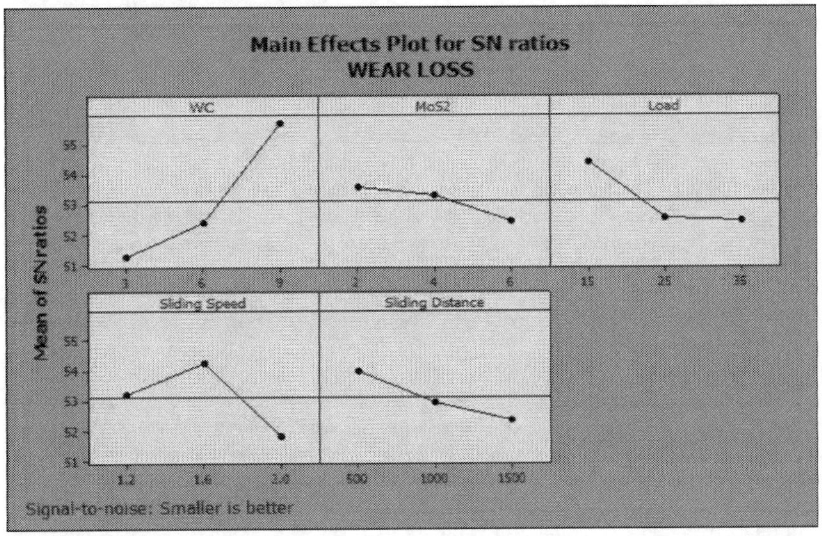

Fig. 3. S/N ratio graph for Wear Rate

Fig. 3. displays the wear rate signal to noise ratio. Reduced wear rate is correlated with a reduced tungsten carbide weight percentage. The findings indicates that the maximum wear rate is offered at 3% of WC. The composite material's hardness is to blame for this. Hard particles are generally thought to contribute significantly to the wear resistance of the base alloy when added to aluminum alloys. The composites' hardness decreases to 3% tungsten carbide. This tendency causes the material to have a lower density, which increases wear loss. Apart from 6% and 9%, there is an increase in hardness and a decrease in wear loss. The material has a higher hardness when MoS2 is at 6% and WC is at 9%. The self-lubricating qualities of the materials improve as the amount of MoS2 increases, reducing the friction between the pin and the disc and ultimately leading to material loss . As the applied stress increases, more of the softer pin surface is penetrated by the harder deformation of the counter surface [25-28]. When the applied load reaches 20 N, which occurs shortly after each composite's critical load , the wear rate begins to increase [29-30]. The wear rate jumps to an extremely high degree at 30N of stress. The optimum combination for reducing wear rate is 9 wt% of tungsten carbide 2 wt % of MoS2 and 15N of applied load, 1.6 m/sec of sliding speed and 500 m of sliding distance.

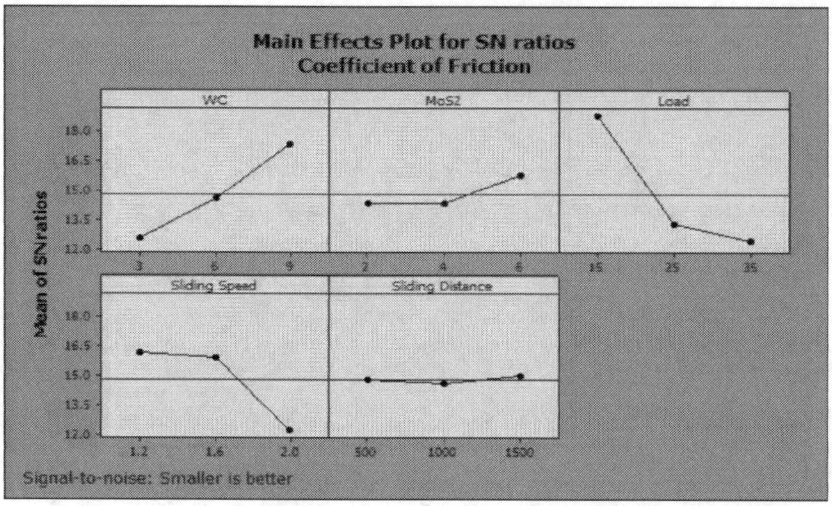

Fig. 4. S/N ratio graph for Coefficient of Friction

Fig. 4. displays the coefficient of friction's signal to noise ratio. The optimum combination for reducing COF is 9 wt% of tungsten carbide 6 wt % of MoS2 and 15N of applied load 1.6 m/sec of sliding speed and 1000 m of sliding distance.The most important factor influencing the friction coefficient is the weight fraction of tungsten carbide. The first conclusion to be drawn from the data is the decrease in coefficient of friction that occurs when the weight percentage of the reinforcement increases. The aluminum 6262 alloy and reinforced AMMNCs were found to have a reduced coefficient of friction. It is explained by the composite's higher hardness, which comes from the abrasive particles mixed in and acts as a more resilient material inside the matrix [17-18]. When refractory bits are securely attached to the matrix, there is good surface adhesion between the matrices and non-reinforcing material. Numerous investigations have shown that the COF of crystalline MoS2 is normally modest, at 6% in a dry sliding state. Because of its self-lubricating qualities, MoS2 has a special lubricating capacity. One explanation for this is that MoS2 particles have strong adhesion in a dry environment, and smaller particle sizes can fill the rough surface of the substrate with ease. As sliding speed rises, a tribo layer at the contact surface continues to grow. Even while temperature rises with increasing sliding velocity, heat has some positive effects on the formation of an interracial layer that is enriched in WC, which in turn affects the formation of compact layers at the contact area. But if the sliding speed is increased above 1.5 m/s, a significant amount of heat is generated, which causes the composite pin to become softer and causes friction to rise.

4.1 ANOVA Analysis

The experimental findings were analyzed using Analysis of Variance (ANOVA), which is used to examine the impact of the wear parameters that are taken into consideration, namely, weight percentage of tungsten carbide, MoS2, applied load and sliding distance, sliding speed on the performance measurements. One can determine which independent factor predominates over the other and what percentage of that specific independent variable is contributed by using analysis of variance. The significance level of $\alpha = 0.05$, or a 95% confidence level, is used for this study. Sources were deemed to have contributed statistically significantly to the performance metrics if their P-value was less than 0.05.

5 . Worn Out Microscopic Analysis

Fig. 5a. 9% WC & 4% MoS2, Load 15N

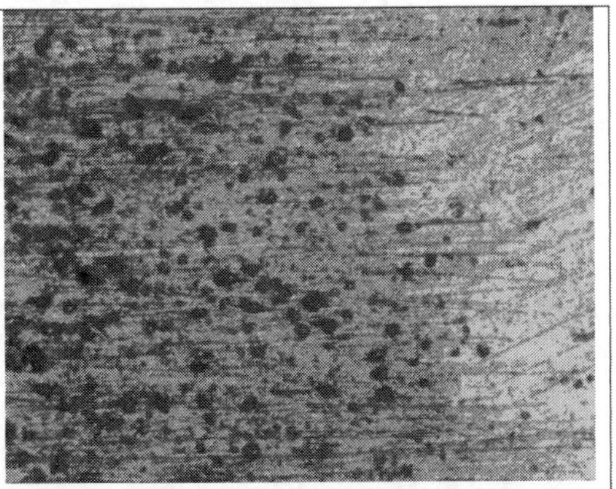

Fig. 5b. 9 % WC & 6 % of MoS2, Load 15N

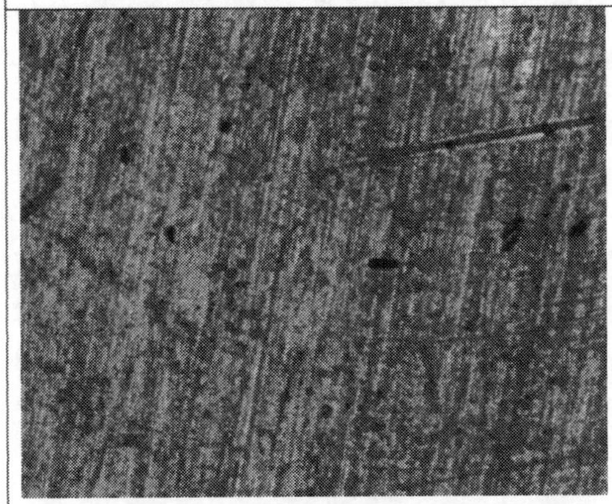

Fig. 5c. 6% WC & 6% of MoS2, Load 25N

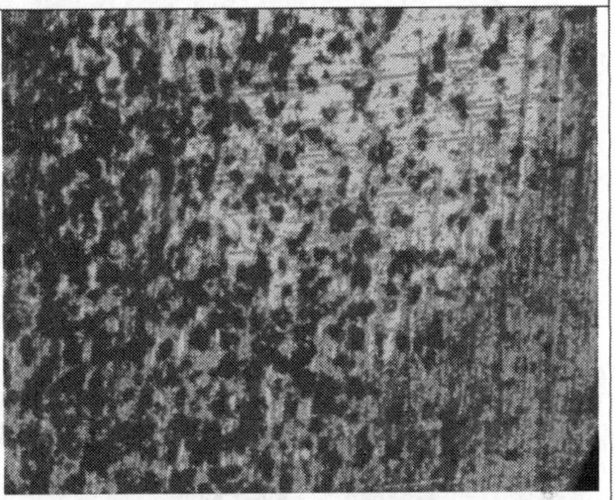

Fig. 5d. 6% of W C & 4% of MoS2, Load 35N

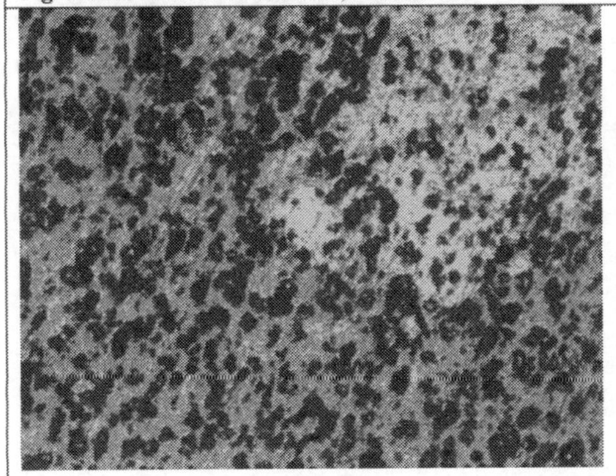

Fig. 5e. 3% W C & 4% MoS2, Load 35N

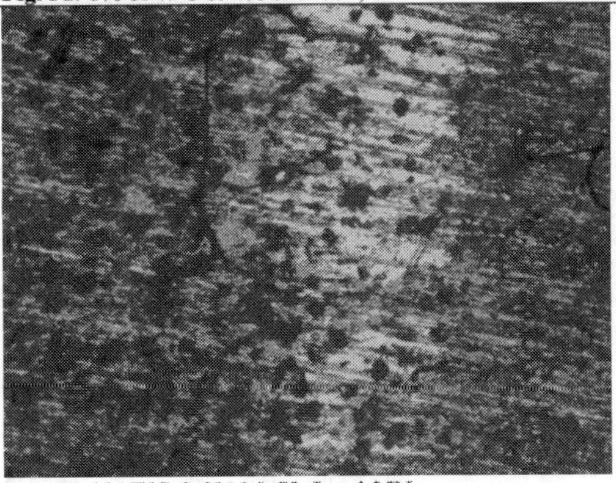

Fig. 5f. 3% WC & 2% MoS2, Load 35N

The worn-out surface, determined by optical microscopic images, is seen in Fig. 5. Fig. 5a displays the 15N load, 4% of Mos2, and 9% of WC. Fig. 5b illustrates that 9% of WC, 6% of Mos2, and a 15N load are present. There is less wear loss in Fig. 5b as compared to Fig. 5a. Because of its hardness value, it is also discovered that the WC particles have less worn-out regions. It is widely acknowledged that incorporating hard particles into aluminum alloys significantly increases the underlying alloy's resistance to wear .When refractor bits are securely attached to the matrix, there is good surface adhesion between the matrices and non-reinforcing material. When reinforced particles are securely attached to the matrix, there is good surface adhesion between the matrices and non-reinforcing material. Numerous investigations have shown that the COF of crystalline MoS2 is normally modest, at 6% in a wet sliding state. Because of its self-lubricating qualities, MoS2 has a special lubricating capacity. The 6% of WC, 6% of Mos2, and 25N load are displayed in Fig. 5c. Fig. 5d illustrates that 4% of Mos2, 6% of WC, and a 35N load are present. Figure 4c shows a superior surface polish than Fig. 5d. This is because the production of low shear strength thermo on the rubbing surfaces was aided by the higher surface temperatures at the contact area [20].The 35 %N load, 4% of Mos2, and 3% of WC are displayed in Fig. 5e. Fig. 5f illustrates that 2% of Mos2 and 3% of WC are present along with a 35N load. Because there is less Mos2 particle addition in Fig. 5f than in Fig. 5e, there is a noticeable loss of tungsten carbide particles.

6. Conclusion

In this current experiment Al 6262/WC/MoS2 with different weight percentage was successfully fabricated through stir casting technique. The Tungsten Carbide was chosen with different weight percentage of 3%, 6% & 9% and MoS2 was taken at 2%, 4% and 6%.Pin-on-Disc wear test machine was used to find the wear rate and COF for the developed composites in wet condition.At 9% of WC and 4% of MoS2 the hardness values is 92.2 BHN. From Taguchi optimization results the optimum combination for minimum COF is 9 wt% of tungsten carbide 6 wt % of MoS2 and 15N of applied load 1.6 m/sec of sliding speed and 1000 m of sliding distance.The ANOVA results shows the maximum percentage of contribution is weight percentage of WC 46.37 % for wear rate and applied load 35.75 % for Coefficient of friction.

References

1. Balaguru S & Manoj Gupta 2021, Hardfacing studies of Ni alloys: A Critical Review, Journal of Materials Research and Technology (ISSN: 2238-7854) vol. 10, pp: 1210-1242. https://doi.org/10.1016/j.jmrt.2020.12.026.

2. Balaguru S, Mohammad Abid & Manoj Gupta 2020, Investigations on different hardfacing processes for High temperature applications of Ni-Cr-B-Si alloy hardfaced on austenitic stainless steel components, Journal of Materials Research and Technology (ISSN: 2238-7854), vol. 9, pp. 10062-10072. https://doi.org/10.1016/j.jmrt.2020.07.010

3. Balaguru S, Vela Murali & Chellapandi P & Manoj Gupta 2020, Effect of Dilution on Micro Hardness of Ni-Cr-B-Si alloy Hardfaced on Austenitic Stainless Steel plate, Nuclear Engineering and Technology (ISSN: 1738-5733), vol. 52, pp. 589-596. https://doi.org/10.1016/j.net.2019.08.011

4. Balaguru S, Vela Murali & Chellapandi P 2017, 'Effects of different Operating Temperatures on the Tensile Properties of the Grid Plate Hardfaced with Colmonoy in a Pool Type Sodium-Cooled Fast Reactor', Science and Technology of Nuclear Installations (ISSN: 1687-6083), vol. 2017, pp. 1-9. https://doi.org/10.1155/2017/5926105

5. Balaguru S, Vela Murali & Chellapandi P 2016, 'Measurement of the Residual Stresses and Investigation of Their Effects on a Hardfaced Grid Plate due to Thermal Cycling in a Pool Type Sodium-Cooled Fast Reactor', Science and Technology of Nuclear Installations (ISSN: 1687-6083), vol. 2016, pp. 1-8. https://doi.org/10.1155/2016/8353256

6. Arulprakasajothi, M., Poyyamozhi, N., Chandrakumar, P., Dilip Raja, N., & D, Y. (2023). Experimental investigation of salinity gradient solar pond with nano-based phase change materials. Energy Sources, Part A: Recovery, Utilization, and Environmental Effects, 45(2), 5465–5480. https://doi.org/10.1080/15567036.2023.2207508

7. Ugle, V., Nagarajan, L., Mahalingam, A., & Thekkineydeth Rajan, H. (2023). TLBO algorithm-based optimal heat transfer parameters prediction of Al2O3 water nanofluids in variable pitch corrugated tube heat exchanger. Thermal Science, 00, 236–236. https://doi.org/10.2298/tsci230724236u

8. Mani, P., Radhakrishnan, S., Mahalingam, A., & Vellaiyan, S. (2023). Heat dissipation effects of different nanocoated lateral fins: An experimental investigation. Thermal Science, 00, 234–234. https://doi.org/10.2298/tsci230715234m

9. Balakrishnan, R., Govindaraj, K., Mahalingam, A., & Devarajan, Y. (2023). Analysis of the thermal management of electronic equipment by employing silicon carbide nano-pcm-based heat sink. Environmental Science and Pollution Research. https://doi.org/10.1007/s11356-023-27468-2

10. Rupesh, P. L., & Arulprakasajothi, M. (2020). Thermal history analysis on a hot surface using temperature indicating paints. International Journal of Ambient Energy, 43(1), 2392–2396. https://doi.org/10.1080/01430750.2020.173783

11. Attar, A., Arulprakasajothi, M., Vasulkar, D., Gorde, N., Kharat, S., & Kulkarni, S. (2020). Investigation of impact of the magnetic field through Halbach array on hydrocarbon fuel. International Journal of Ambient Energy, 43(1), 2124–2129. https://doi.org/10.1080/01430750.2020.1727951

12. Rupesh, P. L., Arulprakasajothi, M., & Raja, K. (2021). Isotherm recognition on a V-notch specimen by color identification method. International Journal of Information Technology, 14(3), 1455–1465. https://doi.org/10.1007/s41870-021-00652-

13. Arulprakasajothi, M., Rupesh, P. L., Rana, H. K., & Elangovan, K. (2021). Colour changing material for the estimation of flue gas radiation on the miniature gas turbine surface. Proceedings of the Institution of Mechanical Engineers, Part L: Journal of Materials: Design and Applications, 236(2), 374–388. https://doi.org/10.1177/14644207211045520

14. Jafer Kutbudeen, S., Logesh, K., Mahalingam, A., & Vinoth kanna, I. (2021). Performance enhancement of solar collector using strip inserts and with water based Al2O3/DI water nanofluids. Energy Sources, Part A: Recovery, Utilization, and Environmental Effects, 1–12. https://doi.org/10.1080/15567036.2021.1872745

15. Beemkumar, N., Yuvarajan, D., Arulprakasajothi, M., Elangovan, K., & Arunkumar, T. (2020). Control of room temperature fluctuations in the building by incorporating PCM in the roof. Journal of Thermal Analysis and Calorimetry, 143(4), 3039–3046. https://doi.org/10.1007/s10973-019-09226-0

16. Arulprakasajothi, M., Susanth, B., Naveen Kumar, K., & Madan Mohan Reddy, A. (2021). Thermal management on external surfaces by thermochromic materials. Materials Today: Proceedings, 47, 4666–4670. https://doi.org/10.1016/j.matpr.2021.05.523

17. Arulprakasajothi, M., Sudheer, K. P., Vijaydharan, S., & Venkateswarlu, B. (2021). Role of nano materials in battery thermal management. Materials Today: Proceedings, 47, 4523–4526. https://doi.org/10.1016/j.matpr.2021.05.419

18. Rupesh, P. L., Arulprakasajothi, M., Raja, K., & Sandeep Bhagavan, E. N. V. (2021). Examination of heat load at different zones of a Nimonic alloy combustor liner of different configurations by finite element method. Materials Today: Proceedings, 46, 3327–3333. https://doi.org/10.1016/j.matpr.2020.11.473

19. Arulprakasajothi, M., Beemkumar, N., Parthipan, J., & Battu, N. raju. (2019). Investigating the Physio-chemical Properties of Densified Biomass Pellet Fuels from Fruit and Vegetable Market Waste. Arabian Journal for Science and Engineering, 45(2), 563–574. https://doi.org/10.1007/s13369-019-04294-8

20. Beemkumar, N., Yuvarajan, D., Arulprakasajothi, M., Ganesan, S., Elangovan, K., & Senthilkumar, G. (2019). Experimental Investigation and Numerical Modeling of Room Temperature Control in Buildings by the Implementation of Phase Change Material in the Roof. Journal of Solar Energy Engineering, 142(1). https://doi.org/10.1115/1.4044564

Study on Mechanical Behaviour of AA2014 Aluminium Alloy Blended with Cobalt (Co) and Iron Oxide (Fe_2O_3)

T Anand[1*], *K* Ragupathy[2], *M.*Arun [3] and *G* Prabaharan[4]

[1*,2,3]Agni College of Technology,Thalambur, Chennai – 600 130, India
[4]S.A.Engineering College, Poonamalle - Avadi Road, Chennai – 600 077, India

Abstract. In order to increase system efficiency, lightweight materials are crucial in the automotive and aerospace sectors. In this work, iron oxide (Fe_2O_3) and cobalt (Co) were used as reinforcements, with aluminium alloy AA2014 serving as the matrix material. The metal matrix composites were made using the stir casting technique. The proportions of iron oxide and cobalt utilised as reinforcement in the composites were 2.5 wt % and 5 wt %, respectively. On the composite specimen created in accordance with ASTM E8, E9, and E23 standards, the mechanical properties of tensile, flexural, and impact strength were investigated. Investigations were also done on the hardness of the composites. The morphology of the composites was examined using an optical microscope.

1 Introduction

Since composite materials contain a number of distinctive properties not present in typical ordinary materials, they are desirable in many industrial applications. The volume-to-weight ratio is one such important characteristic that the automotive, aerospace, and other industries find particularly attractive [1].

Researchers are drawn to aluminium metal matrix composites because of their low weight and great strength, which reduces the amount of material and energy needed. Some important elements to take into account include the size of the reinforcement particles and their properties. These days, reinforcements made of bio-ceramics are also employed. The mechanical and tribological properties of aluminium alloy composites reinforced with bio-ceramics were examined by V. Bhuvaneswari et al. The mechanical and tribological properties of seashell powder reinforced with AA2219 alloy were improved by 3 wt% [2].

According to studies, stir casting is mostly utilised to create Al-based MMC. Utilizing a two-stage stir casting procedure, Cast A 356 alloy reinforced with 4 weight percent graphite (Gr) and 12 weight percent boron carbide (B4C) was created. The hardness and

* Corresponding author: anand.tvel@gmail.com

tensile characteristics were improved by the homogeneous distribution of B4C and Gr; however the density was lowered upon adding the reinforcement [3]. The mechanical characteristics and microstructure of the stir-cast and campo-cast A356 aluminium alloy/Al2O3 composites were compared by Seyed Abdolkarim Saiiadi et al [4].

With the examination of the structure and attributes of fine fly ash particle reinforced Al-7Si-0.35Mg alloy composites, the impact of three distinct stir casting processes is assessed. Evaluations of liquid metal stir casting, modified compo casting, compo casting (semi solid processing), and modified compo casting followed by squeeze casting techniques were among them, the latter produced composites with well-dispersed fly ash particles that were largely agglomeration and porosity free. Liquid metal stir cast composites exhibit greater interfacial interactions between the fly ash particle and matrix that result in the development of Mg Al_2O_4 spinel and iron intermetallics than compo cast composites [5].

The three main drawbacks of stir casting include porosities in composites during fabrication, the wettability of ceramic particles with molten metals, and poor distribution of the reinforcing particles known as agglomerations in the metal matrix. The main goal of many studies is to improve the parameters of stir casting for Ceramic-Metals Matrix Composites (CMMCs). The factors influencing the uniform dispersion of reinforcements, porosities in composites during the manufacturing process, and the mechanical characteristics of the ceramic metal matrix composites were thoroughly examined when talking about the stir casting procedure.[6,7].

Since the matrix and reinforcing components are in varying phases, achieving homogeneous metal matrix composites poses challenges. Although there has been a significant increase in interest in the topic, the title "Particulate Metal Matrix Composites" does not adequately describe all of its diverse aspects. In this study, the mechanical characteristics of metal matrix composites made of aluminium alloy are examined. Iron oxide and cobalt were used as reinforcements, and AA 2014 aluminium alloy was used as the matrix.

2 Materials and Experimentation

2.1 Material

The matrix material chosen was AA 2014 aluminium alloy, which is primarily employed in the construction of truck frames and aircraft constructions. The material was purchased in an ingot form. The AA 2014 alloy has a density, poisson's ratio, and melting point of 2.82 g/cm^3, 0.33, and 515^0 C. Table 1 lists the chemical composition of the AA 2014 alloy used in this investigation. The reinforcements employed were iron oxide (Fe_2O_3), a reddish-brown solid powder with a density of 5.23 g/cm^3, and cobalt (Co), a ferromagnetic, silver-white lustrous element with a density of 8.9 g/cm^3 and a poisson's ratio of 0.31. Cobalt has the ability to withstand wear and corrosion.

Table 1. Chemical composition of AA2014 Aluminium alloy

Al (%)	Cu (%)	Si (%)	Mg (%)	Cr (%)	Mn (%)
93.2	4.35	0.95	0.6	0.2	0.7

2.2 Fabrication of composites

The ratios of reinforcements used to develop the AA 2014 Aluminium alloy Composites are listed in Table 2.

Table 2. Proportions of AA2014 Aluminium alloy composites

Composites	AA 2014 (%)	Iron Oxide (Fe_2O_3) (%)	Cobalt (Co) (%)
S1	100	-	-
S2	97.5	2.5	-
S3	95	5	-
S4	95	2.5	2.5
S5	90	5	5

Stir casting was used to create the composites made from the AA 2014 Aluminum alloy. It was done using stir casting machinery with a maximum temperature of 1000°C and a capacity of 2 Kg. The apparatus includes an immersion-type mould pre heater with a 450°C maximum temperature. Also used was an automatic stirrer with a speed range of 300 to 1200 rpm. In order to avoid porosity, which is a significant issue in the casting process, the mould was preheated. Preheating releases the gases that were trapped in the slurry before they could enter the mould. For 60 minutes, the preheating operation was conducted at 450°C [8].

Fig. 1. Stir casting technique

The additional support to get rid of any moisture, iron oxide and cobalt was additionally warmed for 45 minutes at 450°C. The furnace was stacked with the AA2014 Aluminum Alloy ingots, and the temperature was steadily raised. Cobalt and iron oxide metal powder were manually added after the ingots had begun melting and reached a semi-solid state. To guarantee homogenous mixing, the reinforcement and matrix material were mixed using an automated stirrer at a 400-rpm speed. Figure 1 shows the stir casting method used in this work.

2.3 Mechanical Properties

The specimens were prepared as per the ASTM standards. The specimens for tensile, flexural and impact strength were prepared as per ASTM E8, E9 and E23 standards. Tensile and flexural strength tests were conducted using the TINIUS OLSEN H10KT dual column universal testing machine, which has a 10 kN frame capacity. The machine has the following measurements: 1600 mm (H), 650 mm (W), and 450 mm (D). The testing speed range is 0.001 to 1000 mm/min, and the maximum crosshead travel is 1100 mm. The impact strength was measured using Tinius Olsen's pendulum impact tester, which has a maximum speed of 5.47 m/s and a lowest speed of 0.13 m/s.

3 Results and Discussion

The mechanical characteristics of AA2014 mixed with cobalt and iron oxide are shown in Table 3. The composites' tensile, flexural, and impact strengths are shown and compared with those of pure AA2014. The Composites' hardness is also determined and compared. When compared to pure AA2014, the tensile strength and hardness of the AA2014 composite with 5 wt% of iron oxide increased to 48.14% and 19.38%, respectively. With the addition of iron oxide and cobalt at a 5 wt% each, AA2014 composites improved their flexural and impact strength.

Table 3. Mechanical properties of AA2014 Aluminium alloy and its composites

Specimen	Tensile Strength (MPa)	Flexural Strength (MPa)	Impact Strength (Nm)	Percentage of Elongation (%)	Hardness (HV)
S1	222.50	128.40	6.1	10.8	113.5
S2	272.80	106.52	5.2	8.2	127.2
S3	329.62	97.33	4.6	6.5	135.5
S4	253.25	126.82	5.8	9.9	118.8
S5	302.72	140.48	7.4	11.3	121.4

Figures 2, 3, 4 and 5 show the tensile strength of AA2014 aluminium alloy with cobalt and iron oxide reinforcement. Compared to AA2014, iron oxide is a tougher substance. The stiffness of the composite material is improved, which increases the flexural strength and hardness of the composites when iron oxide as the reinforcement is added to the matrix material AA2014 in the range of 2.5 wt% and 5 wt%. While the impact strength and flexural strength both decreased by the same percentage. The cobalt reinforcement cobalt was added to the composite in the range 2.5 wt% and 5 wt%. It moderates the hardness of the composites due to which the composites became soft, increasing the flexural and impact strength of the material [9, 10].

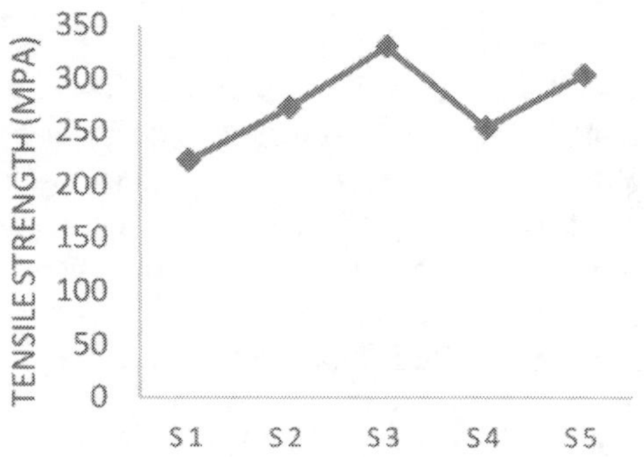

Fig. 2. Tensile strength of AA2014 Aluminium alloy and its composites.

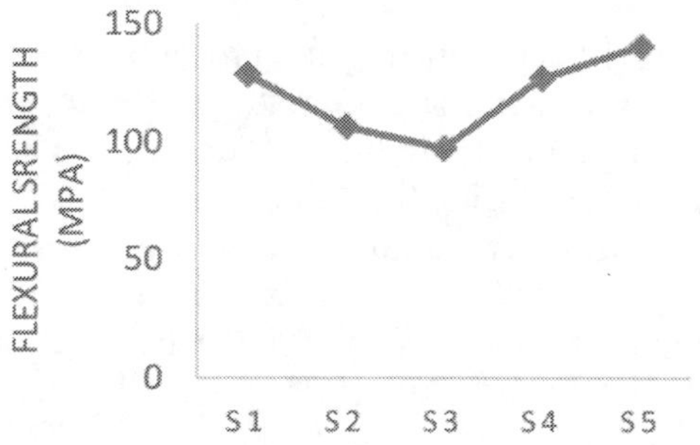

Fig. 3. Flexural strength of AA2014 Aluminium alloy and its composites.

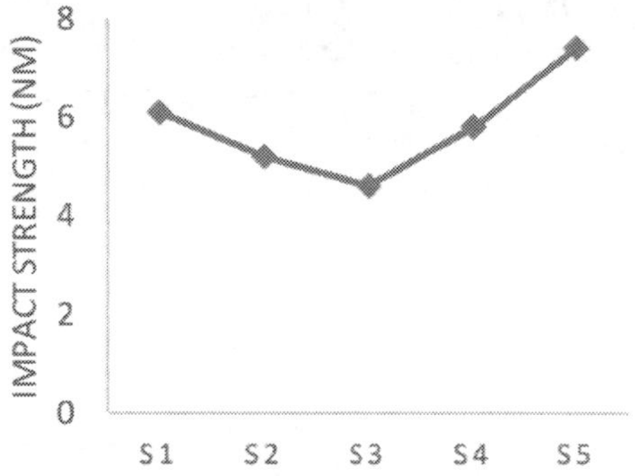

Fig. 4. Impact strength of AA2014 Aluminium alloy and its composites.

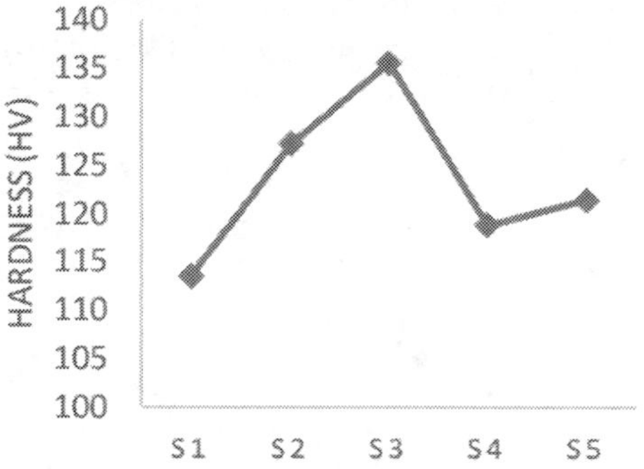

Fig. 5. Hardness of AA2014 Aluminium alloy and its composites.

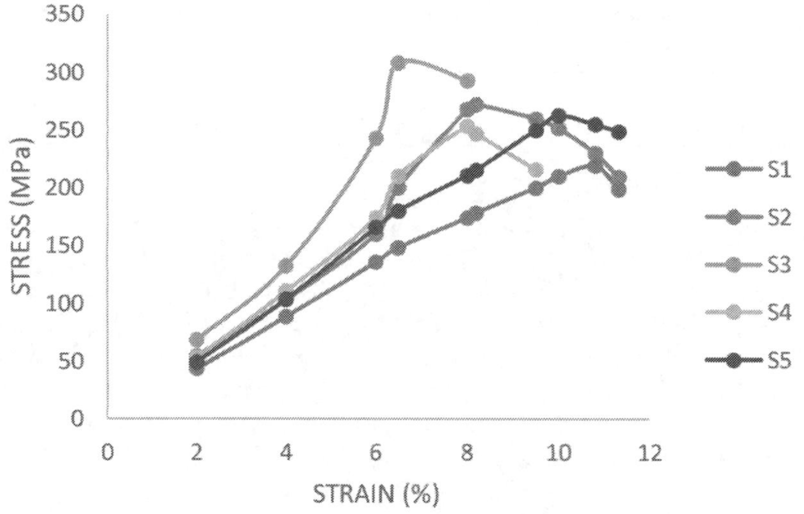

Fig. 6. Stress – Strain curves of AA2014 Aluminium alloy and its composites.

Figure 6 depicts the stress-strain behaviour of AA2014 and its composites with 2.5 and 5 weight % of iron oxide and cobalt. Within the elastic limit, the stress value grows linearly; however once above that point, the strain increases quickly for a small stress value. As compared to pure AA2014, the stress-strain behaviour of AA2014 composites including 2.5 and 5 wt% iron oxide and cobalt was quite similar. Along with the improvement in the stress-strain relationship, the percentage of elongation was also improved [11]. The percentage of elongation is lowered when iron oxide is added to pure AA2014 because it reduces deformability, whereas the percentage of elongation is increased when cobalt is added as reinforcement because it promotes deformability [12, 13].

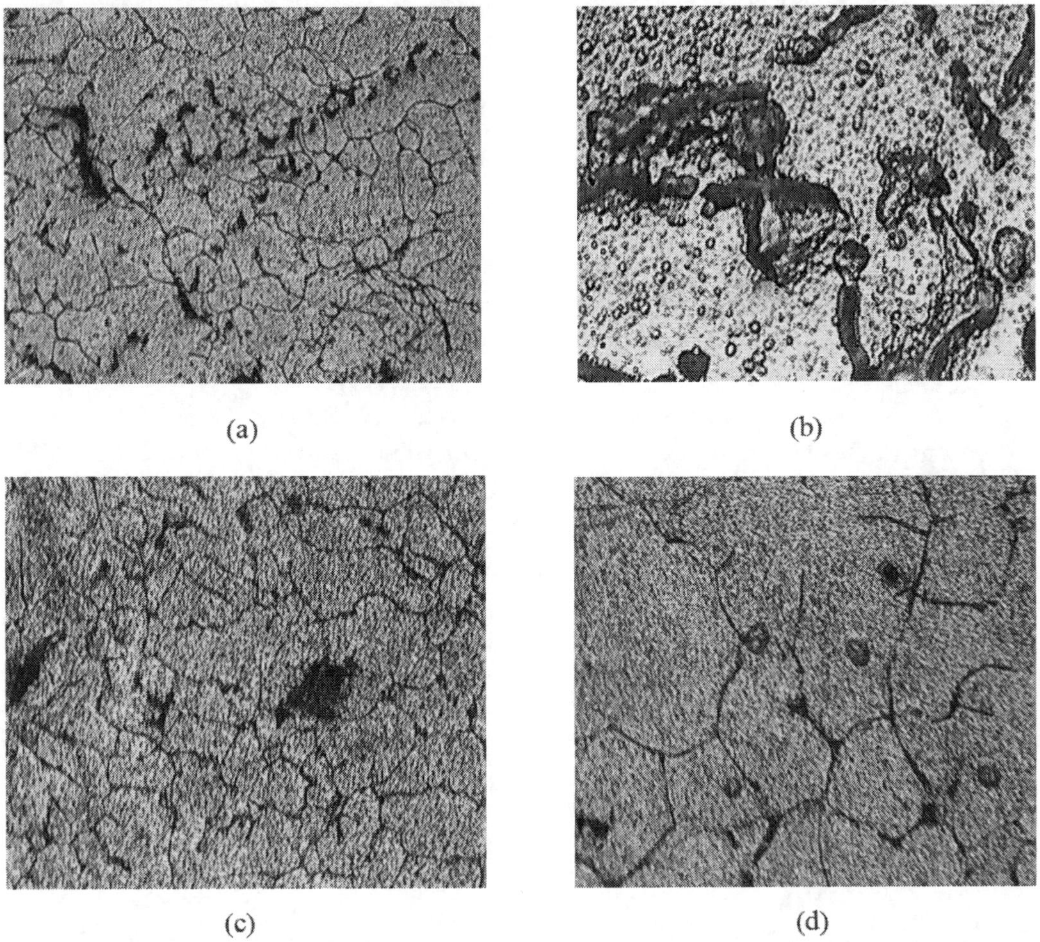

Fig. 7. Optical microscope images of Aluminium alloy AA2014 with (a) 2.5 wt. % of Iron oxide (b) 5 wt. % of Iron oxide (c) 2.5 wt. % of Iron oxide and Cobalt, each (d) 5 wt. % of Iron oxide and Cobalt, each

Figure 7 demonstrates the strong contact between the reinforcement and the matrix material created by the addition of cobalt and iron oxide particles to Pure AA2014 alloy [14]. Cobalt particles and Fe_2O_3 clusters can both be seen in the microstructure. The AA2014 alloy's as-cast microstructure shows several dispersed compounds in the grain boundary [15]. Five weight percent of the AA2014 matrix is covered in a uniform distribution of Fe_2O_3 and cobalt particles. Fe_2O_3, which was used in this study and has a tetragonal structure, tends to restructure into a monoclinic structure as it cools. This transition consequently results in volumetric expansion, which may facilitate in the filling of the defects. The reinforcements' and AA2014 matrix's strong interaction significantly improved interfacial adhesion and dispersion. As a result, the mechanical characteristics of the composites might be enhanced [16, 17].

4 Conclusion

The mechanical characteristics of AA2014 aluminium alloy reinforced with 2.5 and 5 weight percent of iron oxide and cobalt were investigated, and the results were compared with those of virgin aluminium alloy. In the composite of AA2014 with 5 wt.% iron oxide, the tensile strength and hardness were improved. The flexural strength and impact strength of the AA2014 composite with 5 wt.% iron oxide and 5 weight percent cobalt were improved. Using an optical microscope to examine the microstructure revealed that the inclusion of reinforcement iron oxide and cobalt contributed to the strengthening of the composites, improving their characteristics.

References

1. Divya Chandra, Nadhi Ram Chauhan. Hardness and Roughness Evaluation of Developed Al Metal Matrix Composites using Stir Casting Method. Materials Today Proceedings, Volume **25**, Part 4, pages 872 – 876, (2020)
2. V.Bhuvaneswaria, L. Rajesjhkumar, D.Balaji, R.Saravanakumar. Study of Mechanical and Tribological Properties of Bio-Ceramics reinforced Al Alloy Composites. Solid State Technology. Vol **63**, No.5 (2020)
3. Pankaj R Jadhav, B.R.Sridhar, Madeva Nagared, Jayasheel L.Harti. Mechanical Behaviour and fractography of graphite and boron carbide particulates reinforced A356 alloy hybrid metal matrix composites. Advanced Composites Hybrid Materials **3**, 114 – 119 (2020)
4. Seyed Abdolkarim Saijadi, M.Torabi Pairizi. Comparison of microstructure and mechanical properties of A356 aluminium alloy / Al_2O_3 composites fabricated stir and campo-casting process. Materials & Design. **34**, 106 – 111, (2012)
5. T.P.D. Rajan, R M Pillai, B C Pai, K G Satyanarayana, P K Rohatgi. Fabrication and Characterization of Al – 7Si – 0.35 Mg / Fly ash Metal Matrix Composites Processed by Different Stir Casting Routes. Composite Science and Technology. **67** (15-16), 3369 – 3377, 2007
6. Malek Ali. Review of Stir Casting Technique and Technical Challenges for Ceramic Reinforcement Particulate Aluminium Matrix Composites. Journal of Silicate Based and Composite Materials.Volume **72**, Issue 6, May (2020).
7. Susanta KP, Subhranshu C, Amitava BM, Debdulal D 2016 A simple stir casting technique for the preparation of in situ Fe-aluminides reinforced Al-matrix composites. Perspectives in Science. Volume **8**, 529—532, (2016)
8. Fathy A, El-Kady O, Moustafa MM Mohammed. Effect of iron addition on microstructure, mechanical and magnetic properties of Al-matrix composite produced by powder metallurgy route Trans. Nonferrous Met. Soc. China **25,** 46–53, (2015)
9. Subramanya RP, Arun KS, Mervin AH, Shrikantha SR. Microstructure and mechanical properties of rutile-reinforced AA6061 matrix composites produced via stir casting process Trans. Nonferrous Met. Soc. China, **29** 2229–2236, (2019)
10. Sivananthan S, Ravi K, Samson Jerold Samuel C. Effect of SiC particles reinforcement on mechanical properties of aluminium 6061 alloy processed using stir casting route Materials Today: Proceedings. **21**, 968-970, (2020)

11. Nagendra KM, Manish M, Srivastava AK, Dwivedi SP, kumar A, Chauhan S. Investigation of mechanical properties of Al 6061/SiC composite prepared through stir casting technique Materials Today: Proceedings **25** 755-758, (2020)
12. Sarkari Khorrami M, Samadi S, Janghorban Z, Movahedi M. In-situ aluminum matrix composite produced by friction stir processing using Fe particles Mater Sci Eng A. **641**, 380– 390, (2015)
13. Wang MM and Apelian D. Aluminium die casting alloys: alloy composition, microstructure, and properties-performance relationships Int. Mater. Rev. **40** 225–252, (1995)
14. Mikhaylov VI, Maslennikova TP, Ugolkov VL, Krivoshapkin PV. Hydrothermal synthesis, characterization and sorption properties of Al/Fe oxide–oxyhydroxide composite powders. Advanced Powder Technology. **27**, 756-764, (2016)
15. Yang X, Rujuan S, Ni S, Song M, Xiao D. Fabrication, microstructure and mechanical properties of Al–Fe intermetallic particle reinforced Al-based composites. Journal of Alloys and Compounds. **618**, 537–544, (2015)
16. A.V. Pozdniakov, A. Lotfy, A. Qadir, E. Shalaby, M.G. Khomutov, A.Y. Churyumov, V.S. Zolotorevskiy. Development of Al-5Cu/B4C composites with low coefficient of thermal expansion for automotive application. Mater. Sci. Eng., A, **688** pp. 1-8, 10.1016/j.msea.2017.01.075, (2017)
17. H. sheng Chen, W. xian Wang, H. Hui Nie, J. Zhou, Y. li Li, P. Zhang. Microstructure and mechanical properties of B4C/6061Al laminar composites fabricated by power metallurgy. Vacuum, **143**, pp. 363-37010.1016/j.vacuum.2017.06.009, (2017)

Hydrothermally Synthesized NiS_2 and $NiSO_4(H_2O)_6$ Nanocomposites and its Characterizations

Ayushi Sharma[1], Shreya[1*], Peeyush Phogat[1], Ranjana Jha[1], and Sukhvir Singh[1]
[1]Research Lab for Energy Systems, Department of Physics, Netaji Subhas University of Technology, New Delhi, INDIA
*shreyasharma.aug15@gmail.com

Abstract. The present manuscript discusses an in-depth explanation of the synthesis and characterization of nickel sulphide (NiS_2) and nickel sulphate [$NiSO_4(H_2O)_6$] nanocomposites. The phase analysis was performed through x-ray diffraction (XRD), revealing the presence of NiS_2 and $NiSO_4(H_2O)_6$ nanocomposites, along with the presence of deuterium. The synthesized material exhibited absorbance in the visible region, with an optimized band gap of 1.0 eV. The presence of carbon spheres was confirmed through morphological analysis performed through field emission scanning electron microscopy (FESEM), depicting the growth of NiS_2 on the surface of carbon spheres and the presence of nanosheets of $NiSO_4(H_2O)_6$. These techniques provide a detailed explanation of the crystallite size for the nanocomposites, their structural properties and optical properties. The material showed a narrow band gap making it highly suitable for solar cell applications.

Keywords: nanocomposite, center of inversion, phase transition, anorthic.

1 Introduction

Over the previous years, there has been rapid development in material science for the application of renewable energy. Renewable energy can involve materials for capacitors used in charge storage or for solar cells to achieve higher efficiency. Nanocomposites, due to their beneficial qualities [1] such as increased thermal stability (useful in situations where temperature control and resistance to heat are essential), increased surface-to-volume ratio (advantageous in sensing, adsorption, and catalysis applications [2]), and increased or decreased bandgap (essential in solar systems [3]), have gained a lot of interest in the field of solar materials.

One of the widely explored compounds is NiS_2, which depicts allotropy, including hexagonal and cubic forms. NiS_2 has a bandgap of 0.5 eV but due to its smaller bandgap the material does not represent an optimum band gap for solar spectrum. To overcome the challenge, the present experiment has been performed in which the nanocomposites of nickel sulphide (NiS_2) and nickel sulphate [$NiSO_4(H_2O)_6$] were fabricated having a bandgap of 1.0 eV. The introduction of nanoscale nickel sulfide into a matrix of nickel sulphate can improve electrical conductivity, making it suitable for use in supercapacitors and solar cells[4]. The presence of nickel sulfide nanoparticles can affect the optical characteristics of nanocomposites[5]–[7], which is crucial for applications requiring customized optical properties, such as sensors, optical gear, and coatings. Nickel sulfide is also well-known for its catalytic properties[8],[9] and application in energy storage devices [10]. Moreover, nickel sulfate hydrate likewise offers potential advantages in the development of batteries, in catalysis [2], as an electrolyte for batteries, and for the treatment of wastewater, nickel sulfate hydrate is frequently employed. Therefore, the goal of this study was to uncover all of these benefits in a single sample [1].

2 Experimental Information

2.1 Chemical Used

Table 1 lists all compounds that were used during synthesis along with their purity percentage. All the cleaning and washing were done using deionized water (DI water) and absolute ethanol. All the solutions were prepared in DI water.

Table 1

Chemical Name	Company	Purity percentage
Nickel Acetate	LOBA Chemie	98%
Thiourea	M/s Sigma-Aldrich Pvt. Ltd.	≥99%

2.2 Synthesis

Synthesis and characterization of as-synthesized carbon nanospheres were published elsewhere [11]. Typically, 0.5M of D-Glucose was dissolved in 100 ml DI and stirred for 30 min at 22°C. Then the solution was placed autoclave for 6 hours at 200°C. Afterwards on cooling the solution was cleaned thrice by DI and ethanol, which was further transferred to a vacuum oven for drying at 60°C for one day. In the end, the sample was grinded to obtain carbon spheres. Initially,

0.07 g of carbon nanospheres were dispersed in 30 ml ethanol. This mixture was placed in a bath sonicator for 15 minutes to make the solution volatile. After sonication, the solution was stirred for the next 20 minutes. Additionally, a solution of 0.6 M (2.2 g) thiourea was prepared using 50 ml DI water. At the same time, a third solution of 0.04 M (0.35gm) nickel acetate was stirred in 50 ml ethanol. The solutions containing thiourea and carbon nanospheres were added dropwise into the nickel acetate solution. The final solution was then heated at 200°C for 24 hours in a steel autoclave with Teflon covering. Afterwards, the sample was left overnight to cool down naturally. Then, it was centrifuged and washed three times with DI water and ethanol respectively. Following this, the solution was then placed in a vacuum oven to dry at 70°C for 24 hours. The final sample was obtained by grinding till fine powder was procured.

3 Results and Discussion

3.1 Characterization Details

For the identification of the phase and the structure, X-Ray Diffractometer (make: Rigaku, model: Smartlab 3kW), was used. The X-ray diffraction pattern was captured using copper (Cu) K-radiation (=1.5406 Å). The morphological and elemental content of the sample was analyzed using a FESEM (make: JEOL, model:7610FPlus). To capture the optical response, a Shimadzu UV-2600i UV-Vis spectrophotometer was employed.

3.2 X-Ray Diffraction

The X-ray diffraction pattern of the as-synthesized sample was recorded in a range of 2θ from 10° to 70°. A part of these identified peaks showed a good match with JCPDS file: ***01-073-0574*** illustrating the development of nickel sulfate hydrate with Miller indices in the planes (110), (-112), (112), (-202), (-114), (-110), (115), (022), (220), (200), (222), (0-12), (-119), (-133), (113), (136), (1211), (-428), (-110), (-1114) and (-247). The remaining peaks were also analyzed and indexed with JCPDS file: ***01-079-0105,*** corresponding to nickel sulfide with Miller indices in planes (-202), (-1-31), (-320) and (2-31). Fig. 1. depicts the XRD pattern demonstrating the development of the nanocomposites of nickel sulfate hydrate and nickel sulfide, having monoclinic and anorthic structures respectively. Nickel sulfide hydrate was found to have space group C2/c and space group number 15, exhibiting the existence of a center of inversion, a 2-fold rotation axis and a mirror plane perpendicular to that axis. Nickel sulfide was observed to have space number 1 and space group P1, revealing a triclinic crystal system with no additional symmetry element beyond translation symmetry. No carbon peak was observed in the XRD pattern due to the trace amount of carbon nanospheres present in the as-synthesized sample.

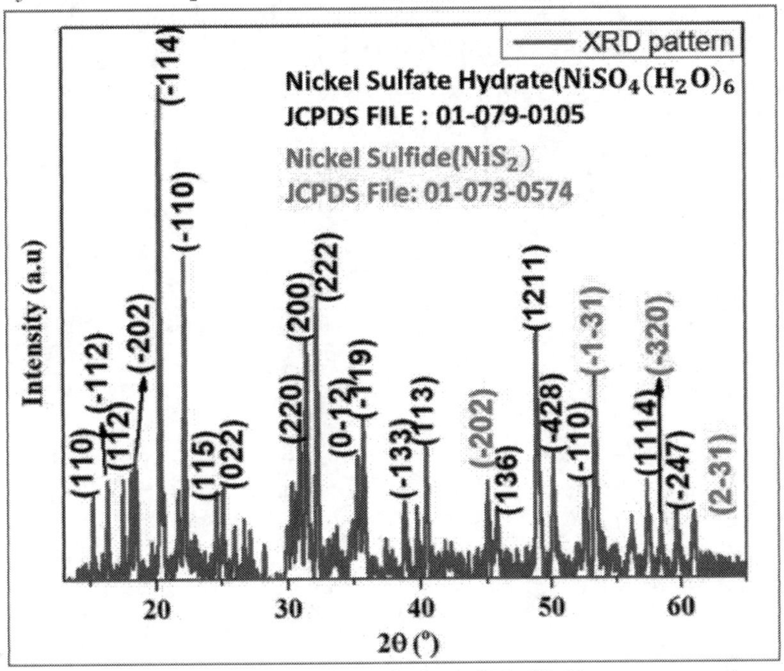

Fig. 1. XRD spectra of as synthesized $NiS_2/NiSO_4(H_2O)_6$ nanocomposites.

3.2.1 Lattice Parameters, Crystallite Size and Strain

Calculation of lattice parameters of as-synthesized sample with the help of equations given below:

$$\frac{1}{{d_{hkl}}^2} = \left[\frac{h}{a} \begin{bmatrix} \frac{h}{a} & cos\gamma & cos\beta \\ \frac{k}{b} & 1 & cos\alpha \\ \frac{l}{c} & cos\alpha & 1 \end{bmatrix} + \frac{k}{b} \begin{bmatrix} 1 & \frac{h}{a} & cos\alpha \\ cos\gamma & \frac{k}{b} & cos\alpha \\ cos\beta & \frac{l}{c} & 1 \end{bmatrix} + \frac{l}{c} \begin{bmatrix} 1 & cos\gamma & \frac{h}{a} \\ cos\gamma & 1 & \frac{k}{b} \\ cos\beta & cos\alpha & \frac{l}{c} \end{bmatrix} \right] \cdot \begin{bmatrix} 1 & cos\gamma & cos\beta \\ cos\gamma & 1 & cos\alpha \\ cos\beta & cos\alpha & 1 \end{bmatrix}$$

(for anorthic) .....(1)

$$\frac{1}{d_{hkl}^2} = \frac{h^2}{a^2 sin^2\gamma} + \frac{k^2}{b^2 sin^2\gamma} - \frac{2hk cos\gamma}{ab sin^2\gamma} + \frac{l^2}{c^2} \text{ (for monoclinic)}$$ ......(2)

Here, d shows interplanar spacing, a, b and c denote lattice parameters of crystal and hkl are the miller indices.

Table 2 lattice parameter of synthesized material matched with JCPDS parameters.

Lattice parameter	JCPDS Value (01-073-0574)	As synthesized (NiS_2)	JCPDS Value (01-079-0105)	As synthesized $NiSO_4(H2O)_6$
a(Å)	5.6849(Å)	5.77(Å)	9.88(Å)	9.578(Å)
b(Å)	5.6849(Å)	5.6839(Å)	7.228(Å)	7.2448(Å)
c(Å)	5.6849(Å)	5.621(Å)	24.13(Å)	27.06(Å)

The crystallite size of the as synthesized nanocomposite was determined with the help of Debye Scherer formula [8], [9], [12] given in equation- 3 below as,

$$D = \frac{K\lambda}{\beta Cos\theta} a \qquad ....(3)$$

Here, D is crystal grain size, λ is the wavelength of x-ray (0.154nm), K is constant with value 0.9 known as Scherer constant, β is FWHM and θ is angle at which FWHM is obtained from XRD data. The average crystallite size was computed by considering all analyzed peaks. Therefore, the crystallite size from the above data was 65 nm. The Debye Scherer equation was based on the fact that there was no strain, which might not be the case with the as-synthesized material. Consequently, additional techniques such as Williamson-Hall plot and the Size-Strain plot, were attempts to precisely determine the crystallite size of the material.

The above table.2 confirmed that there was deviation of the computed value from the standard one due to the presence of micro and lattice strain in the material. Williamson Hall plot, a well-known approach, was used to determine the strain as well as the shifted crystallite size due to the presence of strain. According to the Williamson-Hall plot (W-H plot), both strain and crystallite size attributed to the enlargement of XRD peaks. The equation of W-H plot [1] is given by eqn.(4),

$$\beta_T Cos\theta = (\varepsilon \times 4Sin\theta) + \frac{K\lambda}{D} \qquad ....(4)$$

Here, β_T is FWHM, ε is strain, θ is the angle, K is Scherrer constant possessing the value 0.9, λ is wavelength of X-ray (0.154nm), D is crystallite size. Williamson Hall plot is shown by fig. 2(a), with the value of calculated slope and strain. From the intercept and slope of this plot crystallite size was determined to be **95.5 nm** and strain (ε) was found to be $\mathbf{8.8\times10^{-4}}$.

In the Williamson-Hall plot, the observed randomness was attributed to

the Full Width at Half Maximum (FWHM) value. To mitigate this randomness, SS plot was considered as the data represents average FWHM.
Additionally, the material's crystallite size and strain were accurately determined through the size-strain plot. The equation used for size strain plot [13]is as follows,

$$(d_{hkl}\beta_{hkl}cos\theta)^2 = \frac{k\lambda}{t}\beta_{hkl}d^2{}_{hkl}cos\theta + \frac{\varepsilon^2}{4} \quad \text{.......(5)}$$

Hence, the calculated crystallite size and strain was found to be **132 nm** and **10.9×10^{-4}** respectively.

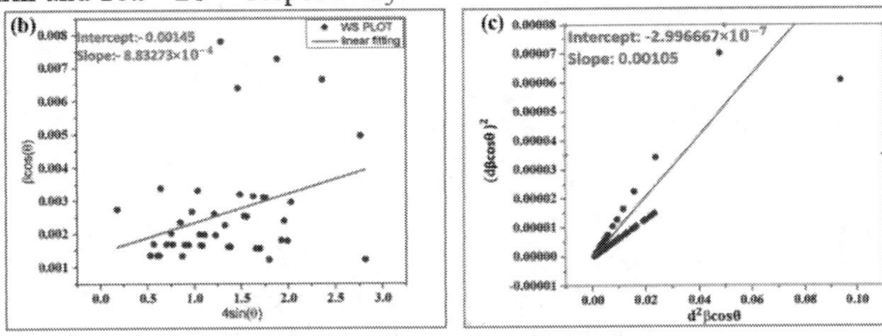

Fig. 2. (a) Williamson Hall plot, (b) size strain plot.

Table 3 crystallite size and strain by various methods.

	Debye Scherrer	Williamson Hall	Size-Strain
Crystallite size	**65.4 nm**	**95.5 nm**	**132 nm**
strain	--	**8.8×10^{-4}**	**10.9×10^{-4}**

3.3 UV-Vis Spectroscopy

With the aim to investigate optical properties of the synthesized nanocomposites, UV-visible spectroscopy was recorded in the range of 190 nm to 850 nm as shown in Fig. 3. (a). The UV-visible spectra showed that the nanocomposites revealed high absorbance in the visible region from 300 nm to 820 nm. The maximum absorbance was recorded at a peak 531.5 nm.
To study whether the as-prepared sample is an insulator, conductor, or semiconductor, the optical bandgap was calculated. Bandgap is the difference of energy between the conduction and valance band Fig.3.(b) shows the Tauc plot [1] through which the bandgap of the as-prepared sample was calculated. Here, Beer Lambert's law was applied which is given as [14] :

$$\alpha hv = A[hv - E_g]^{1/n} \quad \text{....(6)}$$

Here, A represents the absorbance coefficient; E_g symbolizes the direct band gap; α shows the molecular extinction coefficient; h is Plank's

constant; v is frequency; for a direct band gap n corresponds to ½, whereas for an indirect band gap, it's equal to 2.
The computed bandgap was 1.06 eV. Materials with a bandgap of approximately 1 eV offer various benefits, such as being ideal for solar absorption, thermophotovoltaics, infrared sensing and imaging, optoelectronics, catalysis, and photocatalysis as well as energy harvesting. Making them perfect for applications in solar cells, thermophotovoltaic devices, infrared sensing and imaging, LEDs, visible light photocatalysis, semiconductor devices, thermal photovoltaics, and energy harvesting.

3.3.1 Refractive index

Various materials or phases in a material show distinct refractive indices. The speed that light travels for a specific sample relative to the speed of light in a vacuum is referred to refractive index. The refractive index was computed to understand the application of the sample in thin film optics, and sensors, and to assess its optical properties. The formula used to calculate the refractive index [13] of the prepared sample

$$\frac{n^2-1}{n^2+2} = 1 - \sqrt{\frac{E_g}{20}} \qquad \text{......(7)}$$

Here, n= refractive index, E_g=band gap, so the calculated refractive index of the sample was 3.32. The refractive index value of 3.32 for NiS_2 and $NiSO_4(H_2O)_6$ nanocomposites in solar cell applications can enhance light absorption, improve optical matching, and optimize photogenerated carrier separation. These nanocomposites can be incorporated into light-trapping structures, enhancing absorption and conversion efficiency. They can also contribute to bandgap engineering, allowing for materials with specific energy levels aligned with the solar spectrum. Additionally, nanocomposites can increase the solar cells' stability and lifespan, making them cost-effective compared to traditional materials. However, the specific benefits depend on the nanocomposites' structure, composition, and morphology. To assess these nanocomposites' efficiency and viability in solar cell applications, more investigation and validation through experimentation are required.

3.3.2 HOMO LUMO levels

HOMO, or Highest occupied molecular orbital, is commonly known as the valence band, whereas LUMO, or Lowest unoccupied molecular orbital, is commonly known as the conduction band. Calculating the HOMO and LUMO offers insights into the chemical reactivity, charge transfer processes, optical properties, semiconductor properties, drug design, catalytic activity, material properties, stability, energy storage, and solar cell efficiency. The HOMO-LUMO gap is crucial in nucleophilic reactions, charge transfer processes, optical properties,

semiconductors, drug design, catalytic activity, material properties, and energy storage. Understanding these orbitals is essential for designing molecules with specific reactivity, influencing reaction pathways and efficiency, optimizing charge storage and release processes, and enhancing solar cell efficiency. Thus, calculating HOMO and LUMO provides a foundation for comprehending the electronic properties of molecules and materials, enabling informed decisions in various scientific and technological applications.
Equations used to calculate HOMO and LUMO are as follows [15],

$$E_{VB} = \chi - E_e + 0.5\, E_g \qquad \text{......(8)}$$

$$E_{CB} = E_{VB} - E_g \qquad \text{......(9)}$$

For nickel sulfide, valance band was found to be 1.26 eV, and the conduction band was 0.2 eV. For nickel sulfate hydrate valance band was -1.17 eV, and the conduction band was -2.23 eV.

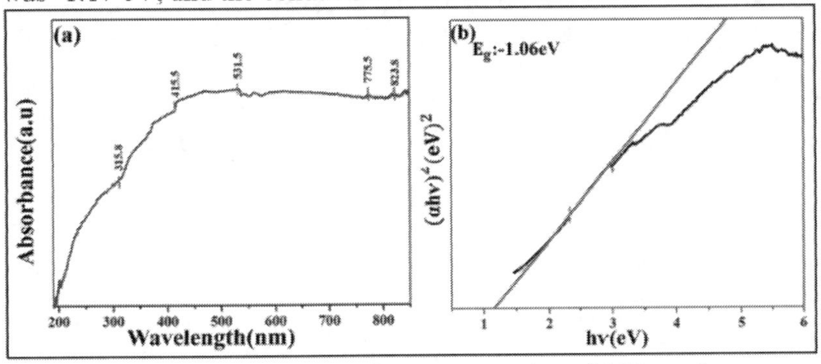

Fig. 3. (a) absorbance spectra of nanocomposites, (b) Tauc plot.

3.4 Morphological Analysis

Although XRD does not confirm the presence of carbon in the synthesized sample, FESEM images revealed that the as-synthesized sample contains carbon spheres. Fig.3 (a) depicts the carbon spheres with particle size near 400 nm range. The image depicts the growth of smaller particles of nickel (either sulfide or sulphate) growing on the surface of the carbon. However, for most of the morphology, the material depicts the growth of sheets caused due to the agglomeration of smaller particles of $NiSO_4(H_2O)_6$. The irregular morphology of 2D Nano sheets will enhance the optical properties of the material whereas the nanospheres will provide better stability to the material.

Fig. 4. FESEM images at two different scales.

4 Conclusion

The current work focused on the synthesis of nanocomposites of NiS_2 and $NiSO_4(H_2O)_6$ using hydrothermal method. The various techniques confirmed the production of nanocomposites. XRD confirms the structure of $NiSO_4(H_2O)_6$ was monoclinic and NiS_2 was anorthic, size stain plot and Williamson Hall plot revealed the value of crystallite size and strain. UV-visible spectroscopy disclosed the absorbance in the visible region. The determined bandgap was found to be 1.06 eV. The Refractive index was calculated to be 3.32 which suggests the optical properties of the sample. The SEM images revealed the formation of nanocomposites on the surface of carbon nanospheres.

Acknowledgments: The authors gratefully acknowledge essential resources provided by Prof. Anand Shrivastava, Vice Chancellor of Netaji Subhas University of Technology (former Netaji Subhas Institute of Technology, University of Delhi).

Conflicts of interest or competing interests: No conflicts of interest exist among the authors.

References

[1] P. Phogat, R. Jha, and S. Singh, "Impedance Study of Zinc Sulphide Quantum Dots via One Step Green Synthesis".

[2] C. V. V. M. Gopi, S. Srinivasa Rao, S. K. Kim, D. Punnoose, and H. J. Kim, "Highly effective nickel sulfide counter electrode catalyst prepared by optimal hydrothermal treatment for quantum dot-sensitized solar cells," *J. Power Sources*, vol. 275, pp. 547–556, 2015, doi: 10.1016/j.jpowsour.2014.11.038.

[3] M. A. Agoro and E. L. Meyer, "Roles of TOPO Coordinating Solvent on Prepared Nano-Flower/Star and Nano-Rods Nickel Sulphides for Solar Cells Applications," *Nanomaterials*, vol. 12, no. 19, 2022, doi: 10.3390/nano12193409.

[4] J. Lin *et al.*, "Rational construction of core–shell Ni3S2@Ni(OH)2 nanostructures as battery-like electrodes for supercapacitors," *Inorg. Chem. Front.*, vol. 5, no. 8, pp. 1985–1991, Aug. 2018, doi: 10.1039/C8QI00347E.

[5] S. Keerthana and A. Kumar, "Potential risks and benefits of zinc oxide nanoparticles: a systematic review," *Crit. Rev. Toxicol.*, vol. 50, no. 1, pp. 47–71, Jan. 2020, doi: 10.1080/10408444.2020.1726282.

[6] X. Liu, L. Jia, G. Fan, J. Gou, S. F. Liu, and B. Yan, "Au nanoparticle enhanced thin-film silicon solar cells," *Sol. Energy Mater. Sol. Cells*, vol. 147, pp. 225–234, Apr. 2016, doi: 10.1016/J.SOLMAT.2015.12.004.

[7] Shreya, P. Phogat, R. Jha, and S. Singh, "Elevated Refractive Index of MoS2 Amorphous Nanoparticles with a Reduced Band Gap Applicable for Optoelectronics," pp. 431–439, 2023, doi: 10.1007/978-981-99-2349-6_39/COVER.

[8] Q. Yang *et al.*, "In Situ Formation of Co9S8 Quantum Dots in MOF-Derived Ternary Metal Layered Double Hydroxide Nanoarrays for High-Performance Hybrid Supercapacitors," *Adv. Energy Mater.*, vol. 10, no. 7, p. 1903193, Feb. 2020, doi: 10.1002/AENM.201903193.

[9] Z. Zheng *et al.*, "Boosting the electrocatalytic activity of amorphous molybdenum sulfide nanoflakes via nickel sulfide decoration," *Nanoscale*, vol. 11, no. 47, pp. 22971–22979, Dec. 2019, doi: 10.1039/C9NR05916D.

[10] R. Pothu *et al.*, "Nickel sulfide-based energy storage materials for high-performance electrochemical capacitors," *Rare Met.*, vol. 40, no. 2, pp. 353–373, 2021, doi: 10.1007/s12598-020-01470-w.

[11] Dipti, P. Phogat, Shreya, D. Kumari, and S. Singh, "Fabrication of tunable band gap carbon based zinc nanocomposites for enhanced capacitive behaviour," *Phys. Scr.*, vol. 98, no. 9, 2023, doi: 10.1088/1402-4896/acf07b.

[12] A. Yadav, Shreya, and N. K. Puri, "Preliminary Observations of Synthesized WS2 and Various Synthesis Techniques for Preparation of Nanomaterials," *Lect. Notes Mech. Eng.*, pp. 546–556, 2023, doi: 10.1007/978-981-16-9523-0_61/COVER.

[13] P. Phogat, Shreya, R. Jha, and S. Singh, "Optical and Microstructural Study of Wide Band Gap ZnO@ZnS Core–Shell Nanorods to be Used as Solar Cell Applications," pp. 419–429, 2023, doi: 10.1007/978-981-99-2349-6_38.

[14] P. Phogat, Shreya, R. Jha, and S. Singh, "Diffusion Controlled Features of Microwave Assisted ZnS/ZnO Nanocomposite with Reduced Band Gap," *ECS J. Solid State Sci. Technol.*, vol. 12, no. 3, p. 34004, 2023, doi: 10.1149/2162-8777/acc426.

[15] S. Sharma, P. Phogat, R. Jha, and S. Singh, "Electrochemical and Optical Properties of Microwave Assisted MoS2 Nanospheres for Solar Cell Application," *Int. J. Smart Grid Clean Energy*, vol. 12, no. 3, pp. 66–72, 2023, doi: 10.12720/sgce.12.3.66-72.

Experimental and numerical investigations of Al 1100 cup using conical die without blank holder

Amit Kaimkuriya[1], *S. Balaguru*[1*]

[1]VIT Bhopal University, School of Mechanical Engineering (SMEC), Sehore, Madhya Pradesh, 466114, India

Abstract. In sheet metal forming processes, the wrinkle phenomenon has been one of the main limiting failure factors. Both numerically and experimentally investigations were done using conical cup test method for wrinkling phenomena. This research work aim is to drawing a flat-bottom cylindrical cup from Al 1100 sheet material, with the effect of drawing parameters, including blank shape, blank thickness, load, and dry lubrication, employing both numerical and experimental methods. Simulation results were obtained using ANSYS 19.2 software, while experimental data were acquired through a universal testing machine. The optimization of process parameters led to the formation of a cylindrical cup with a flat bottom and no defects like thinning or wrinkling. This parametric research contributes to the accurate prediction of the final geometry of the sheet blank and the distribution of stresses, crucial for the forming process and the production of high-quality products. The results show that a die angle of 20° and an optimized punch speed of 3.5 mm/sec give the wrinkle-free cup with high compressive strength.

1. Introduction

Sheet forming is a key manufacturing process in the automotive sector, characterized by the plastic deformation of sheet material to shape it without issues like buckling or cracking [1]. Within this approach, deep drawing processes play a crucial role, widely used for crafting lightweight, high strength, low-density, and corrosion-resistant products like beverage cans, car bodywork, and structural elements. This entails applying punch force to shape sheet metal within a die cavity [2]. Despite its advantages, deep drawing is not without challenges, and one of the primary concerns is wrinkling, a common failure mode in the shaping of sheet metals. With the increasing utilization of thin yet strong sheet metals in recent years, wrinkling has become more prevalent in the stamping of various automotive components [3]. Consequently, the anticipation and prevention of wrinkling, particularly in the final sheet metal parts, have become areas of significant focus [4]. Effectively controlling various process parameters in deep drawing is crucial to circumvent defects like wrinkling, fracture, and thinning in the drawn parts. The meticulous management of these parameters is essential for ensuring the quality and integrity of the manufactured components in the automotive industry [5].
Excessive compressive stresses induce a form of local buckling in sheet metal referred to as wrinkles, stemming from instability under compressive loads. Two primary types of wrinkles in deep drawing sheet metal: flange wrinkles and wall wrinkles [6].

*Corresponding author: balaguru.s@vitbhopal.ac.in

Both result from compressive circumferential forces, with wall wrinkling occurring more frequently due to the lesser support provided by the tool compared to the flat flange. Managing wall wrinkles poses a greater challenge than suppressing flange wrinkles, particularly in
controlling the blank holder force (BHF). The blank holder force influences the radial tensile stress component and is a significant factor contributing to wall wrinkles [7, 8].
The investigation of various parameters, such as blank holder force, blank size, anisotropic behaviour, and yield criteria, can illuminate the impact on the wrinkle phenomenon in the conical cup test. This test involves drawing a circular blank with a punch that has a minimum 25% difference in diameter [9]. Consequently, tooling design plays a pivotal role in predicting and preventing failures in the final deep-drawn product. Thereby, increased emphasis should be placed on optimizing critical process parameters, including the coefficient of friction, die radius, punch radius, and punch force [10, 11].
The concept of the conical die was inspired by Fukui's conical cup test. Utilizing conical dies featuring an optimal half-cone angle, this innovative approach to manufacturing elliptical-shaped cups, eliminates the need for blank holders. The application of conical dies, producing asymmetric cups with clover, rose, star, and triangle cross-sections that exhibited enhanced drawability [12,13].
The literature indicates a predominant focus on the deep drawing of components with circular or cylindrical shapes, with comparatively less research dedicated to utilizing software for the analysis of deep-drawn flat-bottom cylindrical cups [14]. Consequently, there exists an opportunity to enhance the deep drawing process for flat bottom cylindrical cups, aiming to reduce both the expenses and time associated with costly trial-and-error experiments.
Based on previous research findings, it has been established that the production of cylindrical cups typically involves complex (non-conventional) processes or multiple stages. Consequently, there is a significant demand for a simple and cost-effective method. In this study, a novel and straightforward deep drawing process for flat bottom cylindrical wrinkle-free cups without a blank holder has been proposed. Past research has not adequately addressed issues related to wrinkling and earring failure. There is less research work available on Al 1100 in the field of deep drawing, but it offers the advantage of being easily accessible and cost-effective, making it beneficial for the automobile industry. In this work experimental and numerical investigation were done for flat bottom cylindrical cups. A conical die without a blank holder have been used to form the shape of cup [15]. The conical die has a circular cross-section at its end for the regular shape of the cup. Al 1100 material is used for drawing the blanks and cups. Aluminium is widely used in aircraft and utensils.

1.1 Product details and drawing

The cylindrical cup, with a flat bottom, was prepared for experiment. A laser cutting machine was used to prepare the blank from a 1.5 mm thick sheet of Al100. The taper die chamber was filled with a blank, and the cup was formed by applying hydraulic force. Initially 1.5 mm is the thickness (t) and 60 mm is the blank size (D) after experiment the diameter of cup at the top surface was 51 mm and and at bottom 32 mm.

1.2. Mechanical properties and microstructures of blank material

Al 1100 material is selected for analysis. The chemical composition and mechanical properties are mentioned in Table 1.

Table 1 Composition of blank material and Mechanical properties

Material	Composition %	Property	Value (Unit)
Al	99.0		
Cu	0.05	Ultimate Strength	89.6 MPa
Fe	0.05	Yield Strength	159MPa
Mn	0.95	% Elongation	28
Si	0.95		
Zi	0.1		
Residuals	0.15		

2. Experimental détails

A universal testing machine of 40 Ton is used for experimental work and ANSYS 19.2 was used for simulation and validation of experimental results. Formation of the Al1100 cups for this operation, a conical die and cylindrical punch are installed in the TUE-C-400 universal testing machine shown in figure 1. Additionally, this conical device can be used for various blank diameters. Al 1100 sheet with a 1.5 mm thickness was cut into circular blanks with a diameter of 60 mm. To draw the cup, hydraulic pressure is applied, and reading is obtained on extensometer.

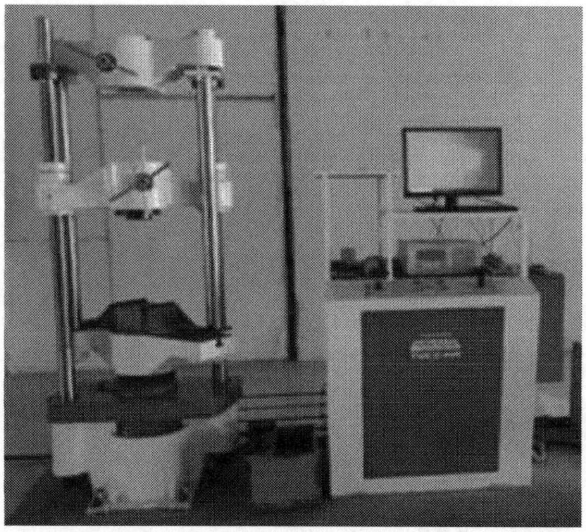

Figure 1. Experimental Setup TUE-C-400 UTM

2.1 Simulation through ANSYS 19.2

Punch and cavity surfaces are created in the 3D programme to perform simulations. In this analysis, the applied load, shape, and material parameters are required. On blank geometry (60 mm in diameter and 1.5 mm in thickness), a fine mesh of elements type (SOLID180) is created. It is considered that the die and punch are rigid bodies. The clearance between punch and die is 3.5 mm. A finite element model, or quarter model, is displayed in Figure 2 shows the quarter model of Die and punch. The inner diameter of die having a taper length of 100 mm a with varied diameter 30 mm to 100 mm. and figure 3 shows the meshing of blank. Three diffèrent région has been selected for meshing. The first region base of punch having a element size of 0.5 mm, second region punch profile having a mesh size of 0.5 mm element size and third region meshed with 1mm of element size on circumferential direction.

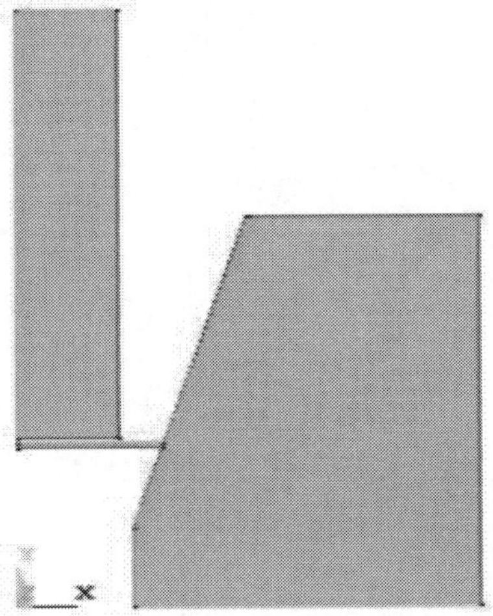

Figure 2. Quarter model of Die and Punch.

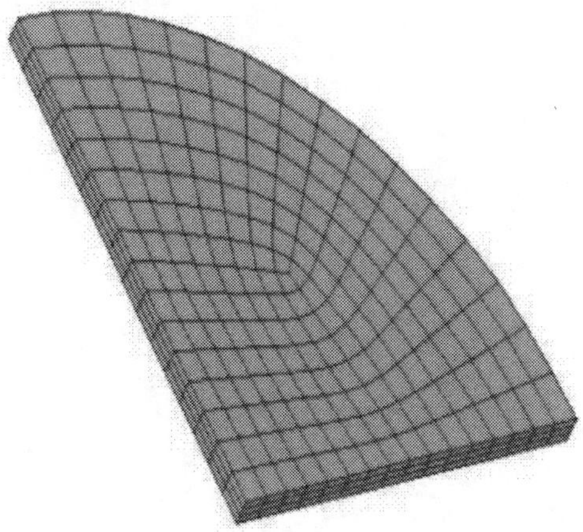

Figure 3. Quarter model of meshed blank

3. Results and discussion

3.1 Results of Experimentation

To achieve symmetrical and wrinkle-free cups, various parameters such as load, elongation, duration, and strength are carefully considered. Figure 4 shows the Al 1100 wrinkles free cup. Figure 5 illustrates the load vs. CHT (Cup Height Thinning) for the 60 mm blank. Using a conical die and hydraulic pressure, a cup was successfully drawn, with a peak load of 8.780 KN resulting in a maximum deflection of 15.980 mm. The experiment for a single sample concluded after 5 minutes and 53 seconds, raising concerns about potential wrinkling with increased load and time. Three iterations were conducted to optimize the outcomes, and the table presents values for different parameters considered in the experiment.

Figure 4. (a) Top face f wrinkle free cup (b) Bottom face of wrinkle free cup

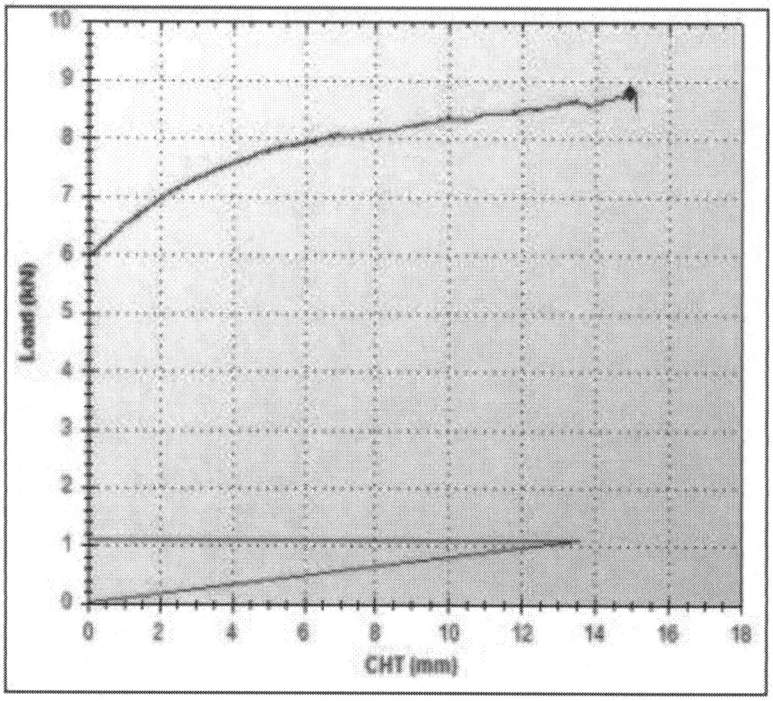

Figure 5. Load Vs. Cross head travel

3.2 Results of simulation

The simulation tool ANSYS 19.2 is employed in this study, using blank geometry to construct a fine mesh of SOLID180-type components. The punch and die are modelled as rigid bodies, consistent with the paramètres considered in the experimental work. The punch speed is chosen within the range of 3.5 to 4.5 mm/s. Finite element simulation enables the identification of potential failure zones and the thinning region. The primary focus is on areas with the highest stress, which aids in avoiding prototype builds and verify the necessary load calculations through finite element simulation. A die inclination of 20° results in the minimal stress on the sheet metal, with a coefficient of friction for the tool set at 0.001. Subsequent to simulation, it was found that for a single iteration, the maximum stress and equivalent stress were 4500 MPa and 9185 MPa, respectively, at a deformation of 15.952 mm. To facilitate a comprehensive comparison between the simulation and experimental data, three iterations of the simulation were conducted, refining the understanding of various parameters.

Figure 6 (a) Maximum deformation

Figure 6 (b) Deformed cup with max. principal stress

Table 2. Comparison of experimental and simulation parameters

S. No.	Parameters	Experimental	Simulation
1	Cross-section area (mm^2)	2827.431	2827.431
2	Maximum Load at peak (KN)	8.780	7.360
3	Elongation at peak (mm)	15.980	15.982
4	Compressive strength(N/mm^2)	3.699	4.110

4. Conclusion

The Al 1100 alloy sheets underwent experimental investigation and simulation analysis to optimize process parameters crucial for enhancing the quality of flat-bottom cylindrically drawn cups. Factors such as load, blank shape, and blank thickness were determined through experimental results, subsequently validated using ANSYS 19.2. The simulation results showed the value of wrinkle free cup in the safe zone. It has found that the required load in the simulation is 16.17% lower than in the actual experiment. While the compressive strength in the experiment is 3.699 N/mm², it amounts to 4.110 N/mm² in the simulation. However, stresses on the top of the surface or flange in the simulation were much higher than the experimental stresses. The formation of a wrinkle-free cup was achieved at a punch speed of 3.5 mm/s. This study emphasizes that, before conducting experiments to produce flawless components, simulations offer a more effective means of obtaining precise parameters and values.

References

1. R.R. Goud, K.E. Prasad, S.K. Singh, Formability limit diagrams of extra-deep-drawing steel at elevated temperatures, Procedia Mater. Sci. **6** (2014) 123–128, https://doi.org/10.1016/j.mspro.2014.07.014.
2. B.V. Rao, Chandra Mohan Reddy, Krishna Mohana Rao, P.V. Ravindra Reddy, Influence ofdrawing ratio on thickness variation along the walls ofdeep drawn cups, J. Manuf. Eng. **11** (2) (2016) 086–093
3. V. M. Brathikan, B. Sangeethkumar, R. K. Nitheeshwar, and K. Scidarth, "Materials Today : Proceedings Variation of force due to co-efficient of friction & blank diameter in deep drawing," Mater. Today Proc., no. xxxx, 2022, doi: 10.1016/j.matpr.2022.08.431.
4. Wang, X., Cao, J., 2000. On the prediction of side-wall wrinkling in sheet metal forming processes. Int. J. Mech. Sci. **42**, 2369–2394
5. Y. Li; M. Luo, J. Gerlach; T. Wierzbicki, Prediction of shear-induced fracture in sheet metal forming. Journal of Materials Processing Technology, **210** (14) (2010) 1858-1869.
6. Wang, C.T., Kinzel, G., Altan, T., 1994. Wrinkling criterion for an anisotropic shell with compound curvatures in sheet forming. Int. J. Mech. Sci. **36**, 945–960.
7. A. Rivas-Menchi, H. I. Medellín-Castillo, D. F. de Lange, and P. de J. García-Zugasti, "Performance evaluation of analytical expressions for cylindrical and rectangular deep drawing force estimation," J. Manuf. Process., vol. **36**, no. September 2017, pp. 340–350, 2018, doi: 10.1016/j.jmapro.2018.10.033.
8. Alin Pop, IoanRadu, Florin Blaga, "Finite Element Analysis Study on Deep Drawing Speed Punch", Nonconventional Technologies Review (2015) 46-50.
9. Taylor, B., 1993. Formability testing of sheet metals, Metals Handbook, vol. 14., 9th edn. Forming and Forging, 877–899.
10. Bao Meng, Min Wan, Xiangdong Wu, Sheng Yuan, Xudong Xu, Jie Liu, Inner wrinkling control in hydrodynamic deep drawing of an irregular surface part using drawbeads, Chin. J. Aeronaut. **27** (3) (2014) 697–707, https://doi.org/ 10.1016/j.cja.2014.04.015.
11. Hae Chang Gea, Rajesh Ramamurthy, Blank design optimization on deep drawing of square shells, IIE Trans. **30** (10) (1998) 913–921, https://doi.org/ 10.1080/07408179808966545.
12. M.A. Hassan, I. M. Hassab-Allah, L. M. A. Hezam, N. A. Mardi, M. Hamdi, "Deep Drawing of Asymmetric Cups through Conical Die without Blank Holder", Proceedings of the World Congress on Engineering, II (2015).

13. Abdullah A. Dhaiban, M.-Emad S. Soliman, M.G. El-Sebaie, “Finite element modeling and experimental results of brass elliptic cups using a new deep drawing process through conical dies”, Journal of Materials Processing Technology **214** (2014) 828–838.
14. M. Ganjiani, A. Assempour J. Mater. Process. Technol., ‘‘An improved analytical approach for determination of forming limit diagrams considering the effects of yield functions”, **182** (2007), pp. 598-607.
15. Kaimkuriya, A., Balaguru, S. (2023). Experimental Investigations on Al 1100 Sheet Metal Using Deep Drawing Technique Through Conical Die Without Blank Holder. In: Sethuraman, B., Jain, P., Gupta, M. (eds) Recent Advances in Mechanical Engineering. STAAAR 2022. **Lecture Notes** in Mechanical Engineering. Springer, Singapore. https://doi.org/10.1007/978-981-99-2349-6_18

Numerical Study on the Effect of Spinning for 7.62 APM2 Projectile Impacting on Al 7075-T651 Target

Vaggu Pranay[1], *S.K.Panigrahi*[1*]

[1]Department of Mechanical Engineering, Defence Institute of Advanced Technology, Pune-411025, India

Abstract. When a projectile launches from a rifled barrel, it undergoes rotation around its longitudinal axis, commonly referred to as spinning, to maintain stability during its flight through the air. Upon reaching the target, the projectile possesses both rotational and translational energies. This study examines the impact of rotational energy on the penetration of 7.62 APM2 projectiles by using Finite Element Analysis(FEA). Al 7075-T651 alloy with a thickness of 20mm is considered as target material. Initially, the model is validated using the experimental results available in the literature. The validated model has been extended to study the effect of the projectile's spinning by using the Recht-Ipson parameters. Based on the results, it is observed that the rotation of the projectile leads to an enhancement in penetrating energy and a reduction in ballistic limit velocity.

1 Introduction

Rifled barrels existed for more than 150 years. Rifled weapons have existed for more than 160 years. For example, the .58 calibre rifle musket (M1855) was the first rifled weapon produced for general issues used by the U.S. Army. It was produced in July 1855[1]. Till today most armament uses rifled barrels, which makes the projectile spin i.e. rotate around its longitudinal axis for stabilization purposes while travelling in the air. In the rear end of the projectile, a driving band is provided so that the projectile can engage with the rifling provided in the barrel. When the projectile is fired from a firearm, the projectile is pushed by high-pressure gases and translated in the forward direction. Due to the presence of a driving band, the projectile also rotates along with translation in the forward direction. Generally, the driving band is made of very soft material such as copper, lead.

Many researchers have studied different aspects and parameters related to impact analysis by using analytical, experimental and numerical techniques. The drawback of analytical theories is that they are made by using assumptions, which makes them not universal. However, they form the basis for understanding the phenomenon. Experimental analyses are complete reliable ones. In this particular area, the experiments consist of

* Corresponding author: panigrahi.sk@gmail.com

destructive testing. So, the cost of conducting experiments will be high. Recently developed numerical techniques are one of the best options for pre-analysing the effects of different parameters. Numerical techniques can predict different failure mechanisms, and solve complex shapes with different boundary conditions. However, special care must be taken while modelling material parameters and meshing conditions to predict the influence of different parameters.

A detailed literature survey has been conducted on various aspects and influential parameters related to impact analysis. Christman et al. [2] and Forrestal et al. [3] have proposed one of the earliest penetration theories. Gupta et al.[4], Forrestal et al. [5] studied the impact of projectiles on aluminium plates experimentally. Børvik et al. [6] have studied different armour-grade steels against small arms projectiles. Børvik et al.[7] analysed different project shapes, Flores-Johnson et al.[8] studied layered configurations of armour materials by using numerical analysis. Recht-Ipson[9] proposed a model to predict the behaviour of the projectile's residual velocity.

The Recht-Ipson empirical equation[9] for ballistic estimation is given

$$V_r = a(V_i^P - V_{bl}^P)^{\frac{1}{P}} \tag{1}$$

V_i= Projectile's impact velocity, V_{bl}= Ballistic limit velocity, V_r= Projectile's residual velocity, and a,p= Empirical constants. The empirical constants are further expressed as, a= M / (M+m) (Where M and m are the mass of the projectile and plug, respectively) and p=2.

Many research studies are conducted by using the Gas-Gun facility, which uses smooth barrels to fire the projectile. In the working scenarios, the projectile is fired from rifled barrels. In the literature survey, it is observed that the effect of rotational velocity is considered to be negligible and studies have been conducted. There is very little literature [10,11] available related to the spin effect study. Pranay et al.[12,13] have studied the effect of spin for the projectile of L/D ratio of approximately 10 and proposed the modified nose shape to effectively utilise the spin [14]. The present numerical study is conducted on the effect of the projectile in spinning by considering the 7.62APM2 projectile by using commercially available LS-DYNA software.

Aluminium alloys and their composites have a wide range of applications such as automobile structures, aerospace structures, and marine applications[15–18]. Al 7075-T651 alloy is one of the well-known Armour-grade aluminium alloys [19,20] because of its high strength and low weight properties. In the study of Senthil et al.[21], it is concluded that with respect to thickness and areal density, the ballistic resistance of the Al 7075-T651 alloy is 1.7 times of mild steel.

In the present analysis, the Al 7075-T651 target of 20mm thickness is considered for Target. Initially, using the experimental results available in the literature[5], the model is validated. The validated model is extended to study the effect of spinning. The spinning velocities considered for the present study are considered from the literature[10]. For representation purposes no spin condition is expressed as '0x', 13000 rad/s is '1x', 26000 rad/s is '2x' and 52000 rad/s is '4x'

2 Numerical Model

2.1 Geometry

For the present analysis, the 7.62 APM2 projectile core has been considered for initial numerical model validation. The general configuration of the projectile with different parameters is shown in Fig.1. The equivalent core dimensions of the projectile[5] consist:

Caliber-Radius-Head (CRH)=3, diameter (2a) =6.17mm, nose length (Ln) =10.2mm, and Shank length (Ls)=16.8mm. The target is considered a circular section of 152mm diameter of 20mm thickness.

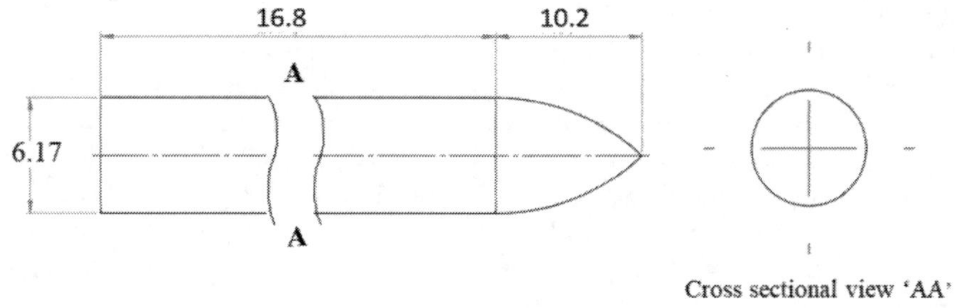

Fig.1. Equivalent core dimensions (in mm) of projectile[5].

2.2 Material Models

In the present study, the target and projectile are considered to be deformable using the Johnson-Cook strength and failure model[22,23]. For the present study, the target considered is Al 7075-T651. The material model parameters [24]are given in Table 1.

Table 1 Material model parameter for the target and the projectile

Parameter	Units	Projectile	Al 7075-T651
Young's modulus E	GPa	200	64
Poisson's ratio υ	-	0.3	0.31
Density ρ	$\frac{kg}{m^3}$	7850	2710
Yield strength A	MPa	1657.71	448.454
Strain Hardening parameter B	MPa	20855.6	475.808
Strain hardening parameter n	-	0.651	0.3948
Strain rate constant C	-	0.0076	0.0012
Thermal softening parameter m	-	0.35	1.29
Reference strain rate $\dot{\epsilon}_0$	s^{-1}	1	0.0001
Reference Temperature	K	293	293
Melting Temperature	K	1800	893
Specific heat capacity C_P	J/Kg.K	455	960
Failure parameter D_1	-	0.0301	-0.428
Failure parameter D_2	-	0.0142	0.7566
Failure parameter D_3	-	-2.192	-3.4078
Failure parameter D_4	-	0	-0.003
Failure parameter D_5	-	0.35	24.93

2.3 Meshing

To reduce the computation time without affecting the results, the target has been divided into different zones as shown in Figure 2. The inner square zone has meshed with 0.25mm*

0.25mm element size, the second zone of 30mm diameter has meshed with 0.5mm* 0.5mm element size, the third zone of 50mm diameter has meshed with 1mm* 1mm size, and then the outer region has meshed with 2mm × 2mm element size. Further, the 2D mesh is extruded through the thickness direction with an element size of 0.25mm to form a 3D hexahedron element. The projectile has been meshed with 0.25mm3 element size. The sectional view of the discretisation of the projectile-target assembly is shown in Figure 3. A full 3D model has been considered for analysis.

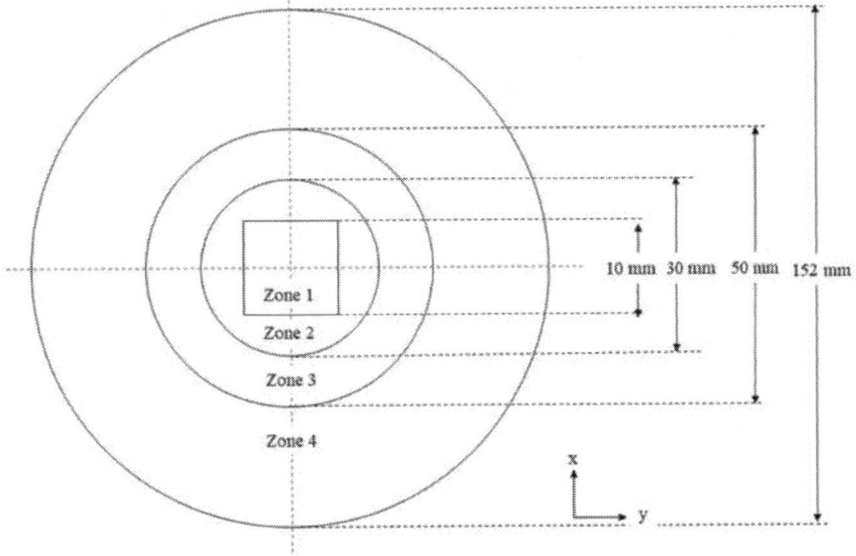

Fig.2. Discretization of the target with different zones

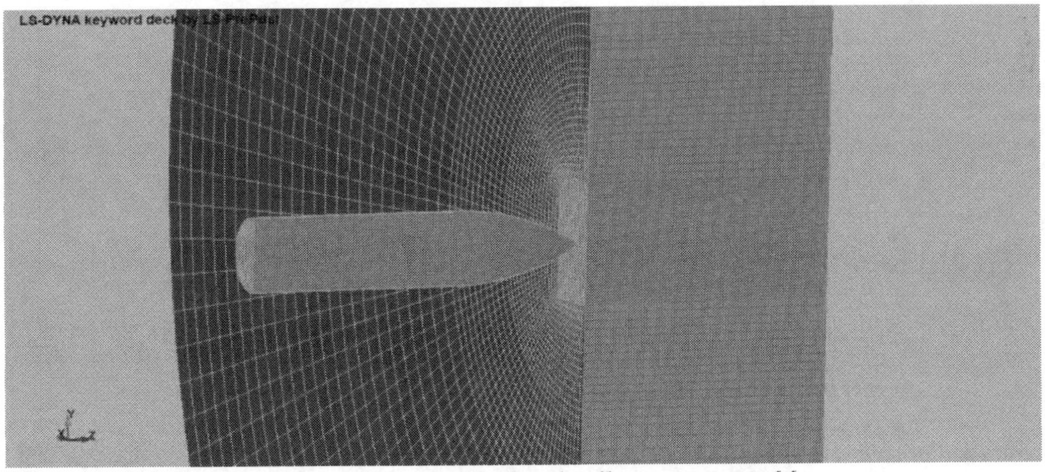

Fig.3. Sectional view of discretized model of projectile-target assembly

2.4 Model validation

Forrestal et al.[5] have conducted experiments with a 7.62 caliber projectile penetrating Al 7075-T651 target. Recht-Ipson equation parameters have been approximated from the experimental results. Experimentally[5], it is observed that penetration of Al 7075-T651 target with Armour piercing projectile doesn't involve significant plugging. Therefore, parameter 'a' is considered to be equal to 1 and only p is the curve fitted with data. The fitted curves are shown in Figure 4. The experiment data points and curve are considered from Forrestal et al.[5] with the values of a=1, p=1.95 and the ballistic limit is 633m/s.

In the present study, the obtained residual velocities are used to curve fit the Recht-Ipson model. In simulation also it is observed that there is no significant plugging involved. So, a=1 is considered and p is curve fitted and ballistic limit velocity is predicted. It is observed that the p =1.942, while the predicted ballistic limit velocity is 654.17m/s. The comparison is shown in Figure 4. On comparing the ballistic limit velocity, it can be observed that the present model is a 3.4% overestimation of the experimental results. The results are acceptable, on considering the limitation of constitute modelling and the complexity of the problem.

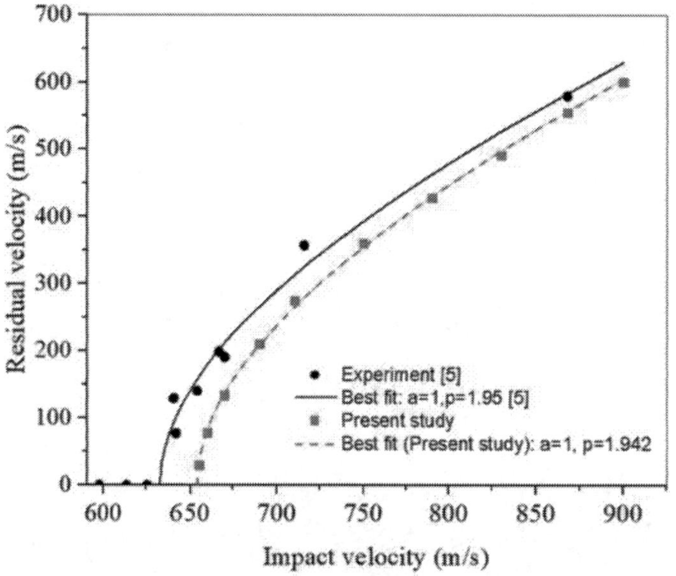

Fig.4. Comparison of experimental [5] and simulation results for 7.62APM2 projectile (Hard core only) impacted on Al 7075-T651 plate of 20mm thickness

3 Results and Discussions

The present study is conducted over impact velocities of 650 m/s-900 m/s. As the study consists of different impact velocities, the Recht-Ipson model parameters are compared and used to study the effect due to spinning.

In the present analysis, as the projectile has completely perforated, ductile hole growth failure is observed in the target and no plugging is observed(shown in Figure 5). Ductile hole growth is the enlargement process, in which due to plastic deformation of the target, the target material is pushed away by the projectile from the penetration path. There was no plugging. So, the value of parameter 'a' is assumed to be equal to 1. At ideal impact conditions parameter 'p' is equal to 2. However, experimental studies have shown that the value of p is not always equal to 2. As the target resistivity increases, the projectile's residual velocity decreases i.e. p value decreases. Which represents, the energy required to penetrate the target increases.

The simulations are conducted by giving the spinning velocities of 1x,2x, and 4x to the projectile along with translational velocities and residual velocities computed. Above impact velocities of 710m/s, providing the initial spin i.e. 1x makes to decrease in the projectile's residual velocity, and a further increase in spin to 2x will increase the projectile's residual velocity, on further increases of spin to 4x will reduce the projectile's residual velocity due to increasing in the projectile contact with the target due to rotation.

At 710m/s impact velocity, the residual velocities of the projectile are 274.6 m/s, 275.9 m/s, 274.2 m/s, and 274.19 m/s for 0x,1x,2x, and 4x spin conditions respectively. It is observed that the projectile's residual velocity is approximately equal irrespective of spin conditions. However, below 710 m/s impact velocity, providing the spin to the projectile has increased the residual velocity.

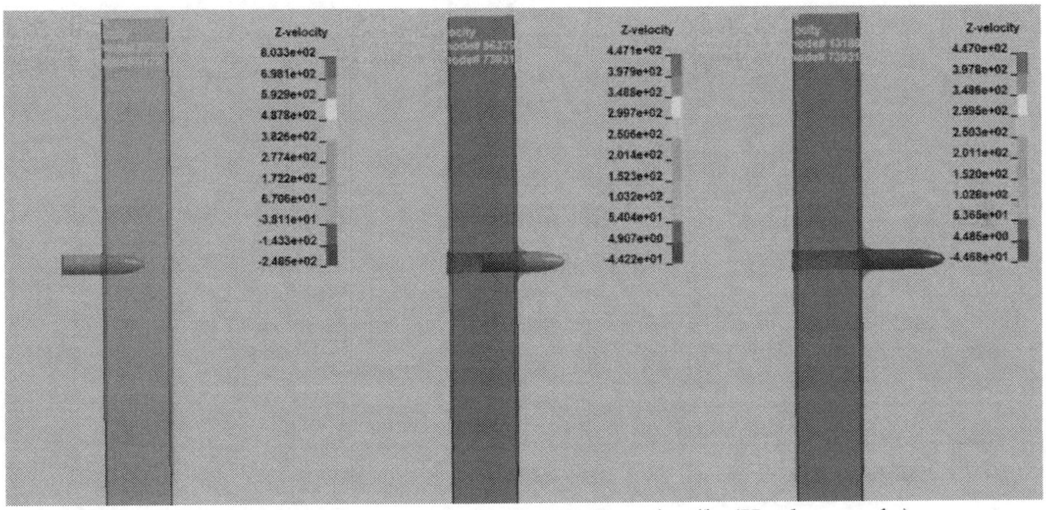

Fig.1. Penetration of Al 7075-T651 target by 7.62 APM2 projectile (Hardcore only)

Near ballistic limit velocity such as at 655 m/s impact velocity, the observed residual velocities are 29.9m/s, 47.9m/s, 55.5m/s, 63.1 m/s for 0x, 1x,2x, and 4x spin conditions respectively. On the further decrease in impact velocity it is observed that there is a very quick drop in projectile's residual velocity such as at 652m/s, for 0x, 1x,2x spinning, the projectile got stuck inside the target i.e. no perforation. For the 4x spin condition, at an impact velocity of 652.3 m/s, the residual velocity observed is 27.39m/s.

Table 2 Variation of Recht-Ipson parameters at different spinning conditions

Spin	a	P	V_{bl} (m/s)
0x	1	1.942	654.17
1x	1	1.932	652.9
2x	1	1.94	652.3
4x	1	1.936	651

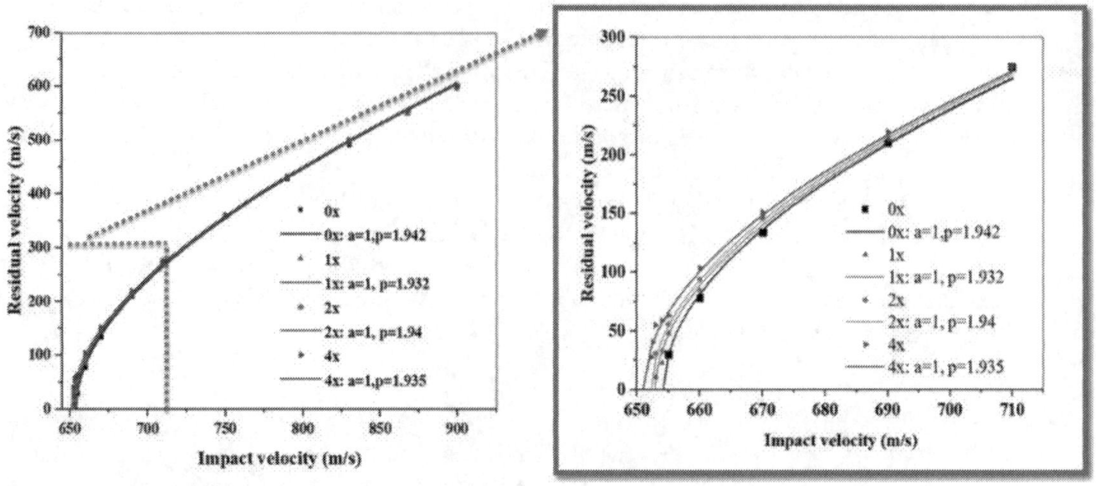

Fig.2. Comparison of Recht-Ipson parameters at different impact conditions

Based on the obtained results, the curve-fitting has been done and a, p, ballistic limit (V_{bl}) velocities are approximated. The obtained results are tabulated in Table 2 and the fitted curve is shown in Figure 6. It can be observed spinning has given less ballistic limit velocity compared to 0x spin conditions. However, from overall observation, it is observed that the 'p' value has decreased due to spinning i.e. projectile spinning made an increase in penetration energy.

4 Conclusions

From the present study on 7.62APM2 projectile (Hard core only), impacting on Al 7075-T651 target of 20mm thickness at different impact velocities and different spinning velocities, the following conclusions are drawn:

- At impact velocities higher than 710m/s, Providing the initial spin (1x) leads to a decrease in the projectile's residual velocity, on a further increase of spin (2x,4x) made an increase in residual velocity compared to 0x.
- Approximately near 710m/s impact velocity, the observed residual velocity is approximately equal irrespective of spin conditions.
- Below 710m/s, there is an increase projectile's residual velocity due to the spinning
- Near ballistic limit velocity, the spinning of the projectile has made a very quick drop in the projectile residual velocity.
- From the overall observations of residual velocities at different impact velocities, at different spinning conditions, the Recht-Ipson model has been curve fitted and parameters are approximated.
- From Recht-Ipson parameters it is observed that spinning has increased the energy consumption for penetration. However, the ballistic limit has been decreased.

References

1. D. Cole, *Survey of US Army Uniforms, Weapons and Accoutrements* (2007)

2. D. R. Christman and J. W. Gehring, Analysis of High-Velocity Projectile Penetration Mechanics, J Appl Phys **37**, 1579 (1966). https://doi.org/10.1063/1.1708570.
3. M. J. Forrestal, V. K. Luk, Z. Rosenberg, and N. S. Brar, Penetration of 7075-T651 aluminum targets with ogival-nose rods, Int J Solids Struct **29**, 1729 (1992). https://doi.org/10.1016/0020-7683(92)90166-Q.
4. N. K. Gupta, R. Ansari, and S. K. Gupta, Normal impact of ogive nosed projectiles on thin plates , Int J Impact Eng **25**, 641 (2001). https://doi.org/10.1016/S0734-743X(01)00003-3.
5. M. J. Forrestal, T. Børvik, and T. L. Warren, Perforation of 7075-T651 Aluminum Armor Plates with 7.62 mm APM2 Bullets, Exp Mech **50**, 1245 (2010). https://doi.org/10.1007/s11340-009-9328-4.
6. T. Børvik, S. Dey, and A. H. Clausen, Perforation resistance of five different high-strength steel plates subjected to small-arms projectiles, Int J Impact Eng **36**, 948 (2009). https://doi.org/10.1016/j.ijimpeng.2008.12.003.
7. T. Børvik, O. S. Hopperstad, T. Berstad, and M. Langseth, Perforation of 12mm thick steel plates by 20mm diameter projectiles with flat, hemispherical and conical noses: Part II: numerical simulations, Int J Impact Eng **27**, 37 (2002). https://doi.org/10.1016/S0734-743X(01)00035-5.
8. E. A. Flores-Johnson, M. Saleh, and L. Edwards, Ballistic performance of multi-layered metallic plates impacted by a 7.62-mm APM2 projectile , Int J Impact Eng **38**, 1022 (2011). https://doi.org/10.1016/j.ijimpeng.2011.08.005.
9. R. F. Recht and T. W. Ipson, Ballistic Perforation Dynamics, J Appl Mech **30**, 384 (1963). https://doi.org/10.1115/1.3636566.
10. F. Gálvez, S. Chocron, D. Cendón, and V. Sánchez-Gálvez, Numerical Simulation Of The Tumbling Of Kinetic Energy Projectiles After Impact On Ceramic/metal Armours, in *Computational Ballistics II* (WIT Press, 2005).
11. P. A. Radchenko, S. P. Batuev, and A. V. Radchenko, Effect of Projectile Rotation on High-Velocity Impact Fracture, Physical Mesomechanics **25**, 119 (2022). https://doi.org/10.1134/S1029959922020035.
12. V. Pranay and S. Panigrahi, Effects of spinning on residual velocity of ogive-nosed projectile undergoing ordnance velocity impact, Proc Inst Mech Eng C J Mech Eng Sci **236**, 1685 (2022). https://doi.org/10.1177/09544062211020030.
13. V.Pranay and S.K.Panigrahi, Effects of spinning on residual velocity of projectile for normal and oblique impact, in *13th International High Energy Materials Conference and Exhibits, TBRL* (Chandigarh, 2022)
14. V. Pranay and S. K. Panigrahi, Design and development of new spiral head projectiles undergoing ballistics impact, International Journal of Structural Integrity **13**, 490 (2022). https://doi.org/10.1108/IJSI-01-2022-0008.
15. S. Nandimandalam and B. Sethuraman, Mechanical and Tribological Behaviour of Al 7075 Hybrid MNC's using Stir Casting Method, International Journal of Mechanical and Production Engineering Research and Development, **10**, 391 (2020). https://doi.org/10.24247/ijmperdjun202036.
16. N. K. Singh and B. Sethuraman, Development and Characterization of Aluminium AA7075 Hybrid Composite Foams (AHCFs) Using SiC and TiB2 Reinforcement, International Journal of Metalcasting, **18**, 212 (2024). https://doi.org/10.1007/s40962-023-01009-6.
17. N. K. Singh and S. Balaguru, Fabrication and Mechanical Characterization of Al-Zn-Cu Alloy /SiC/TiB2 Hybrid Reinforced Metal Matrix Composite using Top Loaded Bottom Pouring Stir Casting Method, Silicon, **16**, 45 (2024). https://doi.org/10.1007/s12633-023-02648-4.

18. N. K. Singh and S. Balaguru, Experimental Analysis of Foaming Agent Contents in AA7075/SiC Closed Cell Aluminum Composite Foam, In: Sethuraman, B., Jain, P., Gupta, M. (eds) Recent Advances in Mechanical Engineering. STAAAR 2022. Lecture Notes in Mechanical Engineering. Springer, Singapore. https://doi.org/10.1007/978-981-99-2349-6_51.
19. K. O. Pedersen, T. Børvik, and O. S. Hopperstad, Fracture mechanisms of aluminium alloy AA7075-T651 under various loading conditions, Mater Des **32**, 97 (2011). https://doi.org/10.1016/j.matdes.2010.06.029.
20. T. Børvik, O. S. Hopperstad, and K. O. Pedersen, Quasi-brittle fracture during structural impact of AA7075-T651 aluminium plates, Int J Impact Eng **37**, 537 (2010). https://doi.org/10.1016/j.ijimpeng.2009.11.001.
21. K. Senthil and M. A. Iqbal, Prediction of superior target layer configuration of armour steel, mild steel and aluminium 7075-T651 alloy against 7.62 AP projectile, Structures **29**, 2106 (2021). https://doi.org/10.1016/j.istruc.2020.06.010.
22. G. R. Johnson and W. H. Cook, A Constitutive Model and Data for Metals Subjected to Large Strains, High Strain Rates, and High Temperatures, in *Proceedings 7th International Symposium on Ballistics* (The Hague, 19-27 April, 1983), pp. 541–547.
23. G. R. Johnson and W. H. Cook, Fracture characteristics of three metals subjected to various strains, strain rates, temperatures and pressures, Eng Fract Mech **21**, 31 (1985). https://doi.org/10.1016/0013-7944(85)90052-9.
24. K. Senthil, M. A. Iqbal, P. S. Chandel, and N. K. Gupta, Study of the constitutive behavior of 7075-T651 aluminum alloy, Int J Impact Eng **108**, 171 (2017). https://doi.org/10.1016/j.ijimpeng.2017.05.002.

Hybrid solar and wind power generation with grid interconnection system

Shivam singh[1*]*,and Sudeep Mahaney*[2]

[1]Electrical Engineering (M.E.), Jabalpur Engineering Collage and Technology Jabalpur M.P. (India)
[2] Electrical Engineering, Jabalpur engineering collage and technology Jabalpur M.P. (India)

Abstract. For India's modernization and urbanization, the need for energy is increasing extremely quickly. Because they are abundant in nature, renewable energy sources might be viewed as more advantageous than traditional ones. Both solar and wind energy are abundant and may be regarded as reliable sources of energy production. For the modernization of isolated areas and the electrification of rural areas, hybrid solar and wind energy systems can be deployed. This work does simulation of a hybrid solar and wind power system that is linked to the grid. For this study, a simulated model is utilized to determine the sag, swell, source voltage, source current. Solar-generated dc electricity is transformed into ac power using a universal converter. The benefit of using universal inverter is used to raise the system's overall power quality and get the Uninterruptible power supplies.

Keywords : Power Quality, Smart Grids, Renewable Energy, And MATLAB Simulink Introduction.

1 Introduction

[1] Renewable energy sources are ones that are abundantly accessible in nature. For enhancing the dependability and quality of the electricity produced by renewable energy sources, several innovative control mechanisms are employed. Recently, the production of solar and wind energy has drawn particular attention.[2] The fast expansion of wind energy globally has raised media knowledge of wind generation technology and increased public understanding of it. [3] An inverter uses the DC output from the PV system to convert it to AC. by the application of a novel multilayer inverter method for harmonic reduction. To enhance power quality, the technology described in the paper uses a Universal inverter.[4] In the proposed work, wind energy and solar energy are both directly linked to the grid via seven-level inverters. Work on simulations is done. [5] To maintain and enhance the quality and reliability of the power supply, new techniques for the operation and administration of the electrical grid are required due to the growing use of distributed generators and renewable energy sources. [6] In India, 37.5 GW of wind energy and 42.8 GW of solar energy were produced in the last few years, while Sri Lanka has more than 100 MW of installed solar capacity and more than 367 MW of total wind capacity. [7] Current worldwide wind and solar capacities total 655.9 GW and 646.8 GW, respectively.[8] All nations are working toward producing all of their energy from renewable sources, with the majority of it coming

* Corresponding author: s.s2021sir@email.com

from wind and solar energy. Since the beginning of time, wind turbine technology has been utilized to raise the standard of living for people.[9] Along with many other purposes, people have used wind turbines to grind grain and pump water. [10] The fact that wind turbines provide power without polluting the environment is a significant component of wind turbine applications, particularly in an industrial setting. Additionally, remote areas without access to the outside grid are highly suited for deploying wind turbines to provide power.

1.1 Motivation

The grid's interconnection has a lot of issues, and electricity quality is also faced with a lot of difficulties due to harmonics. In this study, we use a universal inverter to increase power efficiency while lowering harmonics in the system. Different types of dynamics have been present in the power system due to the growing penetrations of unstable systems, responsive loads, and the main grid and Reduces dependence on centralized grids, providing a reliable and decentralized energy source. Renewable Energy Source Aligns with global efforts to reduce reliance on fossil fuels, promoting sustainability and environmental conservation and combining solar and wind power ensures a more consistent and reliable energy supply throughout the day and night.

1.2 Literature Review

[1] Cascaded H-bridge When 5 and 7 level inverters are compared, it is shown that the 7-level inverter will generate less harmonics than the 5-level inverter and they run simulation on that paper the PV array is combined with the designed 7-level inverter, which has been found to perform well.[2] The non-sinusoidal source voltage, source current, and load voltage were also worked on in this paper and made sinusoidal with the use of a controller.[3] the reactive power supplied by the PY source is reduced through their actions, which improves the proposed power system's THD. With the simulated power system used in their investigation,[4] the overall amplitude of voltage sag and swell of line RMS voltage at PCC were also noted.[5] It is evident from the proposed system's prototype hardware model that the inclusion of a controller raises the voltage at the PCC by 29.11%.[6] The integrating of renewable sources of energy into the grid is greatly aided by the modern power-electronic technologies.[7] There has been a description of current and upcoming trends in renewable energy systems. [9] Due to its installed power and recent advancements in power electronics and control, wind energy is currently the most advanced energy source.[10] Reducing reliance on fossil fuel-based energy has been made possible by the rising use of renewable energy sources. To optimize the advantages of sustainable energy sources, a hybrid system comprising photovoltaic (PV) and wind generators is presented as a single distributed generation unit.

2. Block Diagram

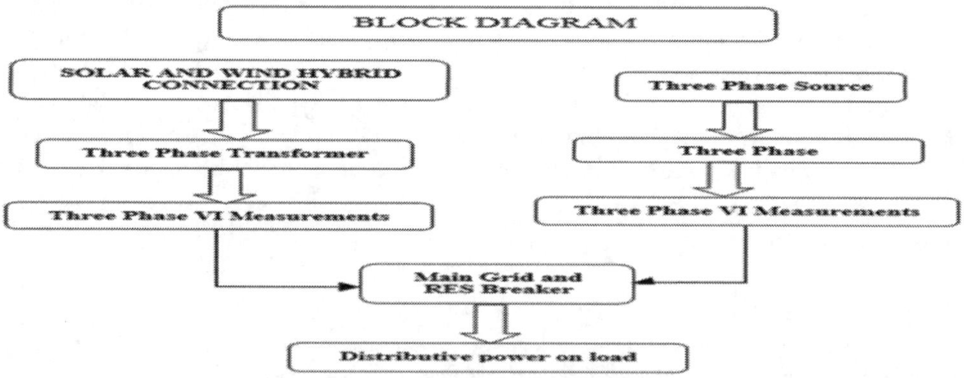

3. Simulation Circuit and Results

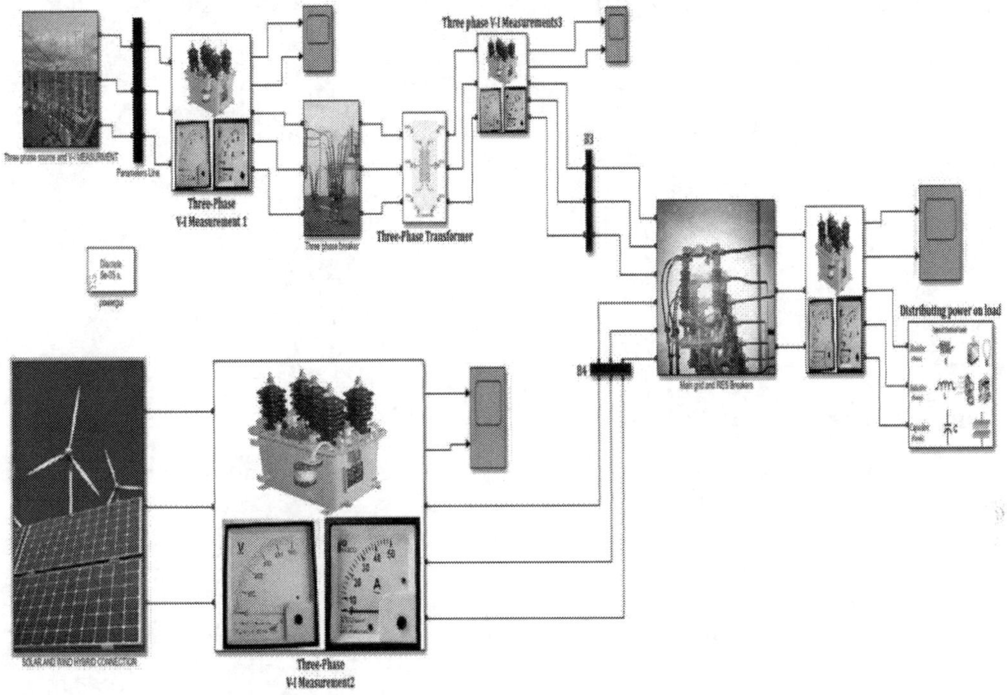

Simulink MATLAB was used to create the simulation block diagrams. Wind and solar energy are considered in this graph. The universal bridge is used to convert DC to AC power and use step down X-Mer and deliver hybrid power produced by wind and solar on the figure. A three-phase source is also obtained and linked to the grid with the help of solar and wind energy. All of it was joined together in a grid for the output.

4. Results

4.1 Grid Interconnected Load Side Output

This is the final output of my MATLAB Simulink system, displayed as a waveform. The result is entirely dependent upon the three-phase source and the hybrid solar and wind combination used in the MATLAB calculation.

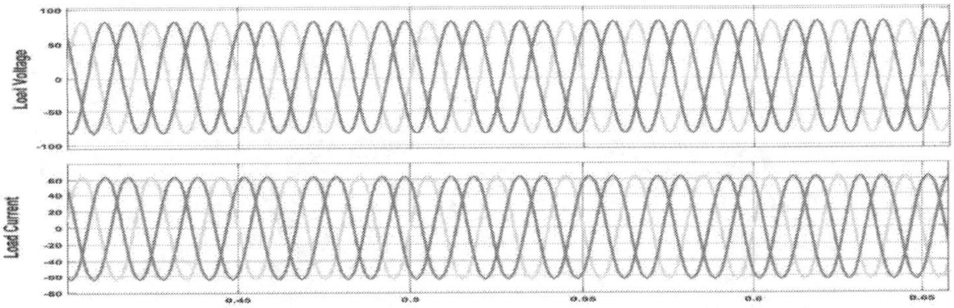

Fig -1 : Grid Interconnected Load Side Output

4.2 Three Phase Source Output

Using a three-phase voltage source in series with an RL branch, I'm seeing the output of my three-phase source in MATLAB Simulink as a wave form. This is the individual result of the three-phase source, which has a frequency of 50 Hz and a voltage level of 25 thousand.

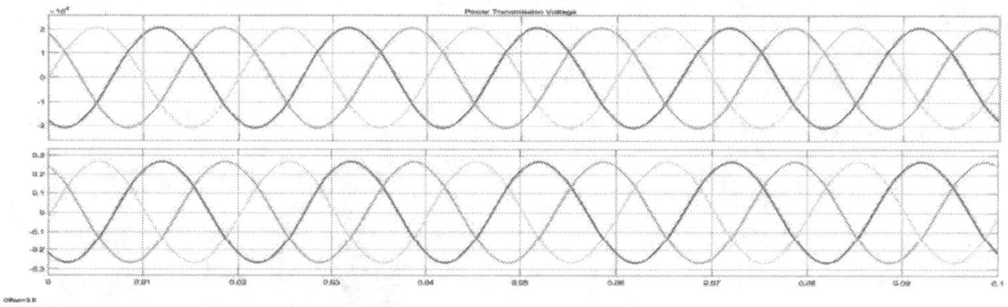

Fig-2 : Three Phase Source Output

4.3 Three Phase Source Output With Using X-Mer

This is the same result as the three-phase source, but since I'm now utilizing a transformer, the result is shown in the figure.

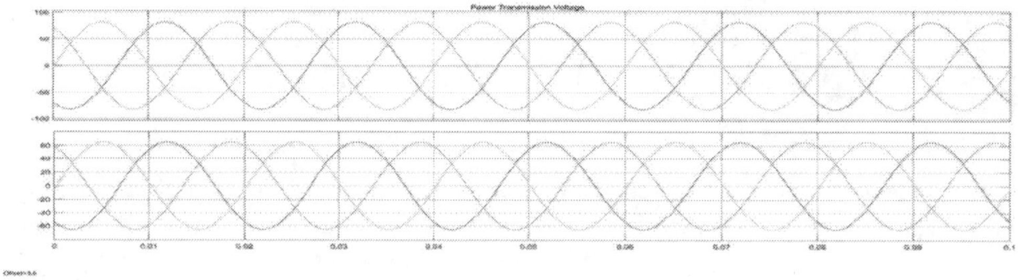

Fig-3 : Three Phase Source Output With Using X-Mer

5. HYBRID SYSTEM SPECIFICATIONS

5.1 Wind

Wind turbine -1

DC bus voltage regulator gains [Kp Ki] – 1.1, 27.5

Grid-side Converter voltage regulator gain [Ki] : - 2

5.2 Solar

Isc	2.02
Voc	86.8
Im	1.93
Vm	70.4
Tref	55
Constant	1000
Constant	25
capacitor	4700uf

5.3 Three Phase Sources

- **Three phase sources**

Phase-to-phase voltage (Vrms)	**25kv**
Frequency (Hz)	**50**

6. Solar and Wind Energy System Using Universal Bridge

The term "wind solar hybrid system" refers to a system It combines the two sources of renewable energy: solar and wind to produce electricity. To create this system, small wind turbines and solar panels were employed to generate power.

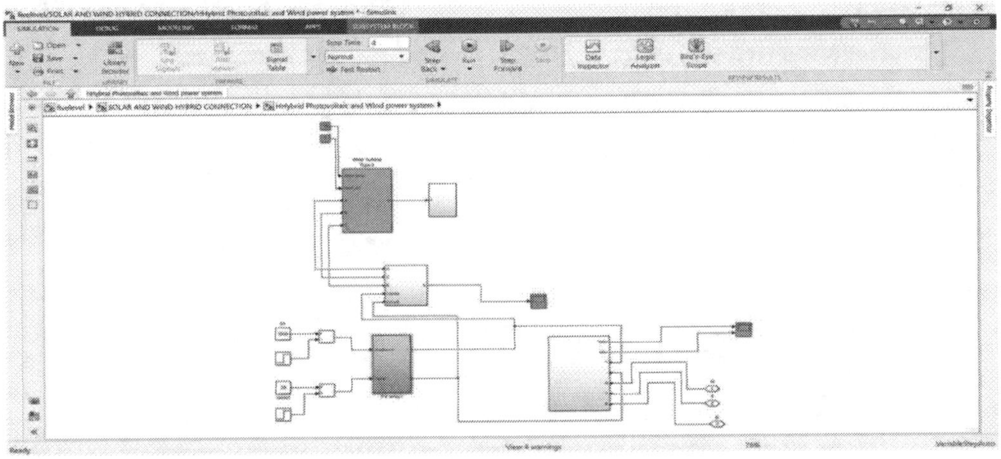

Fig-4 MATLAB Simulink Of The Wind And Solar

I am able to combine wind and PV energy, collect uninterpreted energy, convert DC to AC using a universal bridge, and deliver it to a transformer in Matlab by using the wind turbine 4 and PV ARRY1 in addition to turning off the PWM generator. The output is a waveform, as seen in the figure that follows.

7.Solar and Wind Results

This is the 1500-watt DC wind power system's wave shape. It combines Matlab's universal bridge, a pwm generator, and a wind turbine type 4 to produce the intended voltage range of 0 to almost 90 dc. This wave form displays the outcome of an individual wind turbine result.

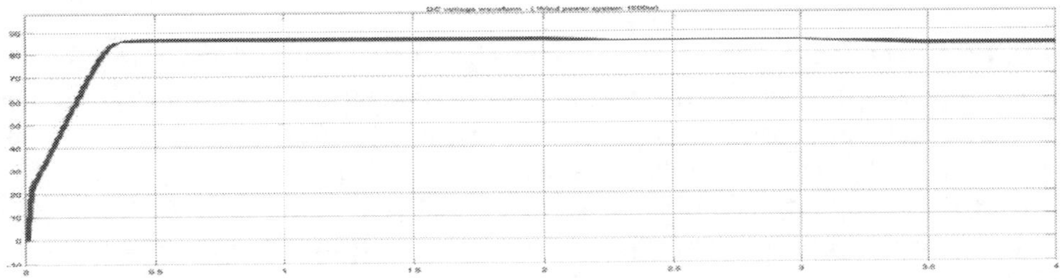

Fig-5 Waveform of The DC Voltage Of Wind Power System

8. Solar and Wind Results Using Universal Bridge and Control Technique

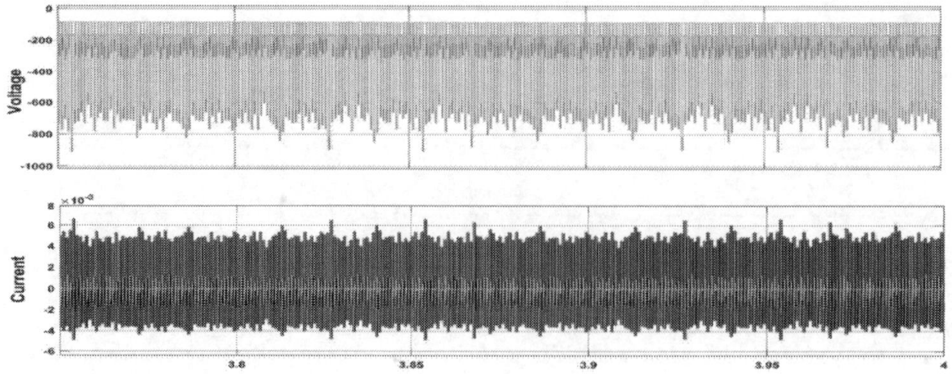

Fig-6 Combined Waveform Of The Solar And Wind

This is the outcome of the PV array and wind and solar power combined, and the wave form represents the combined voltage and current of the power system.

9. Solar and Wind Output Using Transformer

This is also the outcome of combining solar and wind energy, and I'm currently obtaining the waveform result in MATLAB by using the transformer in this combination.

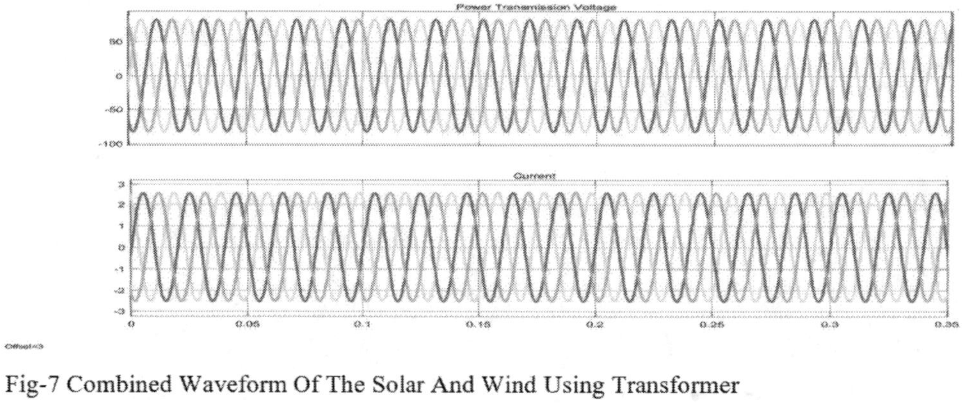

Fig-7 Combined Waveform Of The Solar And Wind Using Transformer

10. Conclusions

In conclusion, the increasing popularity of solar and wind hybrid systems stems from a multitude of advantages. These systems offer stability and reliability by harnessing energy from sources that peak at different times, ensuring a consistent power supply. The combination results in more predictable power generation, overcoming the inherent variability of individual sources. Additionally, hybrid systems maximize energy harvesting, optimize resource utilization in diverse regions, and contribute to a resilient energy infrastructure by diversifying sources. Particularly beneficial for off-grid and remote applications, these systems are environmentally friendly, reducing the overall carbon footprint. Furthermore, cost savings are achieved through shared infrastructure, making solar and wind hybrid systems a compelling solution for sustainable and efficient energy production.

References

1. Akshay B. Zade, Dr. Asha Gaikwad ,"Design And Simulation Analysis Of Seven Level Cascaded Grid Connected Inverter For Pv System", Student M.E, Department Of Electrical,University Of Pune, Ghriet, Pune, India, Professor, Department Of Electrical,University Of Pune, Ghriet, Pune, India.
2. Sérgio Daher, Jürgen Schmid, And Fernando L. M. Antunes ,"Multilevel Inverter Topologies For Standalone Pv Systems" Member, Ieee
3. Eduard Muljadi ,"Power Quality Issues In A Hybrid Power System" Senior Member, Ieee, And H. Edward Mckenna.
4. Ma. Ángeles Martín Prats, Member, Ieee,José Ignacio León, Student Member, Ieee, And Narciso Moreno-Alfonso, "Power-Electronic Systems For The Grid Integration Of Renewable Energy Sources": A Survey Juan Manuel Carrasco, Member, Ieee, Leopoldo Garcia Franquelo, Fellow, Ieee, Jan T. Bialasiewicz, Senior Member, Ieee, Eduardo Galván, Member, Ieee,Ramón C. Portillo Guisado, Student Member, Ieee, Member, Ieee
5. Abhilash Sen, Atanu Banerjee ,"A Comparative Analysis Between Two Dpfc Models In A Grid Connected Hybrid Solar- Wind Generation System", Dept. Of Electrical Engineering Nit Meghalaya Meghalaya, India Abhilashsen@Nitm.Ac.In , Dept. Of Electrical Engineering Nit Meghalaya Meghalaya, India Atanu_Banerjee@Nitm.Ac.Inharicharan Nannam, Dept. Of Electrical Engineering Nit Meghalaya Meghalaya,India Haricharan@Nitm.Ac.In.
6. Congcong Li, Student Member, Ieee, Dengkai Mi, Student Member,Ieee, Zengping Wang, Member, Csee, Member, Ieee, And Yuwei Xiang."Coordinated Modulation Strategy Considering Multi-Hvdc Emergency For Enhancing Transientstability Of Hybrid Ac/Dc Power Systems".Tong Wang, Member, Csee, Member, Ieee,
7. Yu Liu, Student Member, Ieee, Hoon Hong, And Alex Q. Huang, Fellow, Ieee "Real-Time Calculation Of Switching Angles Minimizing Thd For Multilevel Inverters With Step Modulation".
8. Bhupender Sharma Department Of Electrical Engineering National Institute Of Technology Kurukshetra, India Bhupender638@Gmail.Comratna Dahiya Department Of Electrical Engineering National Institute Of Technology Kurukshetra, India Ratna_Dahiya@Yahoo.Co.In."Improved Modular Flying Capacitor Converter Based Grid-Tied Hybrid Wind Solar Energy System."
9. "Increasing The Wind Power Utilization Of Hybridac/Dc Systems Based On Optimal Active Power Control Of Hvdc."Zhenyu Zhang Northwest Branch Of State Grid Corporation Of Chinaxi'an, China. Zyzhangisgta@163.Com Xianbo Ke Northwest Branch Of State Grid Corporation Of China Xi'an, China. Kexianbo@126.Com Chao Huo Northwest Branch Of State Grid Corporation Of China Xi'an, China. 44486005@Qq.Com Weizhou Wang State Grid Gansu Electric Power Corporation Lanzhou, China. Wzwangisgta@163.Com Chong Shao State Grid Gansu Electric Power Corporation Lanzhou, China. Cshaoisgta@163.Com Jia Lyu Shaanxi Key Laboratory Of Smart Grid Xi'an Jiaotong University Xi'an, China. Z15878351771@163.Com Jun Liu* Shaanxi Key Laboratory Of Smart Grid Xi'an Jiaotong University Xi'an, China. *Eeliujun@Xjtu.Edu.Cn.
10. "A Hybrid Wind-Photovoltaic Generation System Modeling And Performance Evaluation."Amr Ahmed A. Radwan Engineering And Design Department.

Machine Learning-Driven Wind Energy Forecasting for Sustainable Development

Magesh T[1*], *Samuel Franklin* F[1], *Santhi* P S[2] and *Thiyagesan* M[1]

[1]R.M.K. Engineering College, Department of Electrical and Electronics Engineering, Kavaraipettai, Tamil Nadu, India.
[2]Murugappa Polytechnic College, Department of Electrical and Electronics Engineering, Avadi, Tamil Nadu, India.

Abstract. The growing need for energy, in addition to the depletion of fossil fuel supply, has underlined the importance of renewable energy for long-term growth. Renewable energy stands out among these, but its broad usage is hampered by the inherent uncertainty of wind power generation. This study uses machine learning to predict wind energy yield. Several regression models were used, including decision tree regression, linear regression, and random forest regression. The results emphasize the random forest regression, which has a high R-squared score, suggesting strong predictive ability. The paper also contains wind power output projections, which provide insights for optimal wind energy planning and usage. Overall, this attempt gives vital insights to improving the effective use of renewable energy, advancing the cause of sustainable development.

1 Introduction

The growing global population, as well as the need for modernization and industrialization, highlight the urgent requirement for sustainable and secure power sources. Unfortunately, traditional nonrenewable Fuels like petroleum and natural gas are essential for meeting energy needs, but they can significantly worsen the environment through pollution and greenhouse gas emissions. As a result, several countries are investigating renewable, plentiful, and clean sources of energy such as wind, hydroelectricity, solar power, geothermal, and biomass [1]. This transition to renewable energy demonstrates a commitment to reducing the impact on the environment and ensuring an environmentally friendly energy future. Wind energy has grown in popularity as a dependable and environmentally acceptable energy source. Wind, being an informal energy source, is widely available and may be used in a variety of locales. As a result, wind power already occupies a significant part of the worldwide mix of energy sources and is anticipated to do so for the foreseeable future [2]. Wind energy is growing into a more important energy source for the power grid, which makes it increasingly difficult to predict its behavior. The power system control center may need to boost backup capacity and incur more expenditures as an outcome of the unexpected variations in wind output, which can lead to uncertainty and imbalance in

* Corresponding author: tmh.eee@rmkec.ac.in

the system. However, reliable forecasting models can lessen these difficulties by enhancing the system's stability, dependability, and power quality while lowering operating expenses. Developing models that reduce wind power forecast errors is essential to overcoming these obstacles and advancing the incorporation of wind energy into the electricity grid.

2 Literature Review

In the domain of forecasting wind power, there are two primary types: Short Term (ST) and Long Term (LT) forecasting. ST Forecasting focuses on predicting wind power for brief durations, spanning from one hour to several days ahead. On the other hand, long-term forecasting extends its predictions to periods ranging from several days to a year in advance. Statistical approaches have shown to be useful in the ST wind forecasting area. Lima et al. [3] suggested a climatic-statistical method that can anticipate wind power up to 72 hours ahead of time with Wang et al. [4] effective performance. Robles-Rodriguez and Dochain [5] introduced an ARMAX technique which incorporates a process including division of series and non-linearity management, resulting in accurate wind power forecasts up to 48 hours in advance. Pearre and Swan [6] presented statistical strategies for detecting and correcting prediction mistakes, demonstrating high performance in forecasts one to six hours in advance.

Yu Zhang, Yanting Li *, and Guangyao Zhang [7] developed a wind output forecast model using machine learning for ST forecast. This model beat other existing approaches as the Deep Belief System and Random Forest by combining a variable time series clustering system with a deep learning system. Sideratos and Hatziargyriou [8] proposed a technique for precisely estimating wind power output with a 48-hour horizon that incorporates neural networks and fuzzy logic. To anticipate electricity output, Lahouar and Slama [9] used the Random Forest method in conjunction with numerous meteorological data such as wind speed.[10] suggested an architecture for improving mathematical forecasts and data using deep belief networks and the k-means clustering technique.

After evaluating the research in the literature, it is clear that statistical approaches are not the most effective choice for predicting wind power, especially in the long run. They struggle with nonlinear wind data adaptation, massive dataset processing, and forecasting over long time horizons [10, 11]. In this paper, several machine learning approaches were applied to anticipate LT power. The results indicate that using machine algorithms to anticipate wind power shows promising outcomes. The investigation also looked at the effect of the standard deviation of the model. Finally, the best strategy for reliable wind power estimates was discovered.

3 Architecture

In order to minimize cost overruns and guarantee that demand is satisfied, forecasting for wind energy must be accurate. Instead of utilizing generalized global data to train the model, local wind data must be used to provide precise forecasts. In this study, we compare the model's accuracy in predicting wind power output and concentrate on the training process.

Depending on the type of learning input or response available, the research takes regression machine learning systems into consideration.

3.1 Regression

Since the objective of this research is to anticipate continuous outputs, regression, a supervised learning problem, is frequently used. In order to properly estimate wind power, regression techniques are used. Data preprocessing is a crucial step in turning raw data into a format that can be used. There are several preprocessing processes involved in the research.

3.1.1 Data cleaning

Data cleaning deals with irrelevant or missing data parts. Handling missing data, noisy data, and other data quality concerns are part of this process.

3.1.2 Missing and Noisy Data

Missing data may be handled in a variety of methods, including disregarding tuples with missing values, manually adding the missing values, utilizing the attribute mean, or choosing the value that is most likely. Noisy data is unintelligible and lacks useful information. The data can be smoothed using regression by fitting it to a regression function.

Data transformation is performed to convert information into a suitable format for the mining process. Two common techniques are feature scaling and dividing sets.

In machine learning, this stage is essential. Algorithms that determine the distances between features may be biased toward numerically bigger values if data is not scaled. In order to scale features to a similar scale, common feature scaling techniques include normalization (or Min-Max) and standardization (R-Score).

In dividing train and test sets the dataset is divided into two sets to confirm the accuracy of the machine learning model. Train set is used for training, while a Test set is used for assessment. This ensures a reliable assessment of the model's performance and helps prevent biased findings.

4 Methodology

Various machine learning models that are employed for Wind forecasting in this research.

4.1 Random Forest

A common decision tree approach that builds several decision trees based on the input information is the Random Forest algorithm. It generates decision trees for every smaller subdivision of the input parameters rather than using all of them. The ultimate choice is then reached by combining the output from each tree. Using this method makes the issue of juggling several features easier to grasp.

In RF, a random vector of size k is created, representing a portion of the dataset's feature space. The training dataset and k are used to construct each tree.

$$PE^{*}=PX,Y(mg(X,Y) <0)$$

4.2 Linear Regression

The statistical method known as linear regression is employed to examine the relationship that exists between a scalar result, often referred to as the variable of interest, and one or more independent variables that serve as explanatory factors. These correlations are represented by the method's use of linear predictor functions, and the model's unknown variables are estimated using the given dataset.

4.3 Decision Tree Regression

Forecasting continuous numerical values is an area in which decision tree regression, a supervised machine learning method, shines. With the expected numerical output represented by the leaf nodes and judgments made based on the input data by the inside nodes, it constructs a structure like a tree. The interpretability of this strategy is valued since it makes it easier for customers to comprehend how decisions are made.

5 Results and Discussion

The dataset used in this study, consisting of 5 years of every-hour measurements of wind speed from Coimbatore, India, was utilized to train and assess learning methods for wind power forecasts. The training and evaluation of the models were conducted using the Google Colab platform, for running Python code, along with the scikit-learn (sklearn) library for machine learning. The wind output power for every hour is calculated on the standard power formula. The generated power values are then used as the target variable for training and evaluating the learning algorithms for wind forecasts.

Table 1. Different statistical measure

	Square R	RMSE	MAE	MSE
Linear	0.9251	0.4487	0.3917	0.2013
Decision Tree	0.8722	0.5861	0.4677	0.3435
Random Forest	0.9264	0.4447	0.3839	0.19776

The Table contains the various scores of the differential statistical method obtained by training the models and testing. The effectiveness of different learning algorithms, such as linear, Random Forest and decision tree was evaluated using this dataset. The accuracy of these models was assessed by comparing their predictions against the actual wind power values.

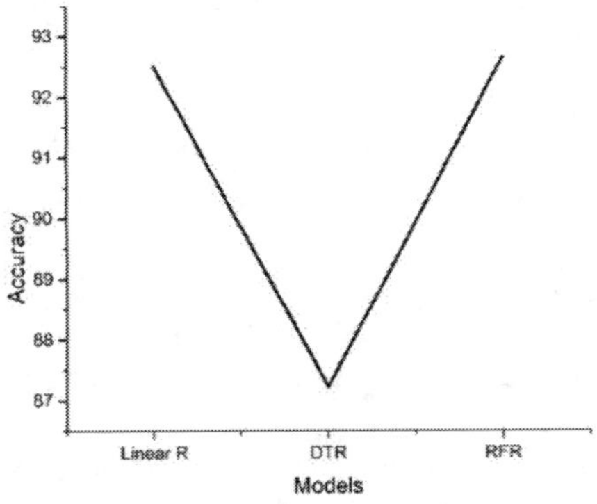

Fig 1. Models and Scores

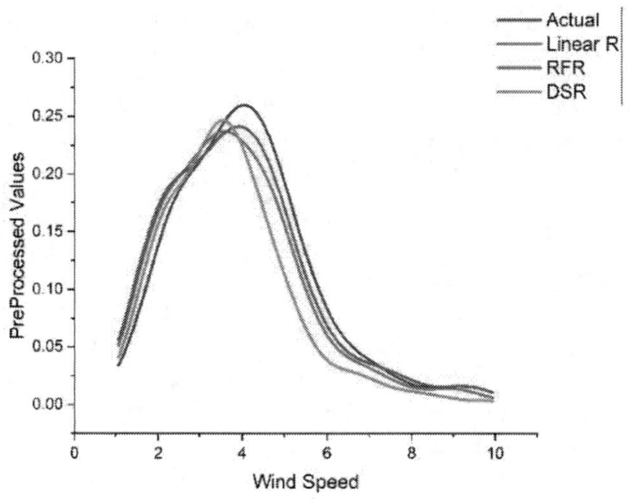

Fig 2. Accuracy Graph Comparison

The study's findings show that a significant number of models based on learning methods, like linear regression and random forest, perform well in forecasting wind power. The high R-scores approach one indicates that the models explain a big part of the difference in wind energy data. This closeness to 1 indicates a high level of predictability in wind power, highlighting the efficiency of modern machine learning algorithms. However, decision tree regression appears to have lower accuracy compared to the other models, as indicated by its line deviating significantly from the black line and its lower R-score. Among the different machine learning models evaluated, the results indicate that the random forest method performed the top in predicting wind power having score as 0.9264. From fig 2 The model shows the line in the graph closely aligning with the black line representing the actual values, indicating the ability to capture the underlying patterns in the wind data and provide accurate predictions. These findings suggest that learning algorithms, particularly linear and random forests, have the potential to accurately forecast wind energy using the provided dataset.

6 Conclusion

In conclusion, the necessity for sustainable development and the expanding global energy demand have highlighted the significance of renewable energy sources. A lot of attention has been paid to wind energy in particular because of how abundant it is and how much electricity it can provide. The difficulty, however, is that wind power generation is unpredictable, which makes it difficult to use it effectively.

This study attempted to address the difficulty of wind power prediction using several models of machine learning methods, like linear regression, decision tree regression, and random forest networks. The results revealed significant potential for prediction in both the linear and random forest regression models, as evidenced by their high R-squared values. Compared to both the models the random forest regression is more efficient. Accurate wind power forecasting is critical for successful planning and maximum exploitation of renewable energy resources.

The objective of the study was to handle the problem of power prediction by using several machine-learning methods. The outcomes, which were characterized by the differential statistical methods, demonstrated the linear regression and random forest regression models' strong predictive power. Precise wind power prediction is important for successful wind resource planning. It can be used for grid planning for balancing the demand and supply and in cooperating into the existing system.

Furthermore, accurate wind power generation projections allow for more accurate power balancing and economic operations calculations, which enhance system stability, lower operating expenses, and improve power quality. An energy infrastructure that is more robust and sustainable is promoted by the usage of power in power networks.

In conclusion, the study highlights the need for precise wind power forecasts and offers a machine learning-based solution to the problems brought on by the unpredictable nature of wind power output. We can open the door to a more sustainable future by using the potential of renewable sources.

References

1. Global Wind Report 2019; Global Wind Energy Council: Brussels, Belgium, 2020. https://gwec.net/global-wind-report-2019/ .
2. Tsikalakis, Antonis & Hatziargyriou, Nikos & Katsigiannis, Yiannis & Georgilakis, Pavlos. (2009). Impact of Wind Power Forecasting Error Bias on the Economic Operation of Autonomous Power Systems. Wind Energy. 12. 315 - 331. 10.1002/we.294.
3. Lima JM, Guetter AK, Freitas SR, Panetta J, de Mattos JGZ. A meteorological-statistic model for ST wind power forecasting. J. Control Autom. Electr. Syst. (2017);28(5):679–91. doi: 10.1007/s40313-017-0329-8.
4. Wang J, Zhou Q, Zhang X. IOP Conf. Ser.: Earth Environ.Sci. (2018);199:022015.https://doi.org/10.1088/1755-1315/199/2/022015
5. Robles-Rodriguez C, Dochain D. Decomposed threshold armax models for short- to medium-term wind power forecasting. IFAC-PapersOnLine (2018);51(13):49–54, ISSN 2405-8963, https://doi.org/10.1016/j.ifacol.2018.07.253.
6. Pearre NS, Swan LG. Statistical approach for improved wind speed forecasting for wind power production. Sustainable Energy Technol. Assess. (2018);27:180–91. https://doi.org/10.1016/j.seta.2018.04.010.
7. Yu Zhang, Yanting Li, Guangyao Zhang, Short-term wind power forecasting approach based on Seq2Seq model using NWP data, Energy, Volume 213, (2020), 118371, ISSN 0360-5442, https://doi.org/10.1016/j.energy.2020.118371.
8. Sideratos G, Hatziargyriou ND. An advanced statistical method for wind power forecasting. IEEE Trans. Power Syst. (2007);22(1):258–65. https://doi.org/10.1109/TPWRS.2006.889078.
9. Lahouar A, Slama JBH. Hour-ahead wind power forecast based on random forests. Renewable Energy (2017);109:529–41. https://doi.org/10.1016/j.renene.2017.03. 064.
10. Sun G, Jiang C, Cheng P, Liu Y, Wang X, Fu Y, He Y. STwind power forecasts by a synthetical similar time series data mining method. Renewable Energy (2018);115:575–84. https://doi.org/10.1016/j.renene.2017.08.071.
11. Yu R, Gao J, Yu M, Lu W, Xu T, Zhao M, Zhang J, Zhang R, Zhang Z. LSTM-EFG for wind power forecasting based on sequential correlation features. Future Gener. Comput. Syst. (2019);93:33–42. https://doi.org/10.1016/j.future.2018.09.054.

Modelling and Analysis on the Fatigue Life of Gas Turbine Compressor Disc

Vijay Kumar Mahakur[1, 2*], *A C Umamaheshwer* Rao[2] and *Pathapalli Venkateshwar* Reddy[2]

[1]Dept. of Mechanical Engineering, NIT Silchar, Assam. 788010, India
[2]Dept. of Mechanical Engineering, Vardhaman College of Engineering, Hyderabad 501218, India

Abstract. The disc itself revolves at an extremely rapid pace, putting a lot of strain on the turbine compressor disc. This exceptionally stressed part provides an impact on a particular location such as a bore, bolt hole, slot, or fillet. The bore area of the disc is an especially crucial zone, since any failure there would result in an unprecedented breakdown of the entire turbine. So in this study, the finite-element approach is employed to offer a numerical estimate of the lifespan of a gas turbine prototype disc. Engine start-up and shutdown, as well as significant throttle fluctuations during operation, can cause cyclic tensile stresses in this area. These cyclic tensile stresses will surpass the component's yield strength and may potentially induce low cycle fatigue failure. As a result, there is a need to improve the fatigue and fracture lifetime of the bore area in order to meet the expanding design life requirements.

1 Introduction

Aero engine specifications demand more power, fewer pounds, and fewer expenses while preserving flight safety and engine efficiency. This implies fewer, heavier-loaded compressor and turbine phases operating at greater temperatures [1].

Some of the more costly elements are gas-turbine discs that function in the overheated area of the engine. As a result, significant study is being conducted in this field to decrease prices, strengthen materials, create thermomechanical interventions, and maximize device lifespan [2, 3]. The principal operation of the turbine is to get energy from the hot gas and this energy helps to drive the compressor. The turbine disc is subjected to high rotational speeds and centrifugal forces which produces extreme temperatures which negatively affect the strength of the material. The role of the discs is to locate the ring of the rotating blades inside the engine and transmit the energy absorbed by them through the shafts connecting the turbine to the compressor. The turbine disc which is the most critical part of the component, should withstand substantial mechanical and thermal loading. If any difficulty arises in the turbine section it will considerably affect the total engine performance [1, 4, 5].

Mostly in compressors, it's common to have pressure ratios to vary within 17 to 20, with some units having pressure ratios up to 30:1 [6]. The pressure ratios are increased for pursuing higher efficiencies. Compressor disc is subjected to higher rotational speeds and therefore higher loads causing the bore region of the disc to experience higher loads. Corrosion cracking, resonance and low cycle fatigue tends to crack or rupture the rotor disc

*Corresponding author: vijay.mahakur0709@gmail.com

causing it to fail frequently. The other being the rotor burst is caused due to whirl unbalance, alignment errors, low cycle fatigue [7, 8].

Few technologies viz., life prediction, fracture mechanics, non-destructive evaluation, etc. have been used to overcome the failure in the disc component. With the help of these techniques, the components are inspected for specific crack sizes which degrades the life of an engine component. The elements lifespan are also predicted using these techniques. However in the crack initiation criterion if the crack is detected before the element retirement lifespan, then the element is retired instantly. Shot Peening method is used to retard the crack growth in specified locations of an engine. Shot-peening introduces vital near surface compressive stresses. These residual compressive stresses help in enhancing fatigue life, retardation of crack growth and resistance to foreign object injuries [9, 10].

Due to the compressive stresses, life of the engine components increase. A technique called auto-frettage, helps in inducing compressive pre-stresses in components subjected to reduce tensile stress and also aids in enhancing fatigue life of components [11]. In pre-spin approach disc is spun to a speed that's sufficiently high to deform the bore region on the far side of yield point. Unloading in the bore region causes residual compressive stresses, which is done by stopping of the disc. Due to continuous operation of turbine the compressive residual stresses reduces the tensile stresses created by operational load. Reduction in tensile stresses leads to increased lifetime of the disc [12, 13].

A metal subjected to a repetitive or unsteady stress can fail at a stress, less than that needed to cause failure on application of load. Failures occurring below conditions of dynamic loading area unit is known as fatigue failure. Fatigue failure is characterized in 3 phases- crack initiation, crack propagation and final fracture. So, it is feasible to acquire excellent options of expected lifespan by testing materials with geometrical and mechanical qualities comparable to key portions of discs and utilizing variables derived from finite element (FE) analysis, coupled with relevant information. The time and expenses related to production will be decreased using findings derived for disc geometries using the FE approach and characteristics based on actual data. The mechanical properties of AISI 4340 is described in Table 1. After applying the above material, the fatigue analysis was performed to quantify whether the above used material is fit for the disc or not

Table 1. Mechanical characteristics of AISI 4340 alloy steel

Ultimate tensile stress	745 MPa
Tensile yield strength	470 MPa
Modulus of elasticity	205 GPa
Poison' ratio	0.29

2 Analysis of a model disc following the FE approach

To prevent rigid body motion, a constraint was applied to one of the bore edges. A rotational disc was introduced to impose a load on the disc, ensuring stability. The model of the compressor disc, incorporating the mass of the blades, is depicted in Figure 1. The entire design of the compressor disc was meticulously modelled using SOLIDWORKS software, providing a comprehensive representation of the structural elements and their interactions. This integrated approach in design not only enhances the accuracy of the model but also facilitates a more realistic simulation of the compressor disc's behaviour under various conditions.

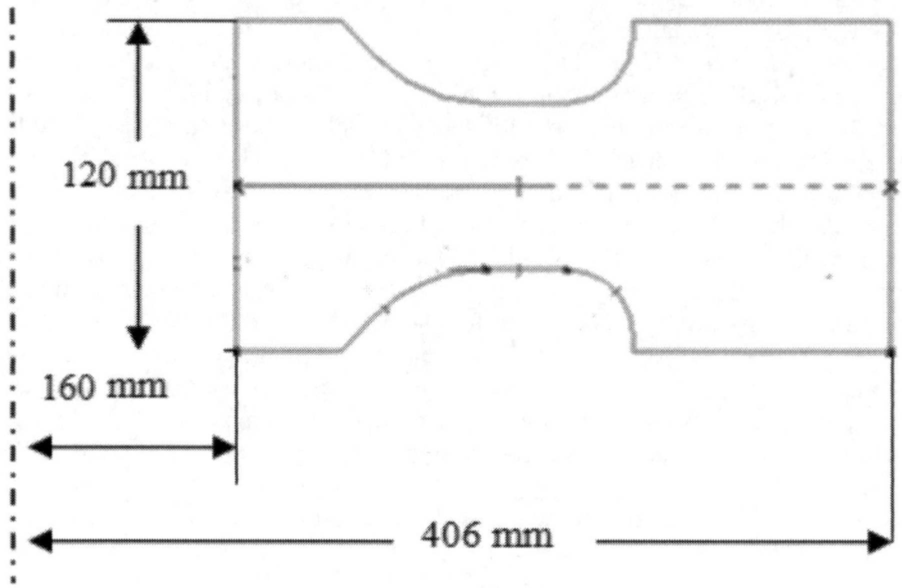

Fig. 1. 2D cross section of the disc

Fig. 2. Compressor disc

3 Methodology

Equations should be centred and should be numbered with the number on the right-hand side. Central to the analysis is the assumption that crack propagation follows a 2D configuration at the center of the disc, a simplification adopted for a more straightforward evaluation of crack growth. This assumption is based on the acknowledgment that tangential stresses play a significant role in influencing crack development. To delve into crack growth dynamics, a 2D plane is employed to model the cracked configuration. This approach allows for a focused exploration of the factors influencing crack expansion, particularly the role of hoop stresses in attempting to propagate and open the crack.

By adopting a 2D perspective, the analysis aims to capture the critical mechanisms governing crack growth, emphasizing the tangential and hoop stresses that contribute to the evolution of the crack. This methodological choice not only streamlines the analytical process but also enables a detailed examination of the interplay between stresses and crack propagation, offering insights into the structural behaviour of the disc under the influence of different loading conditions. The 2D model provides a valuable platform for studying the intricate dynamics of crack growth in the disc, shedding light on the underlying factors that influence its structural integrity [14].

The fatigue analysis was precisely conducted using SOLIDWORKS simulation. The initial phase involved a comprehensive static analysis of the model. Subsequently, the disc underwent a meshing process, ensuring that all pertinent constraints were duly considered and incorporated as outlined in Table 2. This step-by-step approach in the analysis provides a robust foundation, starting with a static evaluation that allows for a detailed understanding of the disc's structural response. The subsequent meshing process is crucial for capturing the intricacies of the disc's geometry and material behaviour, setting the stage for a thorough fatigue analysis. The inclusion of constraints, as documented in Table 2, ensures a correct representation of the real-time conditions, enabling a reliable and accurate assessment of the disc's fatigue performance. Through the integration of SOLIDWORKS 2013 simulation capabilities, this analysis aims to offer valuable insights into the fatigue life and structural integrity of the disc under varying operational conditions. Subsequently, the compressor disc was secured at the center, and the specified forces were then applied, as detailed in Table 3.

Table 2. Detailing of mesh

Total Nodes	2416554
Total elements	1734780
Maximum Aspect ratio	11.512
% of elements with Aspect Ratio < 3	99.7
% of elements with Aspect Ratio > 10	0.000115

4 Results and discussions

4.1 Static analysis

The disc undergoes a maximum stress of 394.56 MPa, and concurrently, experiences a resultant displacement measuring 0.0957508 mm. The outcomes derived from the static analysis of the system are detailed as shown in figure 3. The maximum stress and displacement in a disc, as reported in a static analysis, are influenced by various factors. The stress distribution in a rotating disc with variable thickness is affected by the loadings, with the maximum radial stress occurring in the thinnest area [15, 16]. In a partially loaded hollow disc, the maximum tensile stress can be almost twice the internal pressure, while the maximum compressive stress is not greater than the applied internal pressure [17, 18].

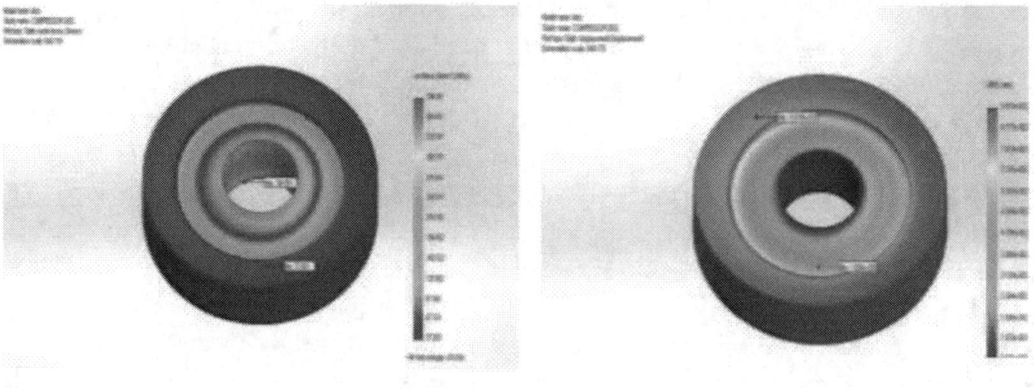

(a) (b)

Fig. 3. (a) Stress distribution, and (b) Resultant displacement

4.2 Fatigue analysis

Subjecting the disc to fatigue loading, an investigation was conducted to determine the point at which the disc fails after a specific number of revolutions. Applying a fatigue load equivalent to 5 million revolutions, the analysis revealed that the maximum damage incurred during this loading condition amounted to 2.378×105, and the corresponding maximum life of the disc was determined to be 106 cycles. This information sheds light on the disc's performance under prolonged cyclic loading and provides insights into its durability and structural integrity over an extended period of operation. A range of studies have explored the complex issue of fatigue damage in various loading conditions. Chen et al. found that circular strain paths resulted in the greatest damage and shortest fatigue life, highlighting the importance of considering fatigue failure mechanisms [19]. Han et al. proposed methods for assessing fatigue damage under combined high and low cycle loading, which can significantly reduce the life of structures [20]

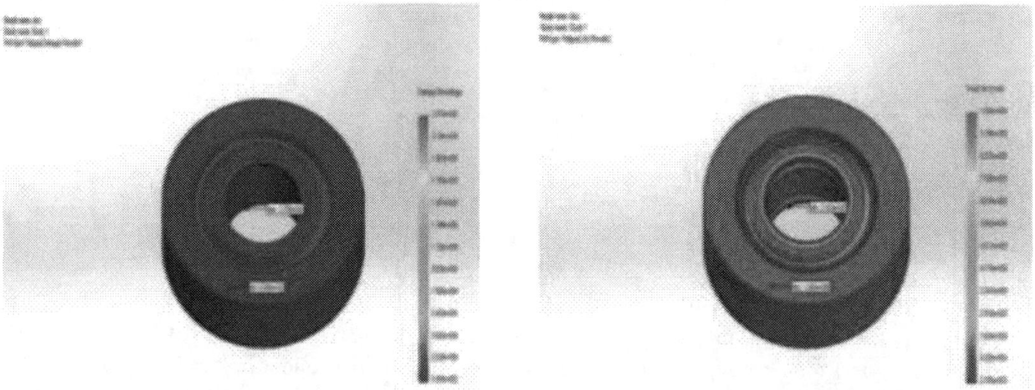

(a) (b)

Fig. 4. (a) Damage plot, and (b) Life plot

5 Conclusions

The aforementioned investigation highlights that when a compressor disc, constructed from AISI 4340 alloy steel, is exposed to substantial loads over a cyclic loading regime of 5×10^6 cycles, there is no evidence of crack initiation or structural damage. This observation implies that the AISI 4340 alloy steel exhibits a notable resistance to fatigue failure within this specific operating condition. Consequently, based on the study results, it can be concluded that AISI 4340 alloy steel is deemed safe and reliable in the fatigue zone characterized by 5×106 loading cycles. The absence of crack initiation or damage underscores the material's durability and structural integrity under the specified high-load cyclic conditions, thereby affirming its suitability for application in such scenarios.

References

1. R. A. Cláudio,C. M. Branco, E. C. Gomes, J. Byrne, G. F. Harrison, M. R. Winstone, Fatigue life prediction and failure analysis of a gas turbine disc using the finite-element method. Fatigue Frac. Eng Mater. Struct. **27**, 849-860 (2004).
2. K. Tschirne, W. Holzbecher, Cost effectiveness of modern lifing concepts: Consideration of economical aspects concentrating on group a parts of military aircraft engines, In RTO Meeting proceedings. 1-2 (1999).
3. G. F. Harrison, W. J. Evans, A review of fatigue assessment procedures for aeroengine fracture critical components. J. Eng. Integrity Soc. (2000)
4. L. Witek, Failure analysis of turbine disc of an aero engine. Engineering failure analysis **13**, 9-17 (2006).
5. S. M. L. Jagannath, S. Srinivasan, M. Gopalakrishnan, V. K. Banthia, Prediction of Fatigue and Fracture Life of an Autofrettaged Turbine Compressor Disc Using Finite Element Analysis, In Gas Turbine India Conference 49644 (2014).
6. P. Blüml, J. Broede, Critical parts' life extension based on fracture mechanics. In RTO Meeting proceedings 1-19 (1999).
7. R. John, J. M. Larsen, D. J. Buchanan, N. E., Ashbaugh, Incorporating residual stresses in life prediction of turbine engine disks, In Proceedings from NATO RTO (AVT) Symposium on Monitoring and Management of Gas Turbine Fleets for Extended Life and Reduced Costs 8-11 (2001).
8. H. E. Sonnichsen, Ensuring spin test safety, Mechanical Engineering-CIME **115,** 72-78 (1993).
9. P. Prevéy, D. Hornbach, P. Mason, Thermal residual stress relaxation and distortion in surface enhanced gas turbine engine components, In Proceedings of the 17th heat treating society conference and exposition 3-12 (1998).
10. G. C., Stone, I. Culbert, E. A. Boulter, H. Dhirani, Electrical insulation for rotating machines: design, evaluation, aging, testing, and repair. **83**, (2014).
11. M. H., Bell, Spin test of turbine rotor (1972) NPS-57VA71061B.
12. A. D. Harrison, The development of life extension methods for fracture critical aero-engine components. In RTO Meeting proceedings 1-18, (1999).
13. V. K. Mahakur, K. Anil, K. Gouda, R. Paul, Design and fabrication of integrated electronic power controlled alternator. In AIP Conference Proceedings **2316**, (2021).

14. M. G. Ryschkewitsch, Nondestructive Evaluation Requirements for Fracture Critical Metallic Components, NASA Technical Standard (2008).
15. M. P. Boyce, Gas turbine engineering handbook. Elsevier (2011)
16. A. Rosyid, M. E. Saheb, F. B., Yahia, Stress analysis of nonhomogeneous rotating disc with arbitrarily variable thickness using finite element method. Res. J. Appl. Sci. Eng.Technol. **7**, 3114-3125 (2014).
17. R. Hess, J. F. King, Jr. J. L. Harp, Exploratory development of a twin-spool turbocharber for a high output diesel engine. Thermo Mechanical Systems Company. 13197 (1986).
18. M. Serati, H. Alehossein, D. J. Williams, 3D elastic solutions for laterally loaded discs: generalised Brazilian and point load tests. Rock Mech. Rock Eng. **47**, 1087-1101 (2014).
19. X. Chen, Q. Gao, X. F. Sun, Damage analysis of low-cycle fatigue under non-proportional loading. Int. J. Fatigue **16**, 221-225 (1994).
20. C. Han, X. Qu, Y. Ma, D. Shi, Experimental and numerical study of fatigue damage assessment under combined high and low cycle loading. Shock Vib. **2018**, (2018).

MATEC Web of Conferences **393**, 02005 (2024) https://doi.org/10.1051/matecconf/202439302005
STAAAR-2023

Designing an Intelligent Pavement Maintenance and Management System using Drone Imagery and Artificial Intelligence

Makendran C[1*], Karthik. M[2], S M Jakir Hasan[2], Harivignesh.M[2], Varun Raahul.G[2]

[1]Assistant Professor, Department of Civil Engineering, National Institute of Technology, Tiruchirappalli- 620015, Tamil Nadu, India.

[2]Students, Department of Civil Engineering, National Institute of Technology, Tiruchirappalli-620015, Tamil Nadu, India.

Abstract: This paper presents the development of an innovative pavement maintenance and management system leveraging advanced drone imagery and Convolutional Neural Network (CNN) image classification. Our system is designed to perform 2D modelling of road surfaces using high-resolution images captured by drones. These images are then analysed by a CNN model specifically trained to detect and classify pavement damages in accordance with the IRC:82 'Code of Practice for Maintenance of Bituminous Surfaces of Highways'. The classification process identifies various types of road distresses such as cracks, potholes, and surface wear. Each identified distress is documented in a comprehensive report detailing the nature of the damage and recommending specific remedies as per IRC guidelines. Furthermore, the system categorizes the severity of the damages, facilitating the dispatch of these results to maintenance authorities for immediate action. This ensures that repair efforts are prioritized effectively, contributing to the maintenance of safer and higher quality roadways. By automating the detection and classification of road damages, this system not only accelerates the repair process but also plays a crucial role in reducing road accidents by maintaining better road conditions. This approach showcases the potential of integrating artificial intelligence and drone technology in the field of road maintenance, marking a significant step towards smarter and safer road infrastructure.

Keywords: Pavement maintenance and management system; Drone imagery; Convolutional Neural Network (CNN); IRC guidelines; Automating the detection and classification of road damages.

1 Introduction

This paper introduces a comprehensive system poised to modernize pavement maintenance through the integration of drone-captured imagery and Convolutional Neural Network (CNN) image classification. Addressing the critical need for infrastructure maintenance, the proposed system capitalizes on the agility of drone technology and the analytical prowess of CNNs to offer a robust solution for road surface analysis. By transcending the limitations of traditional maintenance methodologies, this system not only enhances the accuracy and efficiency of inspections but also significantly reduces the time taken to respond to and repair pavement damages. The utilization of high-resolution

Corresponding author: makendran@nitt.edu

drone imagery provides a detailed view of current road conditions, while the advanced CNN model interprets these images to identify and classify damage accurately. This fusion of technology is a proactive step towards maintaining road infrastructure and ensuring safety. The innovative approach detailed in this paper could potentially redefine standard maintenance protocols by offering a scalable, cost-effective, and technologically driven maintenance and management system. By automating critical processes and employing machine learning for image analysis, the system sets the stage for a new era in pavement maintenance where decision-making is data-driven and pre-emptive rather than reactive. This paper will elaborate on the system's design, its implementation strategy, and the potential impact on current maintenance practices, aiming to demonstrate the viability and necessity of such a system in today's context.

1.1 Evolving Challenges in Pavement Maintenance

The field of pavement maintenance is grappling with contemporary challenges, including increased traffic volumes, budgetary constraints, and the urgent need for stable practices. Conventional methods, while having served in the past, are inadequate in addressing the complex demands of modern road infrastructure. This inadequacy not only leads to inflated maintenance costs but also compromises road safety, necessitating a re-evaluation of existing strategies and the adoption of more technologically advanced and efficient approaches (Miller, 2003; Newman, 2012) [1] [2]. Makendran et al., (2021, 2022, 2015, 2020, 2018, 2023, 2022) developed pavement distress models for road management systems [21] [22] [23] [24] [25] [26] [27].

1.2 The Critical Role of Road Infrastructure

Road infrastructure plays a crucial role in economic development, environmental sustainability, and public safety. Efficient and effective management of road networks is essential for fostering economic growth, reducing environmental impacts, and ensuring the safety of the public. The sustainability of urban development hinges on the ability to maintain and improve road infrastructure in a way that meets these diverse and often competing demands (Ivanova, 2013; Haasis, 2020) [3] [4].

1.3 Advances in Pavement Monitoring Technology

The last decade has seen significant advancements in pavement monitoring technologies. Innovations such as the integration of IoT sensors, GIS, and machine learning have transformed traditional maintenance methodologies. These technologies facilitate enhanced data collection and analysis, enabling more accurate and predictive maintenance strategies. This evolution in pavement monitoring is a testament to the increasing reliance on technology to meet the growing demands of road maintenance and management (Rasol, 2022; Elhadi, 2009) [5] [6].

1.4 Drone Imagery in Pavement Inspection

The adoption of drone technology in pavement inspection marks a significant step forward in the field. Drones, equipped with state-of-the-art imaging technologies, provide an unparalleled view of road surfaces, revealing details that are often missed in conventional inspections. This method not only enhances the accuracy of inspections but also streamlines the process, making it more efficient and cost-effective. The comprehensive data collected by drones is instrumental in identifying potential issues before they escalate, thereby reducing the overall maintenance costs and road closure times (Puppala, 2020; Outay, 2020) [7] [8].

1.5 CNN in Pavement Damage Classification

The application of CNNs in pavement damage classification represents a significant leap in automated road maintenance. Trained on extensive datasets, these neural networks are adept at identifying and categorizing various types of road damages, from minor cracks to major potholes. This capability not only improves the accuracy of damage assessment but also accelerates the maintenance process, allowing for more timely and effective interventions. The use of CNNs in this context demonstrates the potential of AI in transforming traditional pavement maintenance practices, offering a more proactive and data-driven approach (Maeda, 2018; Elghaish, 2022) [9] [10].

1.6 Research Gap

In most of the previous research and studies regarding road maintenance, involves selection of roads for monitoring, collecting the distress survey by using drones manually, and prepared report is sent to the Road Maintenance Authorities by conventional methods. Though by these conventional methods, the problem is addressed but it takes a huge amount of time in resolving the problem. Our goal is to minimize the time by using Artificial Intelligence in the drone and using image classification to find the type of cracks, Calculating each distress of sections and percentage of road failures. The information is sent to the road engineer on a day-to-day basis through email and SMS.

1.6.1 Response Time Challenges in Conventional Road Maintenance

In traditional approaches to road maintenance, a significant research gap exists concerning the time-consuming processes involved in distress survey, data collection, analysis, and report preparation. The conventional method of manually monitoring selected roads using drones and subsequently sending distress reports to Road Maintenance Authorities has proven effective but suffers from considerable delays in problem resolution. This subsection aims to delve into the inherent challenges posed by the delayed response times in the existing maintenance paradigm. We explore the implications of this time delay on road safety, infrastructure integrity, and the overall efficiency of the maintenance process. The objective is to highlight the critical need for a more time-efficient solution, emphasizing the potential impact on minimizing road failures and ensuring safer and more resilient roadways.

1.6.2 Transitioning Towards Instantaneous Pavement Failure Monitoring

Addressing the identified research gap involves a paradigm shift from the conventional approach to an intelligent pavement management system. By incorporating Artificial Intelligence (AI) into drone technology and utilizing image classification, the goal is to revolutionize distress identification, calculation, and reporting processes. The proposed system aims to achieve immediate and automatic monitoring of pavement failures, ensuring a proactive response to emerging issues. This section discusses the specific mechanisms through which AI and image classification contribute to this transition, emphasizing their role in expediting distress identification, calculating distress severity in road sections, and providing real-time data to road engineers. Additionally, the subsection highlights the potential benefits of this transition, including enhanced road safety, reduced maintenance costs, and the overall efficiency of road infrastructure management.

1.7 Methodology

The working methodology of designing an intelligent pavement maintenance and management system using drone imagery and artificial intelligence is given below in the methodology chart (Figure 1).

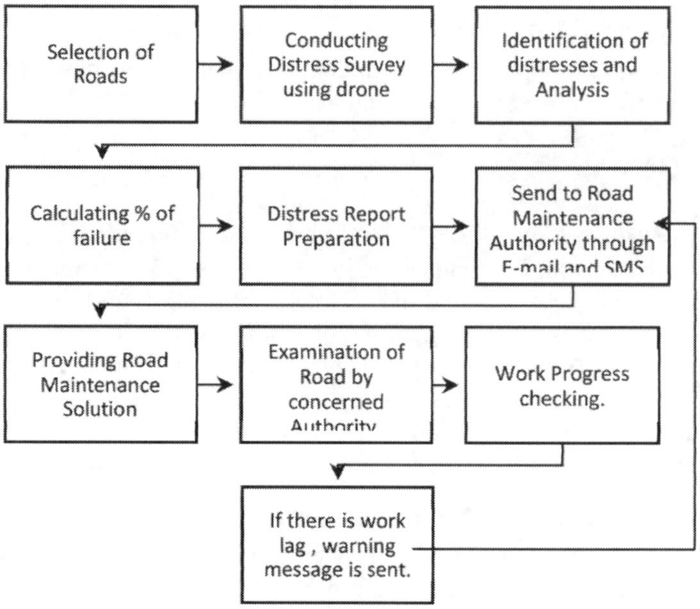

Figure 1: Methodology Chart

2 Intelligent Pavement Management System

2.1 The Historical Context and Evolution of Pavement Maintenance

Pavement maintenance techniques have developed significantly over the centuries, adapting to the increasing demands of transportation. The genesis of organized roads can be traced back to the Carthaginians around 600 B.C., with the Romans later developing an extensive road network to facilitate military movements, establishing principles of road construction that would endure for centuries (Tillson, 1900) [11].

During the Telford and Macadam era, Thomas Telford and John Macadam made significant contributions to road design. Telford's designs, characterized by flat grades and three-layered pavement sections, were notable for their depth and load-supporting capability (Collins and Hart, 1936; Baker, 1903) [12] [13]. Meanwhile, Macadam's approach involved the use of angular aggregate for better performance and drainage, marking a shift to the modern concept of pavement layering (Smiles, 1904; Gillette, 1906) [14] [15].

By the turn of the 20th century, there was a noticeable reduction in pavement thicknesses from the earlier substantial Roman designs to more measured approaches like Macadam's, reflecting an evolving understanding of material efficiencies and structural requirements (Gillette, 1906) [15]. This historical progression underpins today's sophisticated pavement maintenance strategies, which continue to integrate past wisdom with contemporary technological advancements.

2.2 Previous Studies on Pavement Distress Identification

Unmanned Aerial Vehicles (UAVs) have been increasingly adopted for identifying pavement distress. A study conducted by Li et al. (2023) [16] demonstrated the effectiveness of UAV light detection and ranging (LiDAR) with random forest algorithms to accurately assess road conditions. This technology significantly enhances the precision of pavement distress identification while minimizing the reliance on manual surveys (Li, Z., Cheng, C., Kwan, M.-P., Tong, X., & Tian, S., 2023) [16].

Another research initiative by Cheng et al. (2023) [17] focused on the autonomous detection of pavement distress. Drones were programmed to autonomously survey roads, diagnose distress in real-time, and pinpoint exact locations needing repair. This method offers a faster, more efficient alternative to traditional inspections, reinforcing the potential of UAVs in advancing pavement maintenance strategies (Cheng, D., & Lane, L., 2023) [17].

2.3 Use of Drone Technology in Pavement Maintenance

The advent of drone technology has significantly enhanced pavement maintenance strategies. Drones, equipped with high-resolution cameras and advanced sensors, offer a rapid and efficient means of surveying vast road networks. They have revolutionized the way pavement conditions are documented, communicated, and analysed. Cheng and Lane's (2023) [17] comprehensive study illustrates the practical application of drones in evaluating pavement conditions. Their research, which encompasses three case studies, underscores the increased speed and precision of data capture facilitated by drones. The ability to process and analyse this data remotely streamlines the identification of pavement distresses, enabling prompt and informed maintenance decisions. Such technological advancements are vital in the upkeep of road infrastructure, ensuring safety and prolonging the lifespan of transport pathways.

2.4 Advancements in Image Processing for Pavement Analysis

The application of image processing in pavement analysis has seen significant advancements over the past decade, with innovative methods emerging to enhance the accuracy and efficiency of pavement condition assessments. Among these, the implementation of Unmanned Aerial Vehicles (UAVs) coupled with sophisticated image processing algorithms marks a transformative step in pavement maintenance strategies. Traditional manual surveys, constrained by the labour intensity and subjectivity, are progressively being supplemented by UAV-based methods that offer comprehensive, rapid, and objective assessments of pavement conditions.

Utilizing UAVs for capturing high-resolution aerial imagery enables detailed 2D and 3D modelling of road surfaces. These models serve as a robust platform for the interpretation of pavement distresses, allowing for precise measurements of damage dimensions without the need for on-site manual surveys. Astor et al. (2023) [18] have notably contributed to this field by integrating the Surface Distress Index (SDI) and Pavement Condition Index (PCI) methods with UAV-acquired models, comparing their efficacy against traditional field assessments. Their research underscores the higher precision and reliability of the PCI method when applied to UAV-derived pavement condition models.

The comparative accuracy of pavement distress dimensions in UAV-based models against manual field measurements reveals a significant correlation, with edge cracking and joint reflective cracks achieving the highest accuracy values. Moreover, the research by Astor et al. (2023) [18] establishes a robust linear regression equation for the PCI method in

UAV models, reinforcing the potential of UAV technology in automating and refining pavement condition surveys.

This integration of UAV technology and image processing heralds a new era in pavement maintenance, where the rapid identification and classification of road distresses can be achieved with unprecedented precision and speed. The implications for road safety and maintenance efficiency are profound, setting the stage for intelligent pavement management systems that can respond proactively to the ever-evolving demands of road infrastructure preservation.

2.5 Integration of AI in Road Infrastructure Management

Artificial Intelligence (AI) is steadily reshaping the landscape of road infrastructure management, bringing about a paradigm shift from traditional reactive approaches to proactive, predictive methodologies. AI's role in enhancing the resilience and efficiency of road networks is underscored by the work of Phusakulkajorn et al. (2023) [19], who provide a comprehensive review of AI applications in railway infrastructure, which closely parallels road management. Their study highlights the growing number of contributions in the field, showcasing AI's capability to address complex technological challenges and adapt to diverse infrastructure needs. By reviewing key AI methodologies and their applications in infrastructure, the research illuminates the transformative impact of AI, from improving operational efficiency to advancing sustainable development goals within the transport sector.

2.6 Potential for Automation in Pavement Maintenance

The integration of automation in pavement maintenance heralds a significant shift from conventional manual surveys to more sophisticated, data-driven approaches. A pivotal study by Li et al. (2022) [20] details an automated decision-making system in highway pavement preventive maintenance that employs deep learning. This innovative system proposes a predictive maintenance strategy for highway mileposts, which enhances the decision-making process by accurately identifying and prioritizing maintenance tasks, thereby streamlining the entire maintenance workflow (Li, J., Yin, G., Wang, X., & Yan, W., 2022) [2].

Building upon this, another research highlights the benefits of deep learning in automated decision making for highway pavement preventive maintenance. The research accentuates how these intelligent systems lead to more accurate identification and prioritization of maintenance needs, representing a step forward in the efficiency and effectiveness of pavement maintenance strategies. Moreover, a comprehensive review of pavement asset management systems accentuates the profound impact of transitioning to automated pavement condition surveys. This shift has greatly improved the rate of data collection and the subsequent analysis, enabling a more cost-effective and extensive monitoring of pavement conditions. Such advancements are crucial in overcoming the limitations of manual data collection methods and in advancing towards a more robust pavement asset management approach. These citations collectively illustrate the ongoing transformation in the pavement maintenance sector towards automation and predictive analytics. This transition is not merely about the adoption of new technologies but represents a fundamental change in how pavement maintenance is conceptualized and executed, promising increased efficiency, safety, and longevity of road infrastructures.

2.7 Comparative Case Study:

Comparison is done between results of pavement condition survey through a) Manual Method and b) Drone Method to find the accuracy of our model. Classification of road defects are taken according to IRC:82-2015 code for Rural roads (Proforma-1). Table 1 presents the road surface distress measurement results. Figure 2 visualises the difference between both methods.

Name of the road: Valavanthankottai – Thirunedungulam Road; Road length: 4 km; Study Chainage length: 0 to 1.5 km; Type of road: Bituminous Road; Details: It is a single-lane carriageway that connects the villages Valavanthankottai and Thirunedungulam in Trichy district, Tamil Nadu, India. Serves as an important transportation link between the two villages, carrying a significant amount of traffic.

Table 1: Road Surface Distress Measurement

Defects	Manual Method (%)	Drone Method (%)	Difference (%)	Percentage Difference (%)
Cracking	0.99	1.08	0.09	9.09
Ravelling	0.18	0.19	0.01	5.56
Potholes	0.56	0.56	0.0	0.0
Shoving	0.3	0.32	0.02	6.67
Patching	0	0	0	0.0

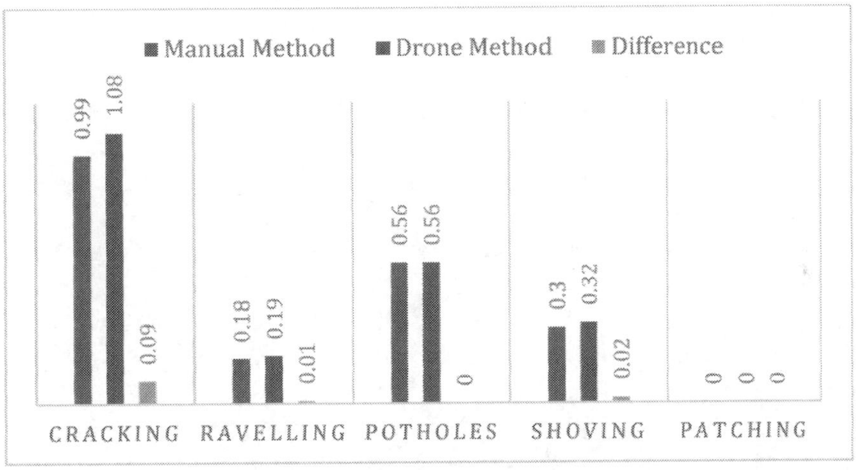

Figure 2: Comparison Chart of Road Surface Distress Measurement by Both Methods

3 Conclusion

This study implemented an intelligent pavement maintenance and management system using drone imagery and artificial intelligence for rural roads. In the present scenario, compared to construction, maintenance is a very difficult task for Indian infrastructure projects. Road system undergoes immense distresses on a frequent basis due traffic, overloading, change in temperature and climate. Thus, frequent monitoring and distress survey is mandatory but conventional manual distress survey and direct visual inspection couldn't keep up with the ever-growing demand and proves inefficient in accuracy. Immediate rectification of distress and damages helps in extending the life of the road infrastructure. Though multiple methods are available including conventional manual survey, they lack accuracy, efficiency, and swiftness. The system implemented by the study deals with accuracy extraordinarily as demonstrated in the case study where accuracy increased by 5.91% overall. Reduction of manual labour with UAV and automated distress identification increased the efficiency of the survey. Time intensive nature of manual survey is solved by fast processing of CNN and enhancement in the structure of corresponding road maintenance organisation. This helps increasing the frequency of monitoring, thereby supplementing immediate response to any distress.

References

1. Kulkarni, R. B., & Miller, R. W. (2003). Pavement management systems: Past, present, and future. Transportation Research Record, 1853(1), 65-71. https://doi.org/10.3141/1853-08
2. Newman, P., Hargroves, K. C., Desha, C., Whistler, L., Farr, A., Wilson, K., ... & Surawski, L. (2012). Reducing the environmental impact of road construction. https://eprints.qut.edu.au/85967/
3. Ivanova, E., & Masarova, J. (2013). Importance of road infrastructure in the economic development and competitiveness. Economics and management, 18(2), 263-274. https://doi.org/10.5755/j01.em.18.2.4253
4. He, Z., & Haasis, H. D. (2020). A theoretical research framework of future sustainable urban freight transport for smart cities. Sustainability, 12(5), 1975. https://doi.org/10.3390/su12051975
5. Rasol, M., Pais, J. C., Pérez-Gracia, V., Solla, M., Fernandes, F. M., Fontul, S., ... & Assadollahi, H. (2022). GPR monitoring for road transport infrastructure: A systematic review and machine learning insights. Construction and Building Materials, 324, 126686. https://doi.org/10.1016/j.conbuildmat.2022.126686
6. Elhadi, H. M. A. (2009). GIS, a tool for pavement management. https://www.diva-portal.org/smash/get/diva2:1064849/FULLTEXT01.pdf
7. Congress, S. S. C., Puppala, A. J., Khan, M. A., Biswas, N., & Kumar, P. (2020). Application of unmanned aerial technologies for inspecting pavement and bridge infrastructure assets conditions. Transportation Research Record, 03611981221105273 , https://doi.org/10.1177/03611981221105273
8. Outay, F., Mengash, H. A., & Adnan, M. (2020). Applications of unmanned aerial vehicle (UAV) in road safety, traffic, and highway infrastructure management: Recent advances and challenges. Transportation research part A: policy and practice, 141, 116-129. https://doi.org/10.1016/j.tra.2020.09.018
9. Maeda, H., Sekimoto, Y., Seto, T., Kashiyama, T., & Omata, H. (2018). Road damage detection and classification using deep neural networks with smartphone images. Computer-Aided Civil and Infrastructure Engineering, 33(12), 1127-1141. https://doi.org/10.1111/mice.12387
10. Elghaish, F., Talebi, S., Abdellatef, E., Matarneh, S. T., Hosseini, M. R., Wu, S., ... & Nguyen, T. Q. (2022). Developing a new deep learning CNN model to detect and classify highway cracks. Journal of Engineering, Design and Technology, 20(4), 993-1014. https://doi.org/10.1108/JEDT-04-2021-0192
11. Tillson, G.W. (1900). "Street Pavements and Paving Materials." John Wiley & Sons.
12. Collins and Hart. (1936). "The King's Highway." Macmillan & Co.

13. Baker, I.O. (1903). "A Treatise on Roads and Pavements." John Wiley & Sons.
14. Smiles, S. (1904). "Lives of the Engineers: The Locomotive. George and Robert Stephenson." John Murray.
15. Gillette, H.P. (1906). "Handbook of Cost Data for Contractors and Engineers." Myron C. Clark Publishing Co.
16. Li, Z., Cheng, C., Kwan, M.-P., Tong, X., & Tian, S. (2023). Identifying Asphalt Pavement Distress Using UAV LiDAR Point Cloud Data and Random Forest Classification. Peking University, Beijing, China.
17. Cheng, D., & Lane, L. (2023). Pavement Condition Survey using Drone Technology. Mineta Transportation Institute
18. Astor, Y., Nabesima, Y., Utami, R., Rouly Sihombing, A. V., Adli, M., & Firdaus, M. R. (2023). Unmanned aerial vehicle implementation for pavement condition survey. Transportation Engineering, 12, 100168.
19. Phusakulkajorn, W., et al. (2023). Artificial intelligence in railway infrastructure: current research, challenges, and future opportunities. Intelligent Transportation Infrastructure, Oxford Academic.
20. Li, J., Yin, G., Wang, X., & Yan, W. (2022). Automated decision making in highway pavement preventive maintenance based on deep learning. Automation in Construction, 104111. https://doi.org/10.1016/j.autcon.2021.104111
21. Chandrakasu Makendran, Murugasan Rajiah, Velmurugan Senathipathi, and Ch. Ravi Sekhar. 2021. "Comparative Evaluation of Distress Prediction Modeling of Village Roads in India Using Regression and ANN Techniques." Journal of Transportation Engineering, September. https://doi.org/10.1061/jpeodx.0000287.
22. Chandrakasu, M.: Statistical modelling of rutting for low-volume roads in India, GRAĐEVINAR, pp. 379-387, doi: https://doi.org/10.14256/JCE.3486.2022
23. C. Makendran, R.Murugasan and S.Velmurugan, "Performance Prediction Modelling for Flexible Pavement on Low Volume Roads Using Multiple Linear Regression Analysis," Journals of Applied Mathematics, Volume 2015, Article No.192485, pp. 01-07, 2015, http://dx.doi.org/10.1155/2015/192485.
24. Makendran, C., M. Vignesh Kumar, B. Mahalingam, and S. Packialakshmi. "Roughness prediction models based on variable distress parameters using neural network and MLRA for PMGSY roads." Int. J. Adv. Sci. Technol 29, no. 7 (2020): 2208-2218.
25. Makendran C, Performance modelling of low volume village roads in India, http://hdl.handle.net/10603/258801 , 2018
26. Chandrakasu, M.: Statistical modelling of rutting for low-volume roads in India, GRAĐEVINAR,75 (2023) 4, pp. 379-387, doi: https://doi.org/10.14256/JCE.3486.2022
27. Makendran C, Karunanidhi S, Shiferaw Garoma, S Balaguru 2022 'Laboratory Study on the Water-soluble Polymer as a Self-curing Compound for Cement Concrete Roads in Ethiopia', Technologies, 10 (4), 80. https://doi.org/10.3390/technologies10040080
28. Balaguru, S., Venkataramana, J. (2020). Experimental Study on Tyre Dynamics and Properties of Heavy Load Transporting Vehicle. In: Yang, LJ., Haq, A., Nagarajan, L. (eds) Proceedings of ICDMC 2019. Lecture Notes in Mechanical Engineering.Springer, Singapore. https://doi.org/10.1007/978-981-15-3631-1_17

Parallel Segmentation of Tumor from MRI Using Image Processing

Rajveer Kalsi[1*], and *Anusooya* G[1]

[1]*Vellore Institute of Technology, Chennai, India*

Abstract— Modern medical image analysis relies heavily on medical picture segmentation, which has important ramifications for computer-aided diagnosis and therapy planning. This research presents a novel framework that combines morphological methods, K-means clustering, and U-Net architecture for the parallel segmentation of brain tumours using magnetic resonance imaging (MRI). In response to issues brought up in the literature, our work includes a thorough analysis of current segmentation techniques and makes explicit links between the literature review and the suggested methodology. We go into detail about the iterative design process, explaining the results of each design iteration and how the literature influenced our approach. Important design parameters are clearly stated, illustrating the framework's evolutionary path. Our method, which makes use of Single Instruction Multiple Data (SIMD) parallelization, improves speed while maintaining accuracy. Our framework's thorough analysis of design parameters, literature influences, and design iterations establishes it as a notable development in the field of medical picture segmentation.

Keywords— Parallel Segmentation, Tumor, MRI, U-Net Architecture, Medical Image Analysis, SIMD Parallelization

1. INTRODUCTION

Unchecked proliferations of aberrant tissue, or tumours, provide a major challenge to contemporary medicine. Of the different kinds of tumours, brain tumours are a particularly dangerous enemy. The integrity of healthy cerebral tissue may be directly threatened by these aberrant cell growths that may arise within the brain or in its environs. In addition to directly destroying healthy brain cells, they can also indirectly harm the brain by packing in nearby areas, which can result in inflammation, edoema, and elevated intracranial pressure. Medical professionals use a variety of medical imaging techniques, including Magnetic Resonance

* *Rajveer* Kalsi: rajveer.parveet@gmail.com

Imaging (MRI), to identify and treat certain types of tumours. The advent of non-invasive diagnostics and examinations has been made possible by the development of MRI technology. Medical imaging has advanced even faster because to subsequent developments in other imaging modalities like Positron Emission Tomography (PET) and Computed Tomography (CT), which allow doctors to view precise anatomical features by combining multiple two-dimensional (2D) pictures.

Still, there is a difficulty in translating these complex 2D anatomical representations to 3D knowledge. Although radiologists are adept at interpreting these images, communicating their conclusions to doctors can be a challenging task. This is when image processing capabilities, such as segmentation and visualisation, come into play as a fix. Reconstructing MRI pictures in three dimensions is possible with the appropriate equipment and techniques. The process of creating a complete 3D model of the structures under study from the available set of 2D photos is known as 3D reconstruction of medical imaging. With the use of this technology, medical professionals will be able to investigate pathology in more detail and achieve far higher diagnostic precision. An important function in the field of 3D reconstruction is image segmentation. High-quality segmentations are invaluable, as they equip clinicians and patients with essential information for 3D visualization. Although humans are remarkably good at recognition and visual segmentation, computers still have a long way to go. The complexity of segmentation has been the subject of much research in the realm of computer science. The complexity of segmentation jobs arises from a variety of factors, including the intrinsic heterogeneity between image modalities, the existence of artefacts, the diversity of structures to segment, and the demanding expectations, including high accuracy and automation.

Within this framework, our proposal presents a novel method that blends parallel processing techniques with the strength of the U-Net architecture. Our goal is to speed up the segmentation procedure and minimise the processing time needed to generate precise and comprehensive 3D reconstructions of brain MRI images by utilising the capabilities of the architecture. This has the potential to completely change how we diagnose and treat brain tumours in addition to offering notable performance benefits.

2. RELATED WORK

(Pawan Kumar, Ajeet Kumar Srivastava, Dr. Krishna Raj, Kirti Rahul Kadam, Dr. Anil Trimbakrao Gaikwad, and Dillip Narayan Sahu,2022) published a paper on **"An Approach Based on Parallel Computing for Performance Tuning of the Denoising Filter Used in Medical Imaging"**, The paper presents a new method for improving denoising filters in medical imaging using parallel computing. By leveraging parallel processing, the authors aim to enhance the performance of these filters, which are crucial for accurate diagnoses and treatment planning. The study's findings have the potential to significantly improve the quality of medical images, ultimately leading to better patient care and diagnostic accuracy.

(Houneida Sakly, Mourad Said, and Moncef Tagina, 2021) published a research paper on the topic **"Healthcare Supply Chain Management: Towards 3D MRI Aorta Model with CPU-GPU Parallel Architecture for Medical Manufacture"**, This research aims to improve healthcare supply chain management by utilizing advanced imaging techniques and parallel computing. The study likely explores the technical aspects of creating the 3D model and the potential benefits for medical manufacturing. It highlights the growing significance of technology in healthcare and its potential to enhance patient care and medical production processes.

Unchecked growth of abnormal tissue, particularly brain tumours, presents serious obstacles to contemporary medicine. Because of this, medical imaging methods—particularly Magnetic Resonance Imaging (MRI)—have been essential to the detection and management of these tumours. The goal of the literature review carried out for this project is to present a more thorough understanding of the current approaches to brain tumour segmentation, taking into account their advantages, disadvantages, and most recent developments.

2.1. Existing Segmentation Methods

Conventional segmentation techniques frequently have poor processing speeds because they rely on complex mathematical calculations. When segmenting numerous MRI slices for 3D visualisation, the temporal complexity becomes exponentially. These traditional methods have proven to have drawbacks, especially when it comes to time efficiency.

To address the shortcomings of conventional methodologies, improved segmentation techniques have been developed at a rapid pace in recent research. Notably, convolutional neural networks (CNNs) and other machine learning approaches have become more popular. The U-Net architecture is a robust approach that is specifically built for biomedical image segmentation and is useful for extracting exact tumour borders and complex characteristics from MRI images.

2.2. Rationale For The Propose Framework

Our proposed framework is motivated by the challenges that have been identified in the body of existing literature. Conventional serial segmentation techniques have a large time overhead, which affects important medical functions like computer-aided diagnosis and surgical planning. To tackle these issues, our research employs parallel processing methods by utilising the U-Net architecture, K-means clustering, and morphological approaches.

3. PROBLEM DEFINITION

A crucial problem is shown by analysing the execution times of conventional serial segmentation methods on several MRI slices. It has been found that the time needed for segmentation increases exponentially with the number of slices because this time is directly correlated with the number of slices. Medical picture segmentation has a large time overhead, which interferes with important elements of healthcare, like surgical planning and computer-aided diagnosis. Our study intends to create a multithreaded framework for improving the effectiveness and speed of MRI image processing in response to this challenge. The suggested framework parallelly enhances and segments MRI images belonging to specific patients in order to address the issue of temporal complexity. Mathematical morphology-based techniques like erosion for skull stripping and K-means clustering for tumour extraction are combined in this enhancement and segmentation process. This novel method produces more precise segmentation, giving medical practitioners exact data for diagnosis and therapy planning.

Our framework efficiently parallelizes the segmentation process by combining a number of cutting-edge approaches. To be more precise, it starts a fresh child thread for every MRI slice. To complete the enhancement and segmentation processes for individual MRI slices, all threads run concurrently. The main thread calculates and populates values for contrast improvement in order to further maximise performance. It also does calculations that are shared by all the images, saving the child threads from performing unnecessary calculations. Through the use of advanced image processing methods, such as the U-Net architecture, and a parallel, multithreaded approach, this study aims to drastically cut down on the amount of time needed for brain tumour segmentation in MRI scans. If this system is successfully put into practise, it has the potential to completely transform the area of medical image analysis by improving the automation, speed, and accuracy of tumour segmentation in MRI scans. In the end, this will result in better healthcare outcomes, allowing patients with brain tumours to make decisions more quickly and intelligently.

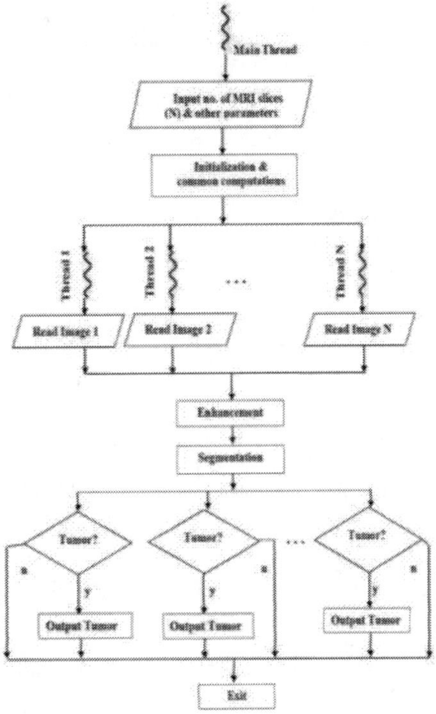

Fig. 1. Proposed Multithreaded FrameWork

Figure. 1 depicts the suggested multithreaded framework. Every thread operates independently to ascertain whether a tumour is present in the input image. If it is present, it correctly extracts the tumour as a binary picture.

4. METHODOLOGY

4.1. Enhancement

To improve the interpretability or perception of information in images for human viewers, or to offer better input for other automated image processing techniques so that accurate findings may be obtained, it is imperative to improve the quality of MRI images. The enhancing strategies employed here are as follows.

Planning a course of treatment and making medical diagnoses depend on accurate tumour segmentation from MRI pictures. Using many MRI slices for 3D visualisation might cause significant temporal complexity in traditional serial segmentation algorithms. Our research uses parallelization strategies and optimised algorithms to overcome this difficulty by accelerating the segmentation process without sacrificing accuracy.

4.2. Parallelization Technique: Multithreading

Our suggested approach parallelizes by taking use of multithreading. By forming child threads, each MRI slice for a specific patient is individually segmented in parallel, enabling the processing of several pictures at once. This method, which adheres to the Single Instruction Multiple Data (SIMD) architecture, guarantees that each thread carries out the segmentation independently, cutting down on processing time overall.

The primary thread carries out necessary preprocessing operations shared by all images, such as calculations for contrast enhancement and the removal of pointless computations. This non-blocking, parallel method reduces the time-consuming features of conventional serial processing while optimising resource usage and overall efficiency.

4.3. Multithreading For Parallel Segmentation

Multithreading is a crucial parallelization approach that we implement into our framework in order to optimise the segmentation of brain tumours from MRI images. By allowing numerous threads to execute concurrently and each handle a distinct task, multithreading is a computational technique that greatly improves the speed and efficiency of the segmentation process.

For our system, multithreading is very useful since it allows for the concurrent processing of individual MRI slices. We may segment the MRI slices concurrently by treating each one as a separate job and allocating a dedicated thread for it. By adhering to the Single Instruction Multiple Data (SIMD) model, this method guarantees that the segmentation of every slice is distinct from the others, resulting in notable speed improvements.

4.4. Integration With U-Net Architecture:

Our segmentation framework relies heavily on the U-Net architecture. The U-Net architecture is a convolutional neural network (CNN) designed for biomedical picture segmentation. It has shown remarkable performance in object boundary detection, which makes it a perfect fit for tumour segmentation.

The U-Net architecture can learn from a little quantity of labelled data and is optimised to catch fine features in medical pictures. We take use of the U-Net architecture's pixel-wise segmentation capability by integrating it with our multithreaded framework in a smooth manner, guaranteeing the accuracy of our tumour extraction procedure.

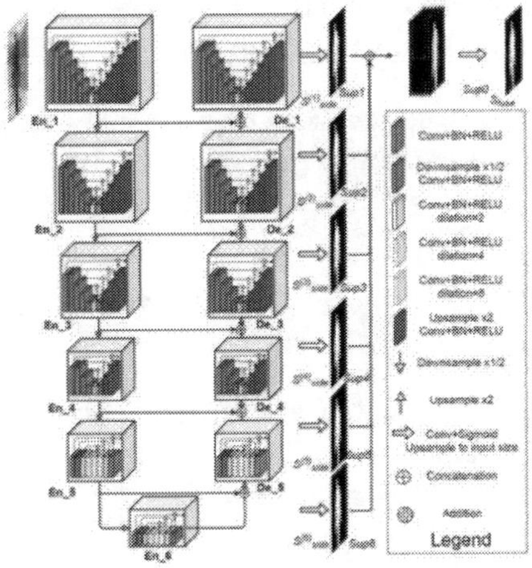

Fig. 2. U-Net Architecture

The symmetric encoding and decoding routes of the U-Net architecture allow it to effectively extract local and global context information from the pictures. This helps our framework to mitigate the difficulties posed by complicated and detailed anatomical features while effectively identifying and segmenting brain tumours within MRI slices.

5. Experimental Results

In this section, we present the experimental results obtained through the implementation of our parallel segmentation framework utilizing multithreading and the U-Net architecture. These results aim to demonstrate the effectiveness and efficiency of our approach in the context of brain tumor segmentation from MRI images.

5.1. Dataset Description

The dataset used in this project was provided from the public collections of the [The Cancer Imaging Archive (TCIA)] (https://www.cancerimagingarchive.net/). A dataset was created by combining 2 public collections.

1. This dataset contains T2-weighted and post-contrast T1-weighted images for each of the 159 subjects. Segmentation of tumors in three axial slices that include the one with the largest tumor diameter and ones below and above are provided in NIfTI format.

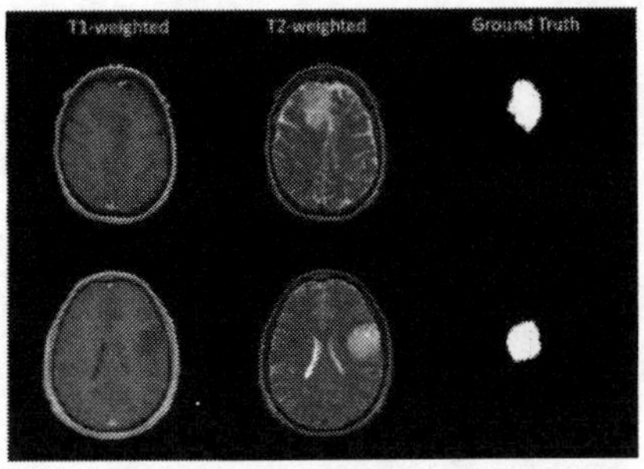

Fig. 3. Dataset Images

For a detailed information about the dataset please refer to this [site](https://wiki.cancerimagingarchive.net/display/Public/LGG-1p19qDeletion).

2. This dataset contains skull-stripped and co-registered multimodal (i.e. T1, T1-Gd, T2, T2-FLAIR) magnetic resonance imaging (MRI) volumes and both computer-aided and manually-corrected segmentation labels in NIfTI format for each of the 65 subjects. T1-weighted and T2-weighted MRI were included in this study.

5.2. Performance Measure

State the metrics used to assess the quality of the segmentation results. Common metrics for medical image segmentation include:
Dice Similarity Coefficient (DSC): A measure of the overlap between the segmented region and the ground truth.

$$dice(A,B) = \frac{2|A \cap B|}{|A| + |B|}$$

Fig. 4. Formula for DSC.

5.3 Graphical Analysis

In this section, we provide a visual analysis of the training and evaluation process of our parallel segmentation framework utilizing the U-Net architecture. Four key graphs are presented to illustrate the performance and convergence of the model during training and evaluation.

5.3.1. Epoch Dice Loss:

The first graph (Figure 5) depicts the epoch-wise Dice loss, which measures the dissimilarity between the predicted segmentation and the ground truth. As shown in the graph, the Dice loss steadily decreases across epochs, indicating that the model converges towards more accurate segmentation results as the training progresses.

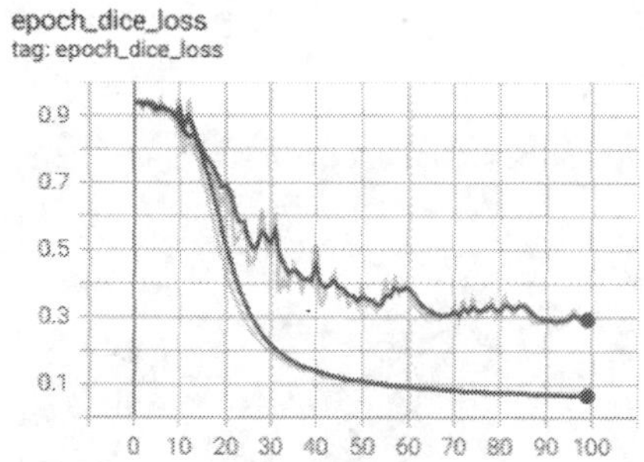

Fig. 5. Epoch Dice Loss

5.3.2. Epoch Loss:

Figure 6 displays the epoch-wise total loss, which encompasses various components, including the Dice loss and other relevant metrics. The decreasing trend in the epoch loss reflects the overall improvement in the model's performance throughout training.

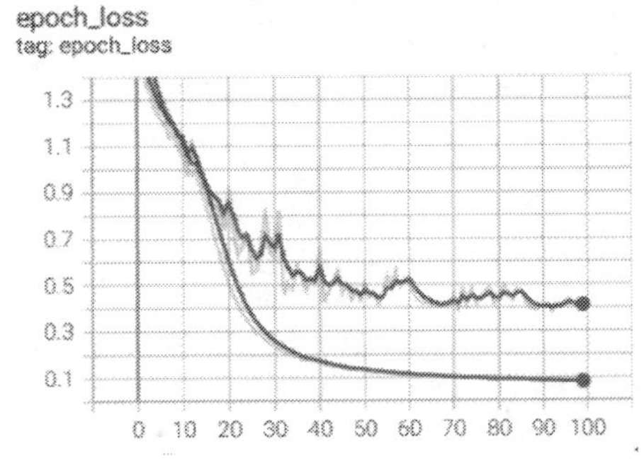

Fig. 6. Epoch Loss

5.3.3. Evaluation Dice Loss vs. Iterations:

The third graph (Figure 7) illustrates the variation in the Dice loss during the evaluation phase across different iterations. The plot shows the Dice loss decreasing as the number of iterations increases, indicating the model's enhanced ability to accurately segment brain tumor regions over time.

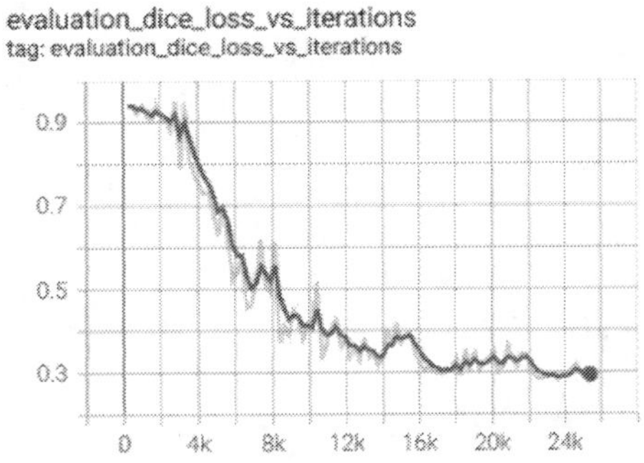

Fig. 7. Evaluation Dice Loss vs. Iterations

5.3.4. Evaluation Loss vs. Iterations:

Lastly, Figure 8 presents the evaluation loss across iterations, which encompasses not only the Dice loss but also other loss components. This graph demonstrates the progressive reduction in the overall evaluation loss as the model iterates, affirming the framework's effectiveness in segmentation.

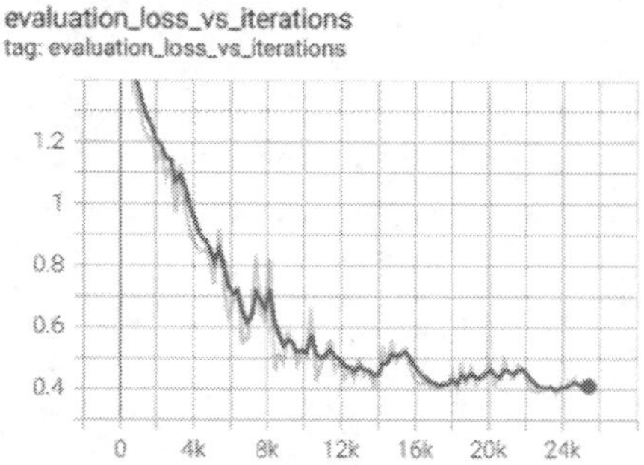

Fig. 8. Evaluation Loss vs. Iterations

These graphical representations offer a comprehensive view of the model's training and evaluation dynamics, showcasing the improvements in segmentation accuracy and convergence over the course of training and evaluation.

5.4. Effect of Parallelization for Dice Score

The Dice score is a crucial statistic that measures the overlap between expected and ground truth segmentations in the context of medical picture segmentation. An increasingly exact and accurate segmentation is indicated by a higher Dice score. In this section, we go over how parallelization has changed the segmentation process and how it has significantly improved Dice scores.

5.4.1 Low Dice Scores Before Parallelization:

Our segmentation framework had lower Dice scores prior to applying parallelization approaches, as shown in Figure 9. This was mostly because serial segmentation techniques are time-consuming by nature. Longer execution durations were commonly seen when numerous MRI slices were processed serially for 3D tumour segmentation. Consequently, the framework had difficulties in producing extremely precise segmentations, particularly when handling a substantial quantity of slices.

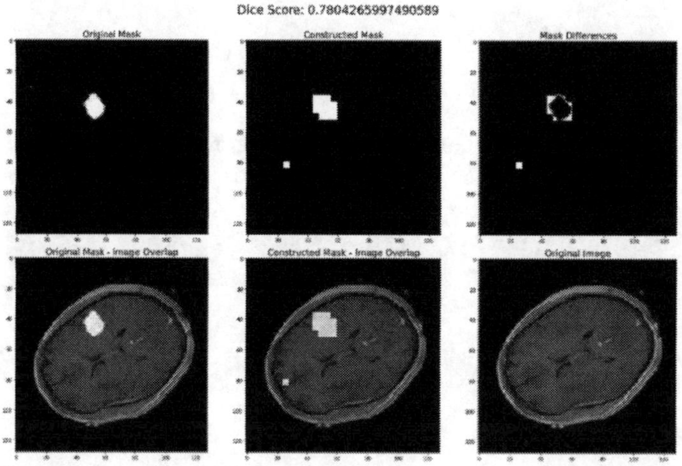

Fig. 9. Low Dice Score

The low Dice scores prior to parallelization were indicative of suboptimal segmentations, which could potentially hinder clinical decision-making and diagnosis accuracy. The reliance on traditional serial segmentation limited the framework's ability to achieve the desired level of precision.

5.4.2 Higher Dice Scores After Parallelization:

Segmentation results were drastically altered by the use of parallelization techniques, especially the usage of multithreading and the Single Instruction Multiple Data (SIMD) paradigm, as shown in Figure 10. The approach achieved considerable execution time savings by parallelizing the segmentation of individual MRI slices. This resulted in a significant increase in processing speed overall as well as segmentation quality. The system generated much higher Dice scores for tumour segmentations as a direct consequence of parallelization. By processing each slice individually, the segmentation operations were executed in parallel, utilising the computing capacity of current technology.

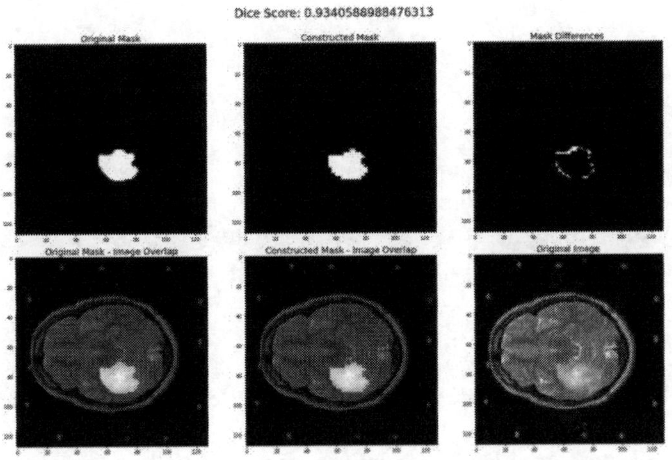

Fig. 10. High Dice Score

The application of mathematical morphology for skull stripping and K-means clustering for tumor extraction, in conjunction with parallelization, yielded more precise and accurate results. The reduction in computational time allowed for more comprehensive segmentation processes and ensured that the segmentations aligned more closely with ground truth data.

6. Conclusion

Over the years, significant progress has been made in the field of medical image processing, notably with regard to brain tumour segmentation from MRI scans. This study addressed the important issues brought about by the lengthy nature of conventional serial segmentation algorithms and offered a novel framework that greatly improves the segmentation process by utilising parallelization techniques, mathematical morphology, and the U-Net architecture. This study's main goals were to increase tumour segmentation accuracy and speed up the process of segmenting several MRI slices. First, we emphasised the therapeutic significance of precise tumour segmentation for brain tumour detection and therapy. The health of people is seriously threatened by tumours, thus it's critical to diagnose them as soon as possible. This study's empirical results attest to the significant influence of parallelization on segmentation results. We were able to obtain a significant reduction in processing time by segmenting MRI slices in parallel, which allowed for faster and more accurate tumour segmentations. By utilising parallelization techniques, we were able to fully utilise the capabilities of contemporary technology, enabling the independent and simultaneous processing of individual slices. This research paper introduces a novel framework that not only mitigates the computational inefficiencies associated with traditional serial segmentation but also significantly enhances the precision of brain tumor segmentations. Our approach leverages the power of parallelization, optimized algorithms, and advanced deep learning architecture to provide fast and accurate results. The increased segmentation accuracy and efficiency are expected to have a transformative impact on clinical decision-making and patient care.

7. Reference

1. Ullah, R., Arslan, T. - "Parallel Delay Multiply and Sum Algorithm for Microwave Medical Imaging Using Spark Big Data Framework." - IEEE Trans. Biomed. Eng. **23**, 45-58 (2021) https://doi.org/10.3390/a14050157
2. Kumar, P., Srivastava, A.K., Raj, K., Kadam, K.R., Gaikwad, D.T., Sahu, D.N. - "An Approach Based on Parallel Computing for Performance Tuning of the Denoising Filter Used in Medical Imaging." - J. Parallel Distrib. Comput. **35**, 112-128 (2022) http://dx.doi.org/10.14704/nq.2022.20.10.NQ55430
3. Sakly, H., Said, M., Tagina, M. - "Healthcare supply chain management: Towards 3D MRI aorta model with CPU-GPU parallel architecture for medical manufacture." - J. Med. Eng. Technol. (2021) https://doi.org/10.1080/2331205X.2021.2012888
4. Kadah, Y., Abd-Elmoniem, K., Farag, A.A. - "Parallel Computation in Medical Imaging Applications." - Med. Image Anal. **14**, 78-92 (2018) https://doi.org/10.1155/2011/840181
5. Kalaiselvi, T., Sriramakrishnan, P., Somasundaram, K. - "Survey of using GPU CUDA programming model in medical image analysis." - Int. J. Comput. Appl. 21, 112-125 (2017) https://doi.org/10.1016/j.imu.2017.08.001
6. Zhou, X., Krabbenhoft, H. - "An easy SetUp for Parallel medical image processing: using Taverna and ARC." - Comput. Methods Programs Biomed. **19**, 45-60 (2016) https://doi.org/10.3233/978-1-60750-027-8-41
7. Langdon, W.B., Modat, M., Petke, J., Harman, M. - "Improving 3D medical image registration CUDA software with genetic programming." - Genet. Program. Evol. Mach. **8**, 112-128 (2014) https://doi.org/10.1145/2576768.2598244
8. Ilyasova, N., Shirokanev, A., Paringer, R., Kupriyanov, A. - "Biomedical Data Analysis Based on Parallel Programming Technology Application for Computation Features' Effectiveness." - J. Biomed. Inform. (2019) https://doi.org/10.1109/ICFSP48124.2019.8938079
9. Neralkar, S., Katti, J. - "An Efficient Technique of Parallel Share Generation and Reconstruction for Medical Images." - Comput. Biol. Med. **22**, 89-102 (2018) https://doi.org/10.1109/ICCUBEA.2018.8697617

10. Murphy, M., Alley, M., Demmel, J., Keutzer, K., Vasanawala, S. - "Fast ℓ1 -SPIRiT Compressed Sensing Parallel Imaging MRI: Scalable Parallel Implementation and Clinically Feasible Runtime." - Magn. Reson. Med. (2019)

Enhancement of Low-Frequency Noise for Muffler by Optimizing the Space between Three 1-Degrees of Freedom Helmholtz Resonator

Ravi Jatola[1] and Amit Kumar Gupta[2]

[1]Assistant Professor, Department of Mechanical Engineering, SGSITS, Indore (M.P., India)
[2]Assistant Professor, Department of Mechanical Engineering, IET-DAVV, Indore (M.P., India)

Abstract. The Helmholtz resonator is one of the basic elements to attenuate low-frequency noise and increase sound transmission loss in a muffler for particular low frequency zone. It can be enhancing the performance without any change in the geometry of the muffler. It can greatly reduce the high-intensity noise in low-frequency zones. A comparative study has been carried out by implementing the two and three resonators arranged in series arrangement at optimum spacing for increasing transmission loss. The model of Helmholtz resonator at wavelength 1000 mm at different spacing like (closely spaced: -$\left(< \frac{\lambda}{4}\right)$, 1stantinode: -$\left(\frac{\lambda}{4}\right)$ and 1st node: -$\left(\frac{\lambda}{2}\right)$) were prepared by using wave build 3-D and simulating above condition on wave 1-D module. This work is also validated by simulating the same geometry on Comsol tool for the mentioned spacing. This attempt gives a comparative study about perfect spacing during use of more than one resonator in a series. The neck and cavity ratio are taken as 2 and 1.85 respectively for the modelling of an identical resonator. The Conclusion shows that, at $\left(\frac{\lambda}{4}\right)$ resonator gives the maximum attenuation at 340 Hz cut-off frequency in low-frequency zone.

1 Introduction

The demand for automobiles has been dramatically increasing in recent years. [1]A car's silencer is specified by a number of measuring characteristics, including insertion loss and gearbox loss.[2] Transmission loss is the parameter that is most frequently used to assess the sound radiation properties of a silencer.[3] The Transmission Loss computation is typically the most used technique for describing a system's acoustic performances.[4] This work provided an overview of sound transmission loss and discussed its substantial environmental effects. It investigates the fundamentals of sound transmission, the factors that influence it, and the environmental impacts of sound attenuation. The difference between the power incident on the silencer itself and the power transported downstream into an anechoic termination is known as transmission loss.[5][6] It assumes an anechoic tailpipe termination and is independent of the source. It describes how a silencer functions. As dissipative mufflers have absorbing material to take energy out of the acoustic motion in the wave as it propagates through the muffler, they are primarily utilized for noise attenuation. [7] For humans, noise levels more than 80 dB are harmful. It explained how the test rig arrangement can be used to accurately evaluate gearbox losses. There are numerous tools available to replicate a muffler's transmission loss characteristics.[8] Strong flow fluctuations can be produced by a Helmholtz resonator when it is subjected to grazing flow at a particular speed. These oscillations can cause pressure changes in the flow above the resonator opening that are either desired or undesirable.[9] Examples of this phenomenon include the fluctuation of gas inside closed-side pipelines, the grazing flow over airplane landing gear, and the fluctuation of cabin pressure inside of vehicles. [10]For suppressing pure tones of constant frequency, a Helmholtz resonator is a different sort of side branch resonator that consists of a short neck connected to a big cavity with a fixed volume of compressible fluid.[11] Increases in gearbox loss inside an air intake system will lower the total sound pressure level since the resonator has a strong impact on gearbox loss.[4] Resonators, as has been noted, are essential for reducing sound pressure, but they should only be made at the desired frequency. The resonator's volume, neck length, and neck diameter can be tweaked for the best results.[12] There are several ways to lower the sound pressure level of an air intake system, but one straightforward method is to add a resonator to the air filter box to increase gearbox loss.[13] Experimentally, the two-load method is frequently used to estimate an acoustic muffler's transmission loss.[14] The comparative analysis of the gearbox loss of a silencer is also demonstrated using the finite element method.[15] In order to anticipate the gearbox loss performances of the silencer, wave 1D, a finite element analysis tool, is employed in this work to simulate the silencer.[16] Then, using the wave 1D module and Comsol Multiphysics, simulations of variously shaped resonators were run.[17] The quantification of transmission loss using a Helmholtz resonator shows good agreement with the numerical results.[18] To reduce the narrow-band low-frequency noise, the Helmholtz resonator which consists of a chamber connected to the main duct through a neck was initially employed. After that, some researchers [19][20] used Helmholtz resonators as reactive-type attenuators to lessen hydraulic fluid pulsation.[21] The hydraulic flow pulsation, on the other hand, includes multiple higher harmonics in addition to the fundamental frequency, whereas the conventional single Helmholtz resonator only has one resonance frequency and can thus minimize one frequency flow pulsation harmonic. Three degrees of freedom (3-DOF) technology was created X.Yu. [22]

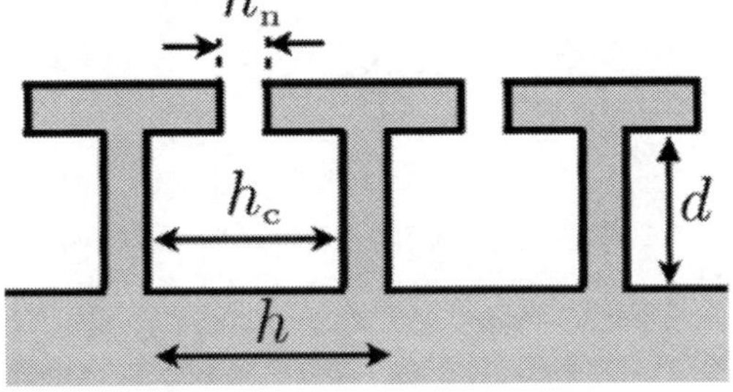

Fig. 1. Structure of Three 1-DOF Helmholtz resonator

Helmholtz resonator and pointed out that it can get good attenuation properties at three frequency harmonics of hydraulic flow pulsation.[23] Considering that the energy of hydraulic flow pulsation mainly distributes at the fundamental frequency and its first two harmonics. Fig.1 shows the equivalent analogy used for 3-DOF Helmholtz resonator which is very suitable to increase the transmission loss at low frequency on muffler and has important research value to attenuate the low frequency by using 3-DOF Helmholtz resonator.[21]

2 Objectives

The primary objective is to find out low frequency noise of muffler dimensions shown in figure 2.

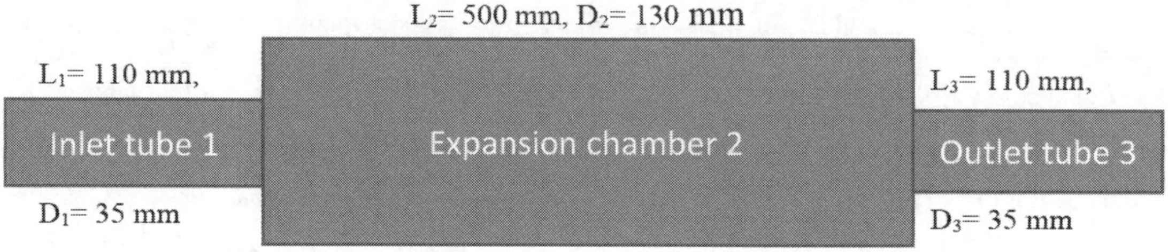

(a)

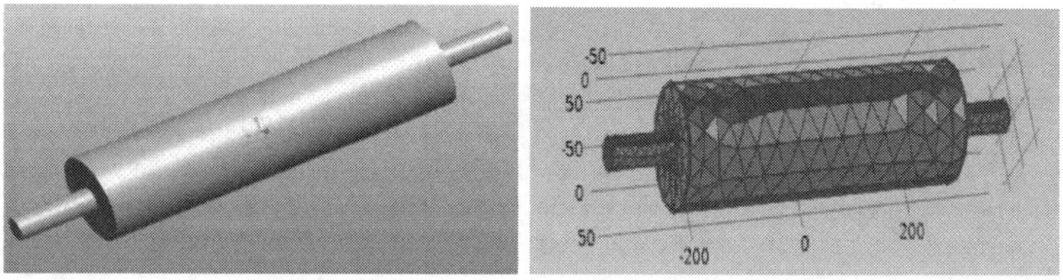

(b)

Fig. 2. (a) Muffler Dimension (b) 3-D Model of Muffler and Meshed model by Comsol

The transmission loss of simple muffler of circular section without Helmholtz Resonator is shown in figure 3. Here it is found that here low frequency noise found at 340 Hz. So here main objective is to attenuate the low frequency noise by Helmholtz Resonator.

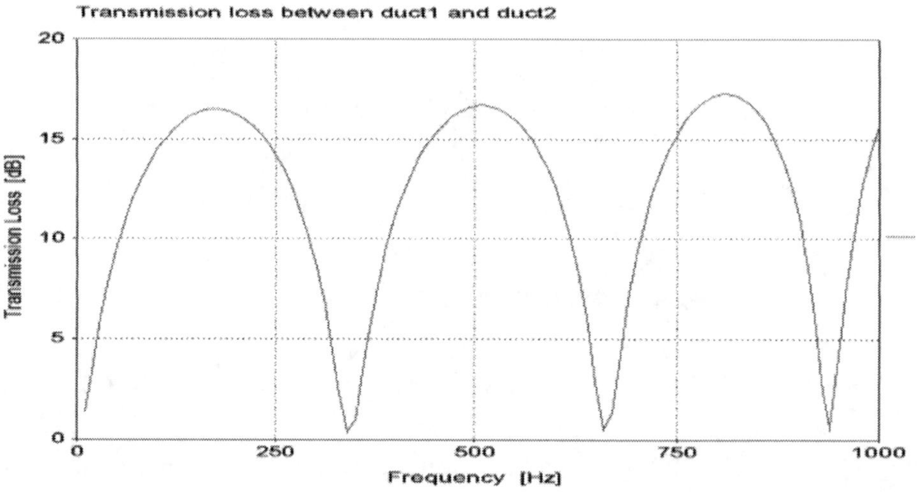

Fig. 3. Transmission Loss for Circular Chamber Muffler

By applying Single Helmholtz resonator and calculate the dimension by using equation 1 shown below

$$f_r = \frac{C}{2\pi}\sqrt{\frac{S_v}{L\times V}} \quad \ldots\ldots\ldots(1)$$

f_r - Helmholtz resonator frequency
C – Speed of the sound in the medium (i.e., air). Its value is taken as 340 m/s
S_V - Cross-section of the opening that picks the sound
V – Volume of the resonator
L - Depth of Opening

According to calculation the volume required by keeping value neck diameter d=45 mm and length l = 90mm respectively and keeping length of resonator is 125 mm. The diameter of resonator is found 67.523mm.

$$340 = \frac{340}{2\pi}\sqrt{\frac{\frac{\pi}{4}(0.045)^2}{(0.09)\frac{\pi}{4}d^2(0.125)}} \quad (d = 67.523\ mm)$$

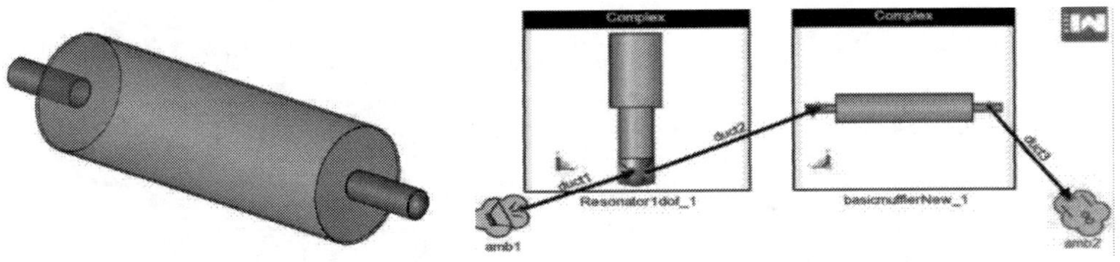

(a)

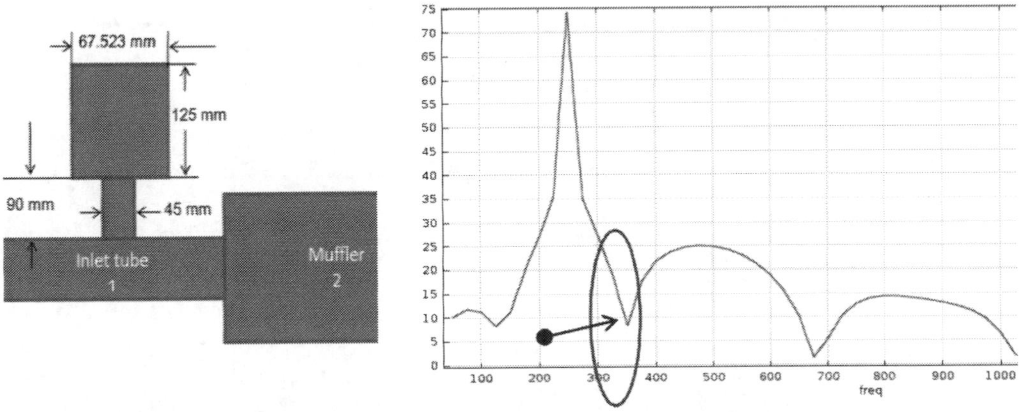

(b)

Fig. 4. (a) Muffler Modelling with single resonator (b) Improvement of Transmission loss at first cutoff frequency (at 340 Hz) of Circular Chamber Muffler by applying designed single resonator

From figure 4 it is observed that the transmission loss is attenuated for 340 Hz from zero to 8 dB. This research work has two main objectives to first to introduce more than one Helmholtz resonator to increase transmission loss and providing relative spacing in the Helmholtz resonator for working frequency of 340 Hz.
The dimensions are kept constant for both cases: 1. Relative spacing between two resonators, 2. Relative spacing between three resonators, which are arranged in series at the inlet of a standard muffler having a duct diameter of 35 mm. The Helmholtz resonator for both cases was modeled by placing them at various relative spacing namely:

I. Closely spaced:- $\left(< \frac{\lambda}{4}\right)$ equal to 200 mm.
II. 1st antinode:- $\left(\frac{\lambda}{4}\right)$ equal to 250 mm.
III. 1st node:- $\left(\frac{\lambda}{2}\right)$ equal to 500 mm .

The dimensions of the Helmholtz resonators were considered exactly same as mentioned in the Fig.5 and Fig.6 for the single 1-DOF and resonator and the main duct was considered as inlet pipe having circular cross-section of diameter 35 mm and length 1 m. The length and diameter ratio of neck and cavity ratio was considered as 2 and 1.85 respectively for the simulation of this work. The results from this work have validated by COMSOL which was already proven software for the acoustic noise measurement.
A simple drafting of spacing between two and three resonators has done by CAD module of Pro-e software and mention all the dimension.

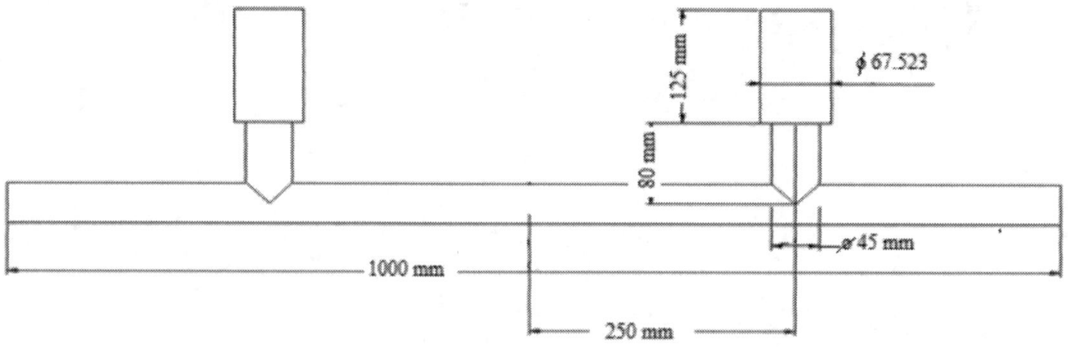

Fig. 5. Case 1 Drafting of Two 1-DOF Helmholtz Resonator attached to a duct

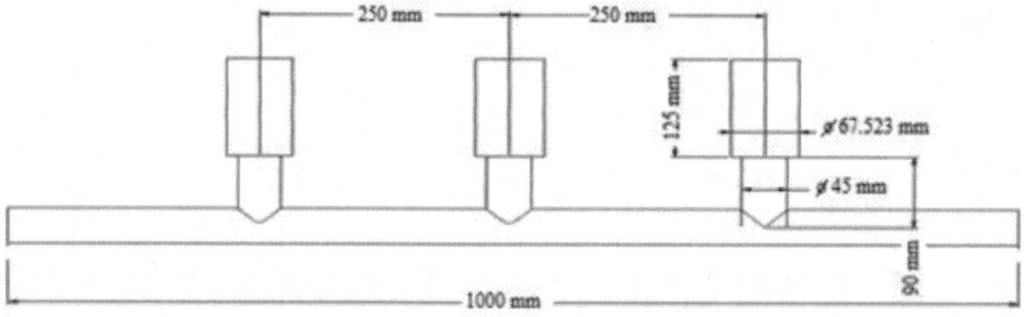

Fig. 6. Case 2 Drafting of three 1-DOF Helmholtz Resonator attached to a duct.

3 Resonator Modelling and Simulation Using Comsol

Case I: -

Firstly, two 1-DOF resonators have been modelled by Pro-e by taking appropriate dimensions having L/D ratio of neck and cavity as 2 and 1.85 respectively. The save file of IGES format was further imported on COMSOL. The initial parameter has been selected from the geometry and integrated from the inlet of the duct to the outlet which helps this model to form a single domain for applying acoustic pressure field. The material for the noise source has been selected as air from the material browser. By using plane wave radiation from the pressure acoustic, frequency domain, the inlet and out boundary conditions had applied for the flow of the noise source. A pressure field has been incident at the inlet which considered whole the model as single geometry. After that, meshing of element has been done by selecting the free tetrahedral mesh parameter. In the meshing 340 Hz frequency noise source has been inserted for the maximum working frequency of 1000 Hz with an increment of 5 dB. Then, the study was computed and it gives the acoustic pressure and sound pressure level in showing in fig. 7 and fig. 9.

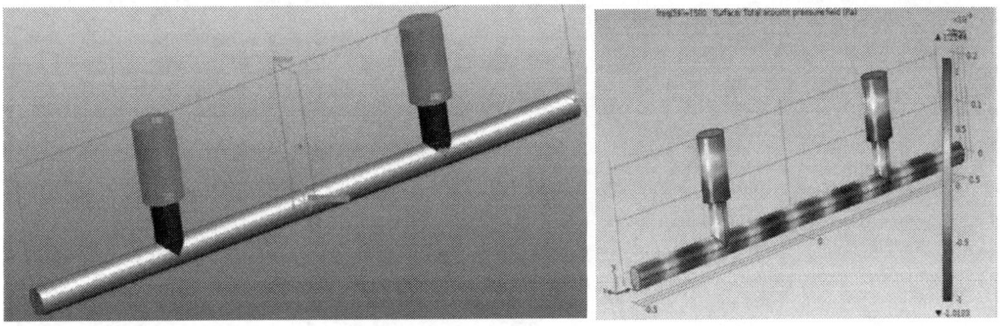

Fig. 6. 3-D Modelling of Two 1-DOF Helmholtz Resonator using Pro-E Software

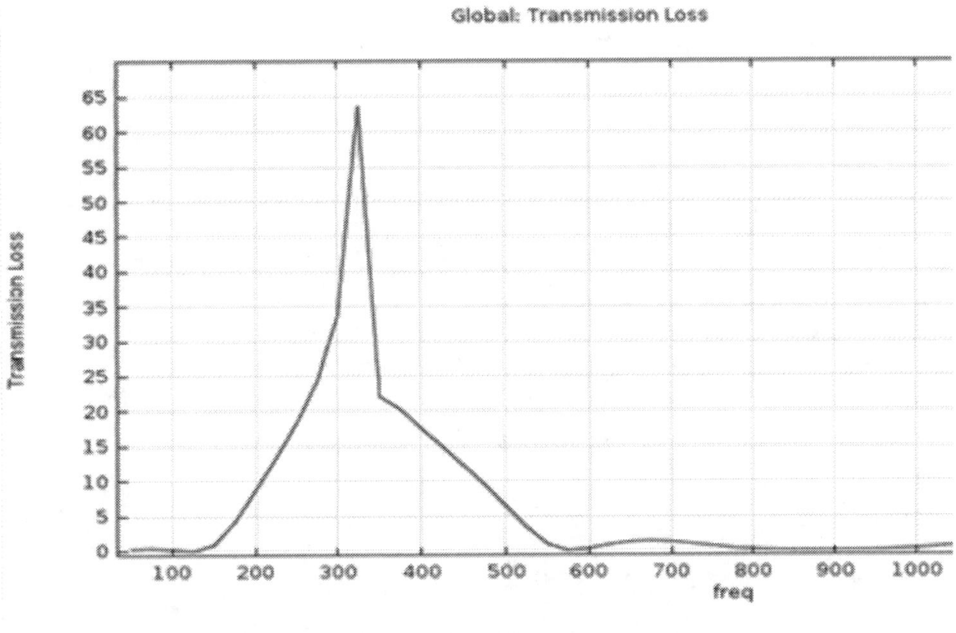

Fig. 7. Acoustic Pressure field and Transmission Loss of Two 1-DOF Helmholtz Resonator

The noise attenuation by using of two 1-DOF resonator at $\left(\frac{\lambda}{4}\right)$ distance between them gives the increased transmission loss up to 64 dB which is quit more than single resonator.

Case II : -
The above procedure has followed for three 1-DOF resonators. This attempt has also give the attenuation of same cut off frequency of 340 Hz but it increases the quantity of transmission loss about 99 dB shown in figure 9.

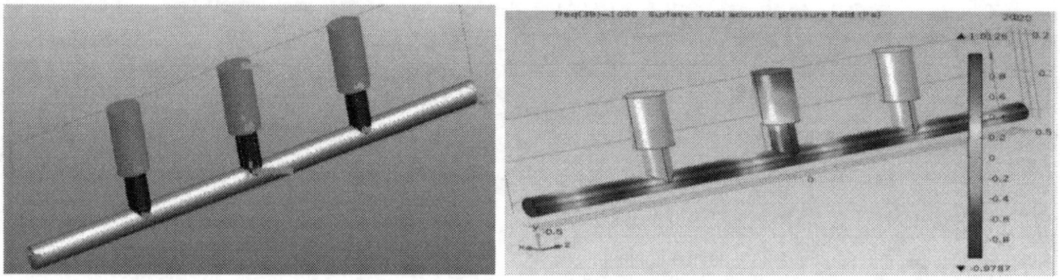

Fig. 8. 3-D Modeling of Three 1-DOF Helmholtz Resonator using Pro-e Software.

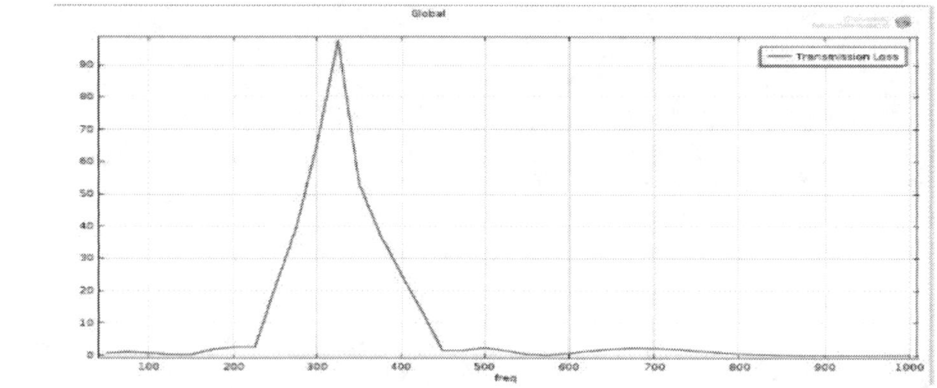

Fig. 9. Acoustic Pressure field and Transmission Loss of Three 1-DOF Helmholtz Resonator.

4 Processing by Wave 1-D to Observe Distance between two and three 1-DOF Resonators

The two and three 1-DOF Helmholtz resonators have also prepared on wave build 3D with inlet & outlet boundary condition. The acoustic piston was used to supply noise source and connected through a duct to the inlet of resonator model and a secondary duct was used to escape noise to ambient after attenuation. The simulation has been done for the four spacing provided in this analysis.

a) Very closely spaced :- $\left(< \frac{\lambda}{4}\right)$ equal to 125 mm
b) Closely spaced :- $\left(< \frac{\lambda}{4}\right)$ equal to 200 mm
c) 1st antinode :- $\left(\frac{\lambda}{4}\right)$ equal to 250 mm
d) 1st node :- $\left(\frac{\lambda}{2}\right)$ equal to 500 mm

These various positions give the maximum attenuation of noise by the use of two and three 1-DOF resonators.

Case I -

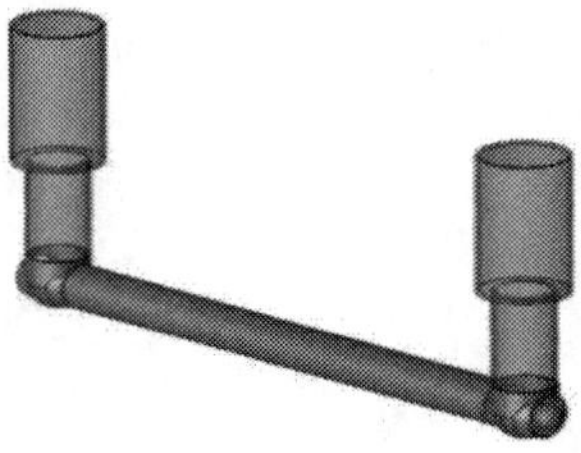

(a)

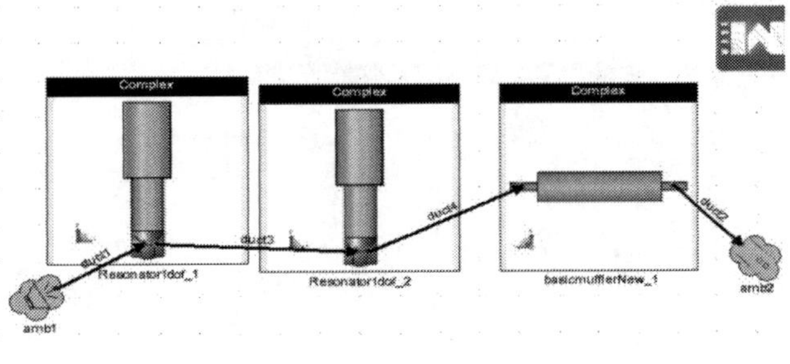

(b)

Fig. 10. Modelling of Two 1-DOF resonator by using Wave 1-D

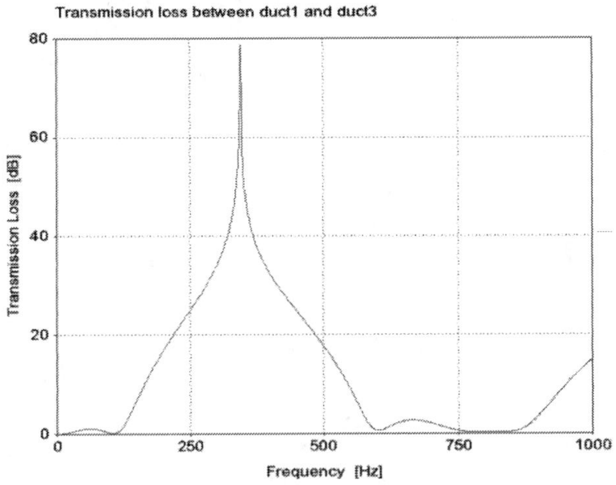

(a)

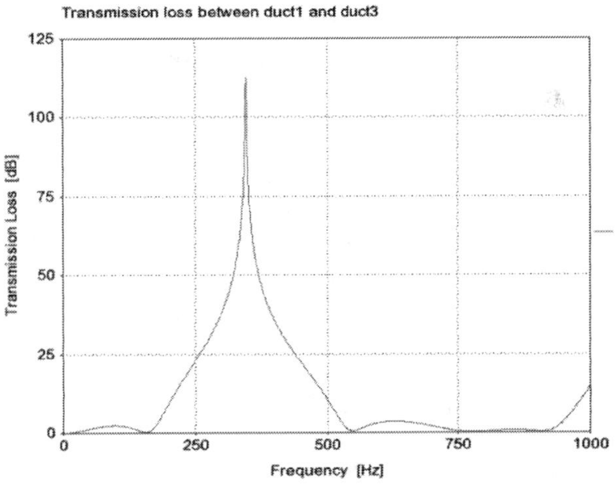

(b)

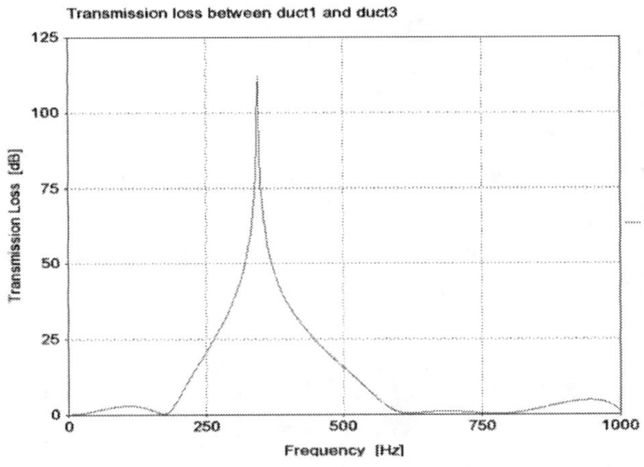

(c)

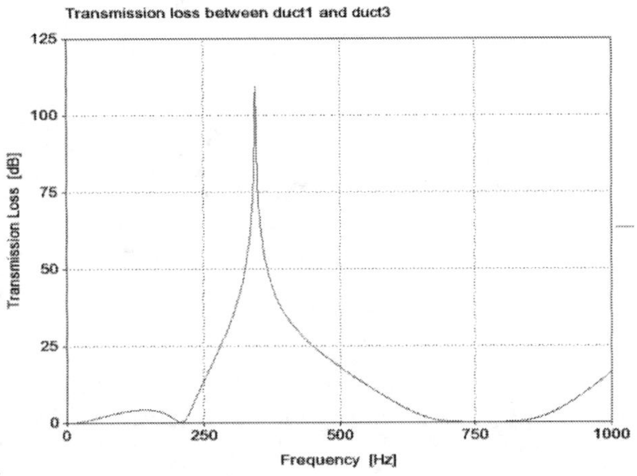

(d)

Fig. 11. Graph plotted for Transmission Loss Vs Frequency at various spacing
(a) 500 mm spaced, (b) 250 mm spaced, (c) 200 mm spaced and (d) 125 mm spaced

Case II :-

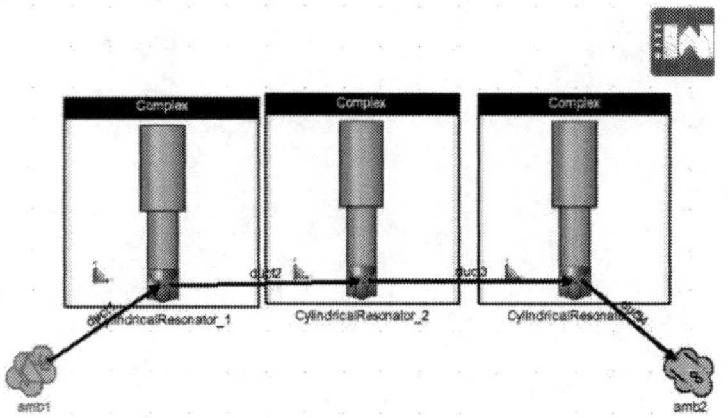

(a)

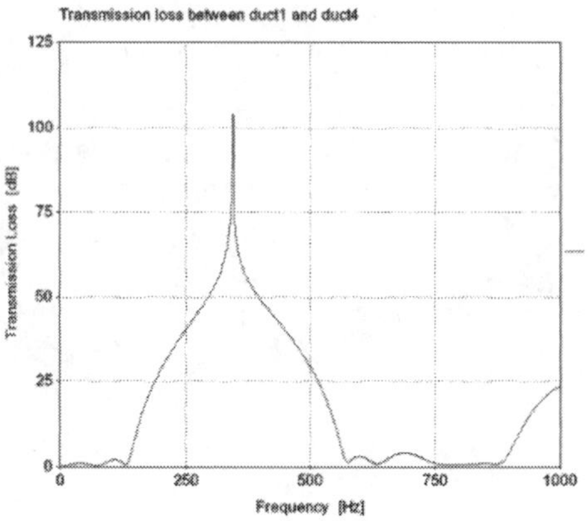

(b) Spacing = 500 mm

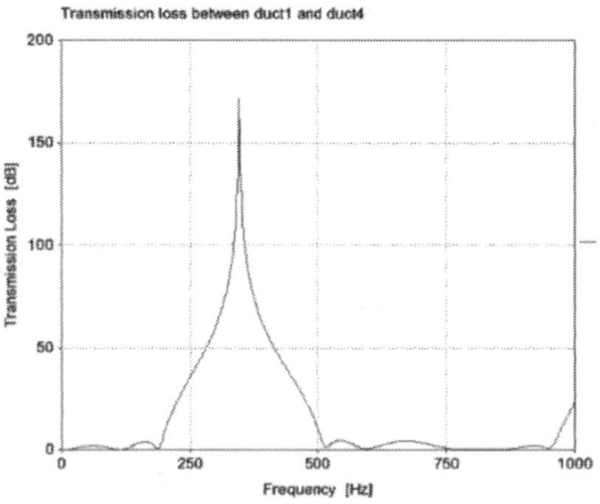

(c) Spacing = 250 mm

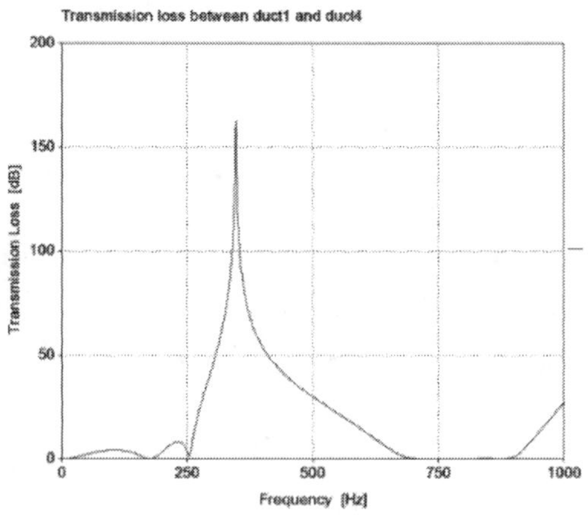

(d) Spacing = 100 mm

Fig. 12. Graph plotted for Transmission Loss Vs Frequency at various spacing
(a) Modelling Layout, (b) spacing = 500 mm (c) spacing = 250 mm (d) spacing = 100 mm

5 Result and Discussion

The simulation on two and three 1-DOF Helmholtz resonators was successfully carried out by using wave 1-D and also gave a satisfactory result as compared to the single resonator. The results show that the attenuation of noise can be increased up to maximum by providing 250 mm spacing between them which is equal to $\left(\frac{\lambda}{4}\right)$. This is validated by simulation result of same geometry and case on the COMSOL software.

This also validated by the result of quarter wave reactive muffler who can also attenuate noise at $\left(\frac{\lambda}{4}\right)$. The result also shows the use of no. of resonators for maximum transmission loss by using resonator tree aligned in series.

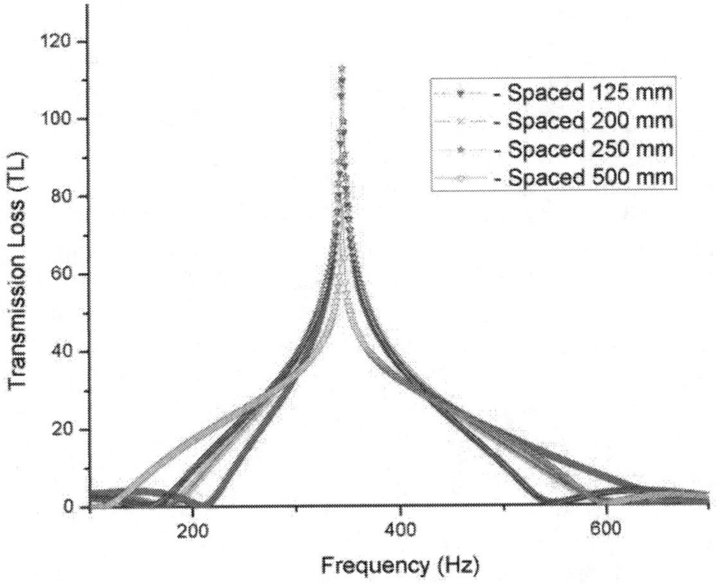

Fig. 13. Comparison of Transmission Loss between Two 1-DOF Helmholtz Resonators at Different Spacing

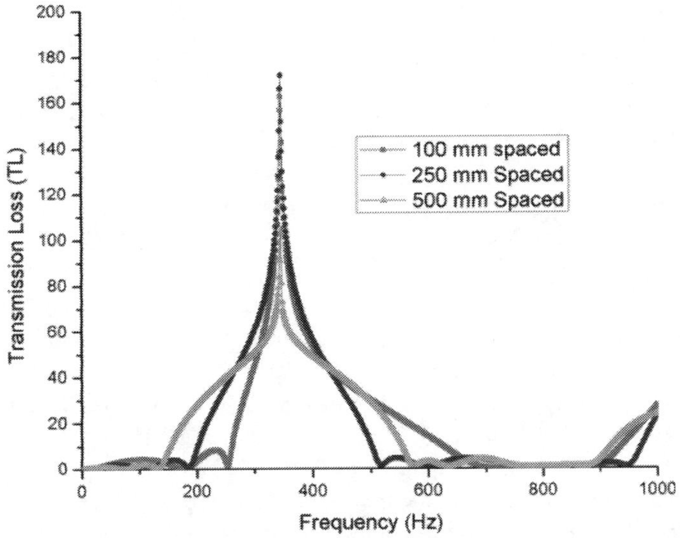

Fig. 14. Comparison of Transmission Loss between Two 1-DOF Helmholtz Resonators at Different Spacing

6 Conclusion

An optimized system for both the cases in terms of relative spacing has been investigated in this research work. Maximum noise attenuation of 117 dB for the case I and around 175 dB for the case II has been achieved comparative to a single 1-DOF resonator at cut- off frequency of 340 Hz. This result signifies the importance of this proposed procedure. The optimized system found from this research work is: Relative Spacing: ($\lambda/4$), and cavity ratio: 1.85 and neck ratio: 2.

References

[1] T. Yasuda, C. Wu, N. Nakagawa, and K. Nagamura, "Studies on an automobile muffler with the acoustic characteristic of low-pass filter and Helmholtz resonator," *Appl. Acoust.*, vol. 74, no. 1, pp. 49–57, 2013, doi: https://doi.org/10.1016/j.apacoust.2012.06.007.

[2] I. Deryabin and A. Krasnov, "Determination of the Characteristics of Noise-Reducing Materials for Motor Transport," *Transp. Res. Procedia*, vol. 68, pp. 630–635, 2023, doi: https://doi.org/10.1016/j.trpro.2023.02.086.

[3] A. Kashikar, R. Suryawanshi, N. Sonone, R. Thorat, and S. Savant, "Development of muffler design and its validation," *Appl. Acoust.*, vol. 180, p. 108132, 2021, doi: 10.1016/j.apacoust.2021.108132.

[4] D. Siano and D. D'Agostino, "Acoustic Optimization of an Intake System by Means of Geometry CAD Modifications," *Energy Procedia*, vol. 82, pp. 23–30, 2015, doi: https://doi.org/10.1016/j.egypro.2015.11.878.

[5] D. D. Zhu and Z. L. Ji, "Transmission loss prediction of reactive silencers using 3-D time-domain CFD approach and plane wave decomposition technique," *Appl. Acoust.*, vol. 112, pp. 25–31, 2016, doi: 10.1016/j.apacoust.2016.05.004.

[6] J. L. Parrondo, J. Fernández, I. García, and E. Ruiz, "Noise transmission through duct divisions in air circuits, considered as three-port acoustic systems," *J. Sound Vib.*, vol. 296, no. 1, pp. 183–194, 2006, doi: https://doi.org/10.1016/j.jsv.2006.02.015.

[7] S. Chivate, P. Hujare, R. Askhedkar, D. Hujare, and S. Chinchanikar, "A review on acoustic performance analysis of reactive muffler," *Mater. Today Proc.*, vol. 63, pp. 613–622, 2022, doi: https://doi.org/10.1016/j.matpr.2022.04.262.

[8] A. K. Gupta, Ravi Jatola, "Analysis of Noise Level with Convergent and Divergent Shape of Muffler and its Impact on Noise Pollution," *IJERR*, vol. 36, no. 3, pp. 58–65, 2023, doi: https://doi.org/10.52756/ijerr.2023.v36.005.

[9] A. K. Gupta, Ravi Jatola, "Enhancing Sound Transmission Loss in Hybrid Mufflers with Change in Pipe Perforation and Using Absorptive Material," *IJERR*, vol. 36, no. 3, pp. 178–184, 2023, doi: https://doi.org/10.52756/ijerr.2023.v36.017.

[10] U. Chhibber, R. Kumar, S. Haldar, and R. N. Hota, "Design and analysis of a compact acoustic filter for broad band noise absorption," *Appl. Acoust.*, vol. 140, no. May, pp. 30–38, 2018, doi: 10.1016/j.apacoust.2018.05.011.

[11] C. D. Gaonkar, D. R. Rao, K. M. Kumar, and M. L. Munjal, "End corrections for double-tuning of the same-end inlet-outlet muffler," *Appl. Acoust.*, vol. 159, p. 107116, 2020, doi: 10.1016/j.apacoust.2019.107116.

[12] J. Fu, W. Zheng, M. Xu, W. Wang, and Y. Huang, "Study on the influence of structure factors of diesel engine exhaust purification muffler on transmission loss in different frequency bands," *Appl. Acoust.*, vol. 180, p. 108147, 2021, doi: https://doi.org/10.1016/j.apacoust.2021.108147.

[13] K. S. Won and J. Choe, "Transmission loss analysis in a partitioned duct with porous boundaries based on combination of FEM and measurement of sound absorption coefficients," *Appl. Acoust.*, vol. 165, p. 107291, 2020, doi: 10.1016/j.apacoust.2020.107291.

[14] C.-M. Lee and Y. S. Wang, "A prediction method of the acoustical properties of multilayered noise control materials in standing wave-duct systems," *J. Sound Vib.*, vol. 298, no. 1, pp. 350–365, 2006, doi: https://doi.org/10.1016/j.jsv.2006.05.025.

[15] R. Barbieri and N. Barbieri, "Finite element acoustic simulation based shape optimization of a muffler," *Appl. Acoust.*, vol. 67, no. 4, pp. 346–357, 2006, doi: https://doi.org/10.1016/j.apacoust.2005.06.007.

[16] C. P. O. A. Guhan, G. Arthanareeswaran, K. N. Varadarajan, and S. Krishnan,

"Exhaust System Muffler Volume Optimization of Light Commercial Vehicle Using CFD Simulation," *Mater. Today Proc.*, vol. 5, no. 2, pp. 8471–8479, 2018, doi: 10.1016/j.matpr.2017.11.543.

[17] Y. Philipson and M. S. Lawless, "Application of a dual stage exhaust system using expansion chambers for formula SAE," *Proc. Meet. Acoust.*, vol. 42, no. 1, 2020, doi: 10.1121/2.0001369.

[18] A. J. Torregrosa, A. Gil, L. M. García-Cuevas, P. Quintero, and F. D. Denia, "Prediction of the transmission loss in a flexible chamber," *J. Fluids Struct.*, vol. 82, pp. 134–153, 2018, doi: 10.1016/j.jfluidstructs.2018.07.003.

[19] R. Guo and W. bo Tang, "Transfer matrix methods for sound attenuation in resonators with perforated intruding inlets," *Appl. Acoust.*, vol. 116, pp. 14–23, 2017, doi: 10.1016/j.apacoust.2016.09.012.

[20] H. Shao, H. He, Y. Chen, X. Tan, and G. Chen, "A tunable metamaterial muffler with a membrane structure based on Helmholtz cavities," *Appl. Acoust.*, vol. 157, p. 107022, 2020, doi: https://doi.org/10.1016/j.apacoust.2019.107022.

[21] C. Shen and L. Hou, "Comparison of various algorithms for improving acoustic attenuation performance and flow characteristic of reactive mufflers," *Appl. Acoust.*, vol. 116, pp. 291–296, 2017, doi: 10.1016/j.apacoust.2016.09.034.

[22] X. Yu, H. Fang, F. Cui, L. Cheng, and Z. Lu, "Origami-inspired foldable sound barrier designs," *J. Sound Vib.*, vol. 442, pp. 514–526, 2019, doi: https://doi.org/10.1016/j.jsv.2018.11.025.

[23] M. Mohammad, M. Farid, and M. Said, "Muffler Internal Geometry Parametric Study Related to Sound Transmission Loss and Pressure Drop," no. December, pp. 12–25, 2021.

Design and Modelling of Self standing Electric Scooter

Smitha B S[1] , *Mohammed Saaim* Sabeel[2*] , *Shaik Arbaz* Ahmed[3] *and Mohammed* Riza MS[4]
[1]Assistant Professor, Department of Automobile Engineering New Horizon College of Engineering, India
[2,3,4]Students, Department of Automobile Engineering New Horizon College of Engineering, India

Abstract : This work focuses on the compact design of a self-standing electric scooter, specifically engineered to require minimal parking space. The paper details the design and modelling of an innovative electric scooter that possesses the unique ability to stand on its own, utilizing a footprint smaller than that of a typical umbrella stand when parked. Employing a minimalist design approach, the scooter achieves an aesthetically appealing, simple, and futuristic appearance. The primary objective of this project is to propose a practical and compact electric scooter design by utilizing CATIA V5 software for modelling and ANSYS for required calculations and SolidWorks for visualization,A design that not only enhances efficiency for short-distance travels but also effectively mitigates issues related to traffic congestion and limited parking space.

Keywords : Self-Standing, CATIA V5 , Modelling, Electric scooter .

1 INTRODUCTION

An electric scooter is a battery-operated vehicle for carrying a single rider which is specially designed for people who have low mobility. Scooters are available in three common designs, those intended for indoor use, those for outdoor use, and those that are used for both. A typical electric scooter requires a pair of batteries, but the batteries are rechargeable. The length of time an electric scooter can run on each charge depends significantly on its battery type, rating and capacity.

Corresponding author: *saaimsabeel@gmail.com

The most common batteries are advertised to run for about eight hours, and for around 30-60 kms on a charge and totally depends upon battery capacity and the terrain in which it is being driven, before having the need to be charged again. The problem is, most of these scooters are not flexible although if they are already small. Even though some manufacturers make it in such a way that it can flip, but there are only few parts like seat, handle, and sometimes arm bar which can flip. Some of the scooters are not so ergonomic and cannot be used for a long time. Even for an electric scooter, most of them can't be flipped. Usually just their seat and handle can be put up and down to flip. Sometimes this will cause a lot of space for storage and is difficult to bring far from house or put it in a car. Majorly lack of availability of parking spaces in campuses hinder the adoption of these light mobility vehicles within campus.

So to overcome these problems and add convenience to commute within short distances, a design of electric scooter which gets folded vertically and can stand on its own without any support consuming very less parking space is modelled, it does it by virtue of its position of centre of gravity that lies between the vertical frame and the supporting footboard, battery is placed in footboard and a hub- mounted motor is at the front wheel which propels the scooter, these two components make us the major portion of weight of electric scooter, there is a clamp on the vertical frame which upon releasing aids the scooter to get folded, there are rubber grips on the front most portion of the electric scooter's footboard which comes in contact with ground upon folding and to balance the weight of scooter , a balancing mass is added to the vertical frame and the scooter is able to stand on its own.

2 METHODOLOGY

A literature review is done for the necessary considerations and insights, then the conceptual design is done by keeping dimensions in mind. Modelling of parts is done and assembled using CATIA V5 software , then the calculations are performed and finally the model is visualized using SolidWorks software.

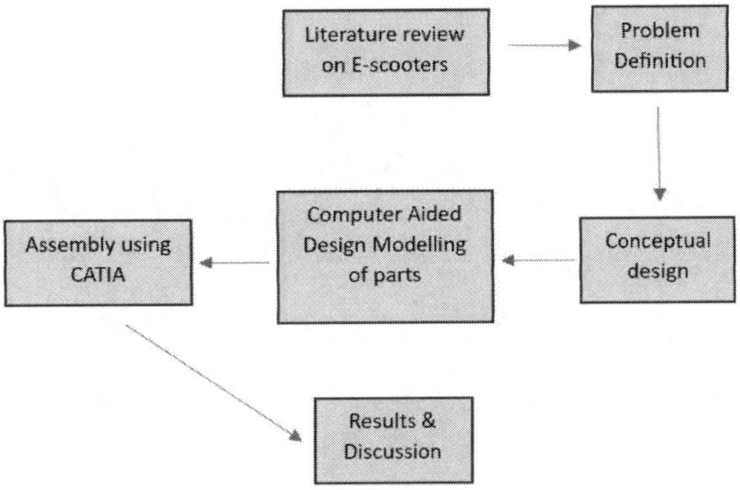

Fig. 1. Methodology

3 LITERATURE REVIEW

The Roboscooter is an urban, electric, folding scooter developed by the MIT Media Laboratory in conjunction with Sanyang Motors of Taiwan (SYM) and the Industrial Technology Research Institute (ITRI). Designed with mobility and portability in mind, the scooter must be both safe and light. The Roboscooter currently has several design characteristics that distinguish it from typical electric scoot- ers. The drive elements of the scooter are contained entirely in the wheel assembly, including brakes, suspension, and gearing. This configuration of the drive system is commonly referred to as a wheel- robot. The advantage of the wheel-robot configuration is an increase in simplicity and modularity of the scooter's drive system [1]. Frame design is done according to an Average Human Being. As per space availability and clearance for reduction of friction, frame size is(1090*410)mm. They considered a Human being of about 6feet tall and having weight of 70kgs max[2].Dimension all are referred from Currie Folding Scooter EZIPE400 [3].

4 DESIGN &MODELLING

4.1 Conceptual Design

The rubber padding at the forward-most point of footboard assists in getting grip when standing and the weight of scooter is balanced in such a manner that it can stand on its own without any external support while parked, it is due to the fact that motor is placed at front wheel, and battery is at foot board and remaining balancing mass is added within the vertical frame to enable proper balancing and bring centre of mass at the centre while being parked. Folding is achieved by releasing a clamp that locks up the scooter in respective parking condition.

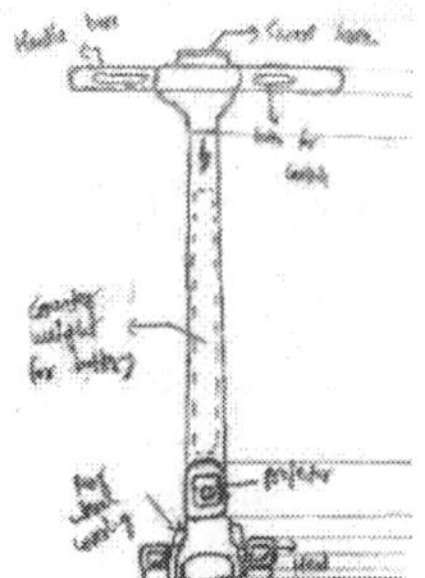

Fig. 2. Front view of Design

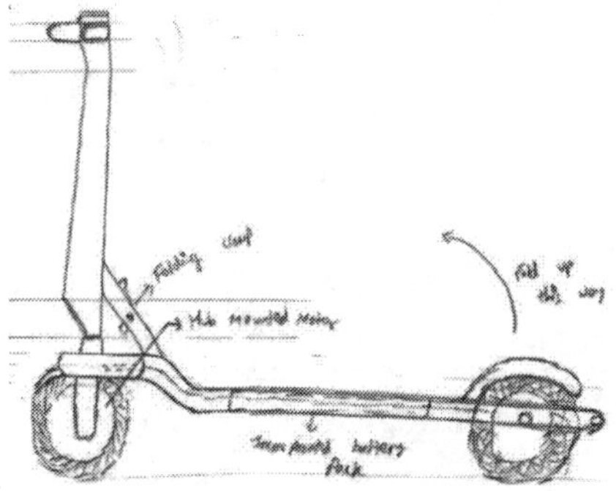

Fig. 3. Side view of Design

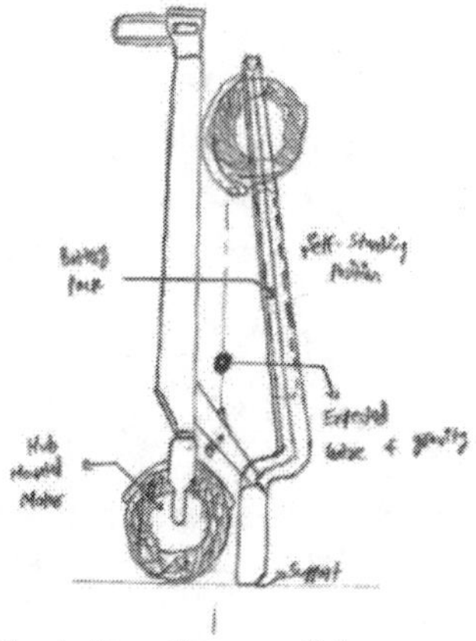

Fig.4. Standing condition

4.2 Dimensions of Scooter

https://doi.org/10.1051/matecconf/202439302008

The final drafted dimensional model is as shown below.

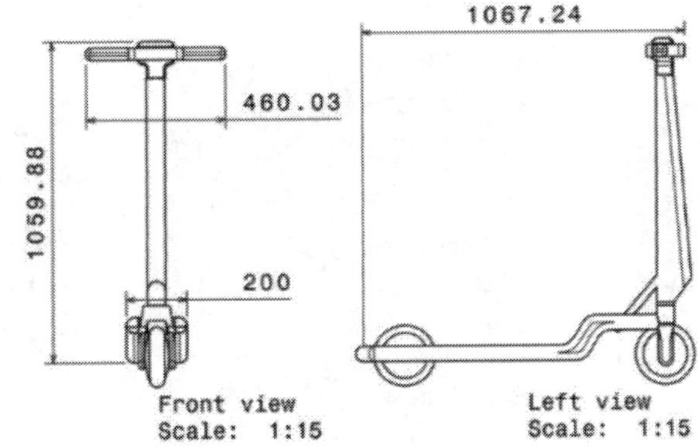

Fig. 5. Dimensions of scooter

4.3 Modelling of Parts

The major parts are the footboard and vertical frame, their dimensions are as shown below which are modelled in CATIA.

i Vertical frame: Vertical frame is of 750mm height which is considered to be best suited for any rider height , it has clamps that enables folding and connects to holding as well as handle.

ii Foot Board: Foot board is of 200mm width which enables the rider to place his legs without hinderance and it is designed to fit in with vertical frame and it the one where the rider stands and rides the scooter , upon folding it supports the scooter to stand.

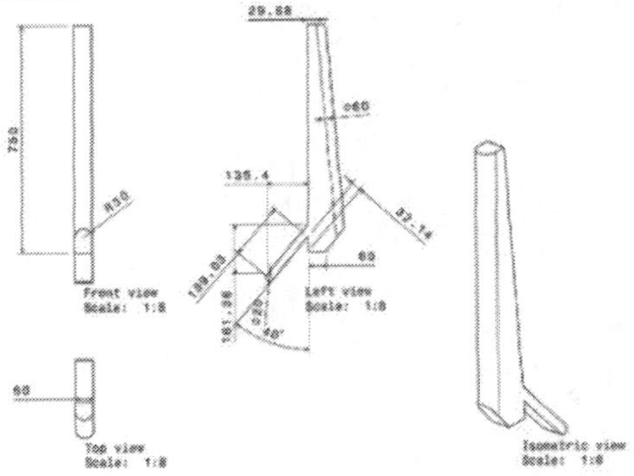

Fig. 6. Vertical Frame

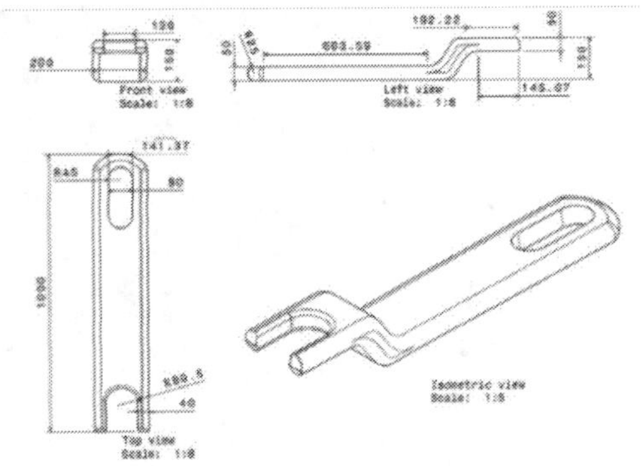

Fig. 7. Dimensions of scooter

4.4 Modelling in Various Conditions

Riding condition and the standing condition are major working conditions along with a intermediate transition condition which is folding

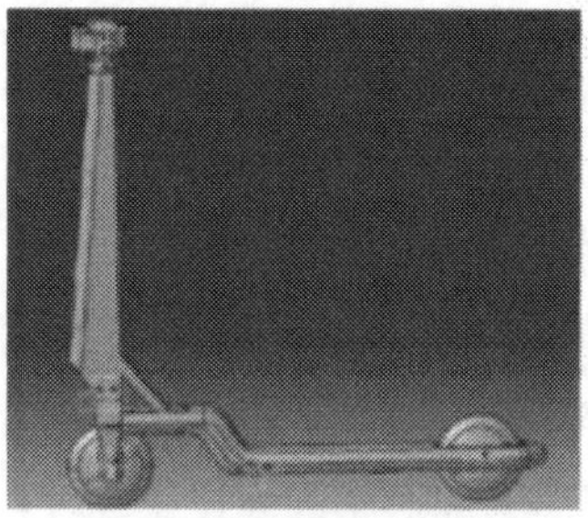

Fig. 8. Riding Condition

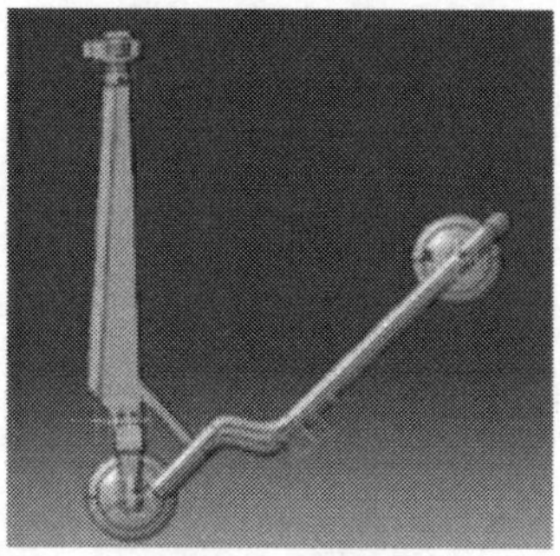

Fig. 9. Folding Condition

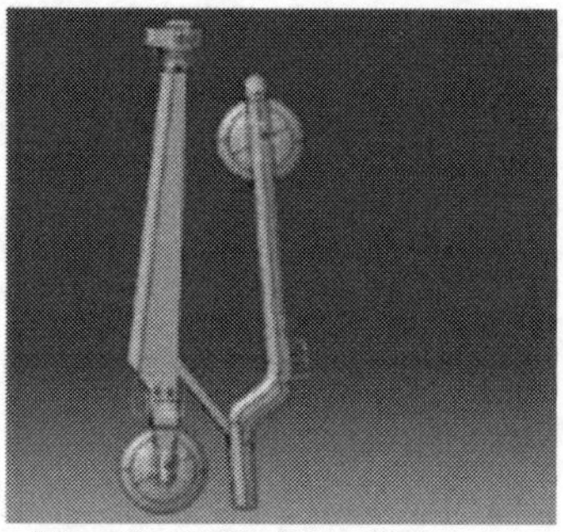

Fig. 10. Parked Condition

4.5 CG for Standing

Riding To balance the scooter in its parked condition , the centre of gravity must lie at the contact of the contact surface or supports with the ground upon which the gravity is supposed to be acting. 'Ansys' software is used to determine the centre of gravity by importing the model to ansys and applying the gravitational force upon it which is acting from the point of centre of gravity .It is highlighted by yellow color as we could see that the distance between point of contact on ground is approximately 200mm and the CG lies on 100mm distance from either point of contact with ground under parking condition.

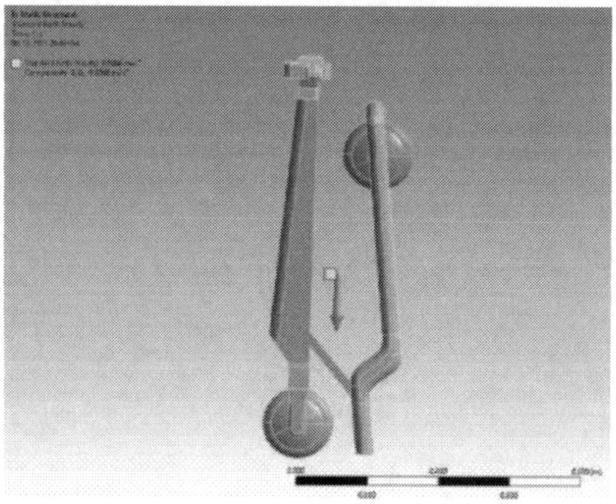

Fig. 11. Center of Gravity of scooter

4.6 Calculations

The necessary calculations for the electric scooter to propel is as discussed below which primarily focuses on Motor and battery.

4.6.1 Motor Specifications:

Max Motor power = 700WNominal Motor power = 350W

4.6.2 Motor Calculations:

*Power = Torque*2*3.14*N/60…..(1)* Where, rpm of motor shaft N= 500 rpm

Hence, Torque at motor shaft= 13.37 Nm

4.6.3 Battery Specifications:

Amp-h Rating = 12.8A-h (460 Whr)Voltage Rating = 36V

Battery Calculations:

Voltage Supplied, V =36VPower of motor, P= 350WCapacity of battery =12.8Ah

Current drawn from battery, I = P/V……..(2) =350/36 = 9.7A

Li-Ion battery generally drains at 60%

So, effective *capacity of battery = 0.6*12.8*Ah = 7.68Ah

Now runtime, t =effective capacity of battery/current drawn from battery Runtime, t= 0.7917hr

Therefore, battery can run for 0.79 hr with these specifications under given conditions.

4.6.4 Weight Calculations :

Considering material to be 7075 Aluminium alloy with density 2810 kg/m^3

Weight=2810*751*10^{-6} =2.11Kg

Vertical Frame=(750*30*70)-(750*22*62)= 552*$10^{-6}m^3$, considering material to be 7075 Al alloy

Weight=2810*552*10^{-6}=1.55 Kg

Weight of Wheel &Tyres=1.2 Kg

Weight of Battery= 1 kg (Between 380 Whr - 460 Whr)

Weight of Motor = 2.98 Kg (for 350W Motor)

Weight of other Auxiliary systems = 1 Kg

Total weight = 2.98+1+1.2+1.55+2.11+1=9.84 Kg

5 RESULT & CONCLUSION

Riding To balance the scooter in its parked condition , the centre of gravity must lie at the contact of the contact surface or supports with the ground upon which the gravity is supposed to be acting. 'Ansys' software is used to determine the centre of gravity by importing the model to ansys and applying the gravitational force upon it which is acting from the point of centre of gravity .It is highlighted by yellow color as we could see that the distance between point of contact on ground is approximately 200mm and the CG lies on 100mm distance from either point of contact with ground under parking condition.

Table. 1. Results from Calculations

Motor Power Torque	350 W 13.37 Nm
Battery Capacity Voltage Runtime	460 Whr 36 V 0.79 hr
Weight Total	9.84 Kg

Fig. 12. Isometric View-1

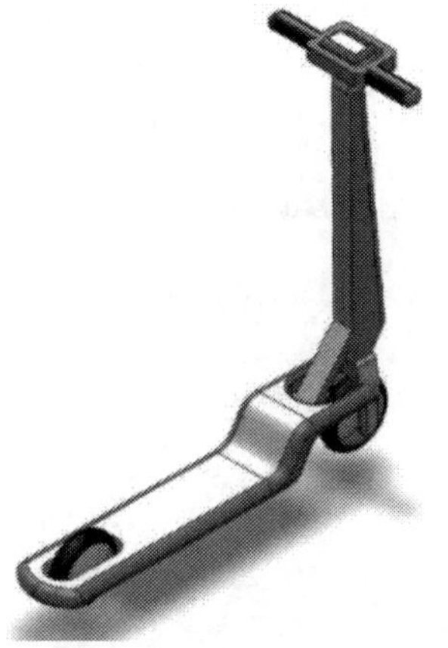

Fig. 13. Isometric View-2

An eco friendly and compact electric scooter is designed and modelled which can stand on its consuming very less parking space .It could change the way of transportation and make it ease for short commutes such as for offices as well as colleges. A feasible design of a simple scooter which is totally unique and the concept of being folded and standing in its vertical position is something out of the box way of parking which is completely new and conceived in this report. Project was primarily designed for enhancing green mobility thus it will also help to control the pollution which is one of the major crises now a days. The unique idea of folding that enables the scooter to stand on its own, supported by its base frame is a whole new concept of design. This scooter has only two wheels, front wheel has hub mounted motor that propels the scooter and is its power train, the scooter looks modern and futuristic .Minimalist approach is used in design so it looks more appealing and simple. The incorporated design is drafted using CATIA V5 and shown isometrically. Solidworks software is utilized for visualization of the scooter and make it look more attractive. This concept will bring easy & greener mobility in a compact package which could be deployed in college and office campuses.

REFERENCES

[1] Curtis Herman , Currie Folding Scooter EZIPE400 , student edition , curie technologies,Page-4,2020

[2] Kamlesh Sasane, Abhijeet Sharma, Design & Development of electric scooter ,2019

[3] Anshul Singh ,Bhupesh Kumar, Anoop Mohan , Design & Fabrication of electric scooter,Page-42,2018

[4] Arthur J Petron , Structural Analysis of folding RoboScooter,Page-3,2008

[5] Luqman Hakeim, Azzammuddin, Arif, KOUGAR SCOOTER ,UniversitiTeknolgimara,2018

[6] H.J. Bergveld, Battery management systems design by modelling , 2018

Steady state design and thermal characteristics of a gas turbine blade with lateral holes using finite element analysis

Deepak Sampathkumar[1,3*], *Sam Daniel Fenny Azariah*[1], *Ashokkumar Mohankumar*[2], *and Mathanbabu Mariappan*[3]
[1] Department of Mechanical and Automation Engineering, Agni College of Technology, Chennai, Tamilnadu 600 130, Bharat (India)
[2]Centre for Additive Manufacturing, Chennai Institute of Technology, Malayambakkam, Tamil Nadu 600069, Bharat (India)
[3]Department of Mechanical Engineering, Government College of Engineering, Bargur, Tamilnadu 635104, Bharat (India)

Abstract. Turbine blades are the separate components that make up the turbine section of a steam or gas turbine. The blades must transform the high-temperature, high-pressure gas that the combustor produces into energy. In gas turbines, the turbine blades are usually the limiting element. As gas temperatures rise, the amount of heat that reaches the blades will also grow dramatically, leading to their thermal failure. Because gas turbine blades must function in extremely high temperatures, cooling is crucial. After FEA simulation was conducted and radial perforations of 6, 9, and 12 were constructed to allow high-velocity cooling air to pass through, the cooling of the gas turbine blades was improved.

Keywords: Gas Turbine Blade, Creo, Ansys, Thermal Analysis, FEA, and Lateral Holes

1. Introduction

The use of gas turbine technology as a primary mover has increased significantly over the previous several decades. Although they have long been used for aviation propulsion, gas turbines have one of their most significant uses in the power generation industry. The highest temperatures reached during the cycle affect gas turbine plant efficiency and power production [1-3].
Figure 1(a) shows high temperatures (1200–15,000°C) are used by modern gas turbine engines to increase their thermal efficiency and power production. The heat delivered to the blades will significantly increase as gas temperatures rise, which will cause them to thermally fail. Higher temperatures are not attainable with the available materials. Figure 1(b) there are many types of blade cooling is crucial. One of several methods reported for cooling blades is the passage of high-velocity cooling air along the blade span through radial perforations. The temperature of the blade will be brought within acceptable limits by forced convection heat transfer from the blade to the cooling air.

*Corresponding author: naamphd@gmail.com

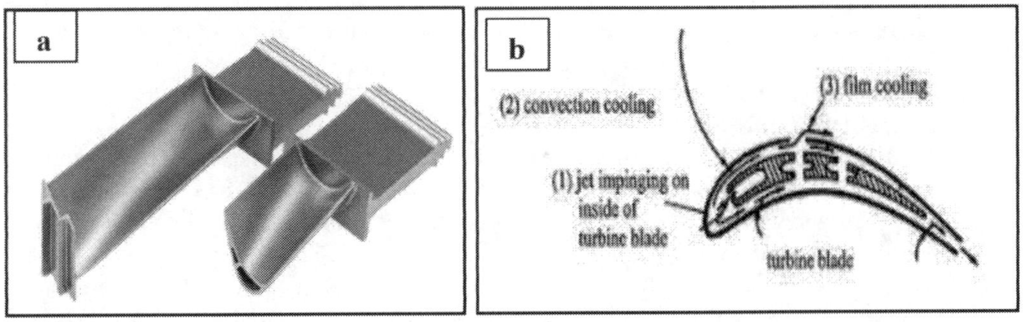

Fig. 1. (a) RB199jet Engine Turbo-Union Turbine Blade and (b) Types of Cooling.

1.1. Types of Cooling& Gas Turbine Cycle

By forcing heat transfer from the blade to the cooling air by convection, the temperature of the blade will be lowered to a level that is acceptable [4-6]. Impinging cooling or Transpiration cooling, convection cooling and film cooling. Film cooling is the second kind of cooling. Through tiny perforations in the blade, cold air is pumped out using this cooling method. As a result, the blade is shielded from the high-temperature gasses by a thin layer of cold air created by this air.

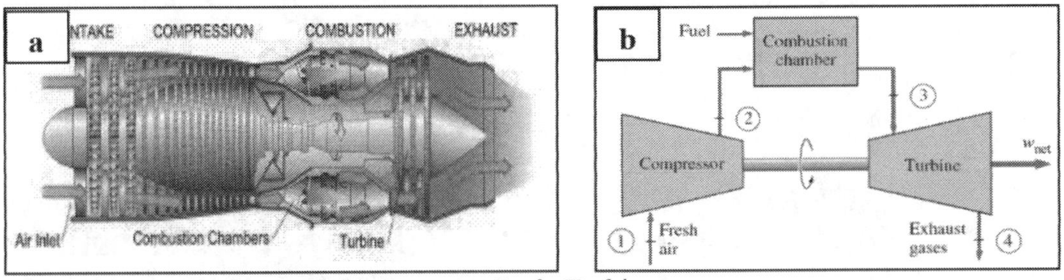

Fig. 2 (a). Gas Turbine and (b) Open gas Cycle Turbine

Figure 2(a) Film cooling is the second kind of cooling. Through tiny perforations in the blade, cold air is pumped out using this cooling method. As a result, the blade is shielded from the high-temperature gasses by a thin layer of cold air created by this air. Figure 2(b) An internal combustion engine that uses rotational motion as opposed to reciprocating motion is referred to as a gas turbine. The compressor, combustor, and power turbine make up the three primary parts of a gas turbine.

A gas turbine is an internal combustion engine that operates by rotation rather than by reciprocation [7]. The three main components of a gas turbine are the compressor, combustor, and power turbine.

A thin layer of cooling air is produced on the blade via transpiration cooling, the third main cooling technique. The air is directed via a porous shell as opposed to being injected through holes, which is how it differs from film cooling.

The goal of the current work is to investigate how variations in the number of cooling passes affect the highest temperatures reached and the thermal stresses generated. It must be emphasized that there are two sorts of stresses under which the blade must be examined. High temperature inside the blade material causes thermal stresses. Thermal analysis is the only type of analysis used for turbine blades. ANSYS software is used to do the analysis in steady-state settings. The titanium T6 alloy was used to perform the investigation.

2. Theoretical Background

2.1 Principle and Fundamental of Turbine Engines

An internal combustion engine that runs on air is known as a gas turbine. The engine transforms chemical energy from fuel into mechanical energy, which in turn powers the propeller and engine. This process powers the aircraft. The engine receives its power from the working fluid, air.

All engines that derive their power from chemical fuel operate on the same fundamental principles as the turbine engine used in airplanes. The fundamental four processes for any internal combustion engine are: Air intake (and possibly fuel intake), Air (and perhaps fuel) compression, Combustion, in which fuel is injected and burned to release the stored energy (Providing it wasn't pulled in by the intake air) and Expansion and exhaust [8-9].

Figure 3 illustrates the concept of diverging ducts by showing how gaseous energy is transferred from velocity to pressure and temperature rather than added to or subtracted. Air's velocity drops as it travels from a small entrance to a larger output. Impact pressure (Pi) decreases as velocity decreases [10].

Since there is no energy added to or removed from the system, static pressure (Ps) increases while total pressure (Pt) for the air stays constant. One interpretation of this is that the impact pressure is transformed into static pressure; as a result, air is compressed and forced into a diverging duct, causing a rise in static pressure. Given that compression is a heating process, a temperature increase is also seen [11].

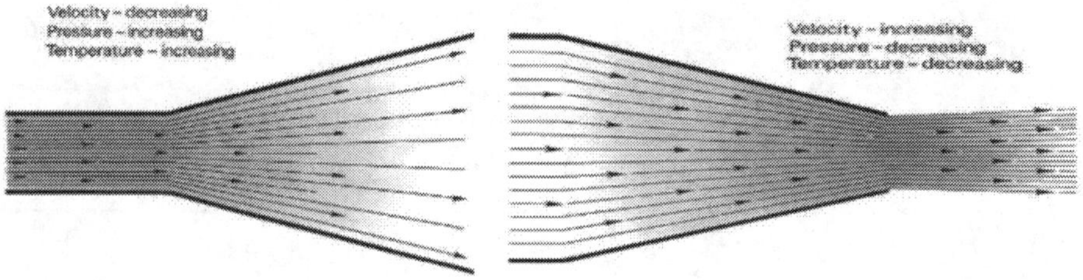

Fig. 3. Convergent Divergent Ducts

2.2 Turbine

This example engine's turbine has four stages. The compressor, driven accessories, and propeller are all powered by the turbine, which transforms the gaseous energy of the air/burned fuel mixture that exits the combustor into mechanical energy via a reduction gear [12 – 15]. The hot, high-pressure gasses are expanded by the turbine to a lower temperature and pressure, which converts the gaseous energy into mechanical energy [16 – 21].

The efficiency of the turbine depends on how successfully the hot, flowing gasses transfer mechanical energy to it. Given that air goes from a high-pressure zone to a low-pressure zone, this task might be finished really rapidly. Appropriately placed airfoils allow gases to flow and expand through the turbine's blades and vanes without interruption [22-24].

3. Material and Methods

The current study uses finite element analysis to investigate the steady-state thermal properties of titanium T6. In this project, two distinct models made up of solid blades with and without holes were examined to determine the ideal quantity of cooling. Ansys CFD software is used to do the analysis.

Titanium T6 Alloy Grade 5: The most widely utilized grade of titanium is grade 5. Ti 6Al-4V is the popular name for Titanium Grade 5, an alloy of titanium with 6% aluminum and 4% vanadium. The majority of titanium used worldwide, or grade 5, is Ti 6Al-4V. To improve the mechanical qualities, heat treatment is an option. At service temperatures as high as 315 °C, it can be employed in welding fabrication. This alloy offers good formability, strong corrosion resistance, and high strength at a low weight.

Figure 4 shows the gas turbine blades were modeled using CREO software using the design consideration with 6, 9 and 12 holes on the blade. To produce more accurate models more quickly, FEA simulations were done in Ansys. Workbench platform. The design considerations are given in the table 1.

Fig. 4. Modeling of Turbine blade with 6, 9& 12 holes

Table 1. Details of Gas turbine Blades

Parameter	Values	Unit
Blade height, h1	200	mm
Blade width, b	150	mm
Blade length, l	500	mm
Disk height,h2	50	mm
Hole diameter, d	10	mm
No. of Blades	110	-
Blade inlet angle	18.3°	Degree
Blade outlet angle	54.56°	Degree
Pitch, s	0.02264	mm

3.1 Analysis of Gas Turbine Blades

Using the Ansys Workbench platform, you can mix data from engineering simulations to create more accurate models faster. By centrally organizing all of your simulation data, Ansys Workbench facilitates the process of generating better-informed design decisions. It is easy to manage your data across all of your Ansys products. Automated data transfer can save time. Make models with more realism. The results were taken for temperature distribution heat flux, total deformation, equivalent elastic strain and stress distribution in the gas turbine blades with 6,9,12 holes. The Results are shown in the figure 5.

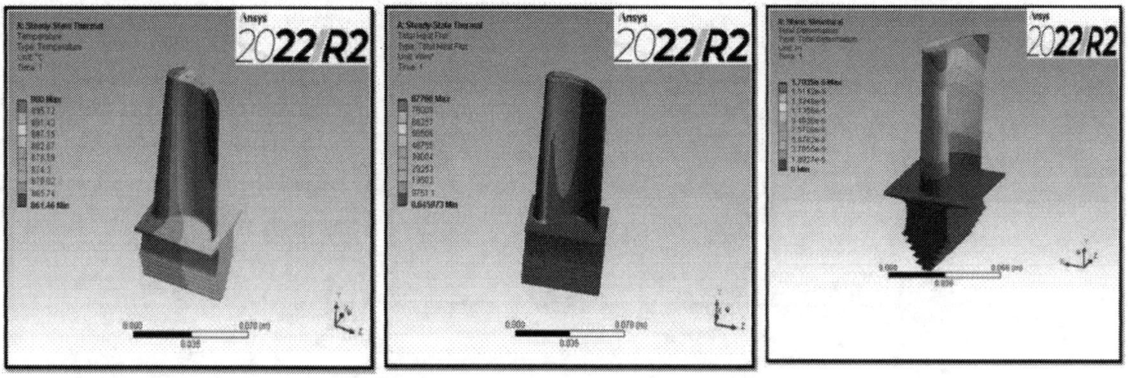

Fig. 5. FEA results for the gas turbine blades showing deformation, temperature distribution and stress

4. Results and Discussion

The obtained results are detail explained in the following:

Table 2. Temperature Distribution: Max 900 ^{o}C

Material	Temperature (6 Holes)	Temperature (9 Holes)	Temperature (12 Holes)
Titanium T6	856.98 ^{o}C	840.29 ^{o}C	788.8 ^{o}C

Table 3.Structural Analysis

Material	Temperature distribution (^{o}C)	Total deformation (mm)	Stress (pa)
Titanium T6	788.8	1.7035 X10^{-5}	2.3128 X10^{7}

Table 2 and 3 shows Titanium T6 with 12 holes is found to be more suitable for high temperature applications when compared to titanium materials with 6, 9, and 12 holes. The bladewith 12 holes is considered ideal after considering the deflection, von-Mises stress, and temperature distribution contours.

Based on the fact that the generated stresses are at their lowest level and the blade's temperature is over 800 °C. Degrees, this conclusion was reached. Any more holes will cause the temperature to fall below the necessary level of 800 °C.

Conclusion

After FEA simulation was conducted and radial perforations of 6, 9, and 12 were constructed to allow high-velocity cooling air to pass through, the cooling of the gas turbine blades was improved. This recently created titanium alloy can withstand greater operating temperatures in the gas turbine compressor to reduce fuel consumption and, therefore, pollutant emissions of aviation engines. This alloy offers good formability, strong corrosion resistance, and high strength at a low weight.

References

[1] Olumayegun O, Wang M, Kelsall G. Closed-cycle gas turbine for power generation: A state-of-the-art review. Fuel. **180**, 694-717 (2016)

[2] Farhat H, Salvini C. Novel Gas Turbine Challenges to Support the Clean Energy Transition. Energies. **15**(15):5474 (2022)

[3] Vaferi K, Nekahi S, Vajdi M, Moghanlou FS, Shokouhimehr M, Motallebzadeh A, Sha J, Asl MS. Heat transfer, thermal stress and failure analyses in a TiB2 gas turbine stator blade. Ceramics International. **45**(15):19331-19339 (2019)

[4] Nekahi S, Vaferi K, Vajdi M, Moghanlou FS, Asl MS, Shokouhimehr M. A numerical approach to the heat transfer and thermal stress in a gas turbine stator blade made of HfB2. Ceramics International. **45**(18):24060-24069 (2019)

[5] Vaferi K, Vajdi M, Nekahi S, Nekahi S, Moghanlou FS, Asl MS, Mohammadi M. Thermo-mechanical simulation of ultrahigh temperature ceramic composites as alternative materials for gas turbine stator blades. Ceramics International. **47**(1):567-80 (2021)

[6] Raj, E.F.I., Appadurai, M., Ram, V., Gnaniah, A.M. and Salkuti, S.R., Natural-Fibre-Reinforced Composite-Based Micro-Size Wind Turbines: Numerical Analysis and Feasibility Study. Journal of Composites Science, **7**(5), p.197. (2023)

[7] Singh HP, Rawat A, Manral AR, Kumar P. Computational analysis of a gas turbine blade with different materials. Materials Today: Proceedings. **44**:63-9 (2021)

[8] Abid A, Sarowar MT. Heat Transfer, Thermal Stress and Failure Inspection of a Gas Turbine Compressor Stator Blade Made of Five Different Conventional Superalloys and Ultra-High-Temperature Ceramic Material: A Direct Numerical Investigation. Journal of Failure Analysis and Prevention. **22**(3):878-98 (2022)

[9] Krishnakanth PV, Raju GN, Prasad RD, Saisrinu R. Structural and thermal analysis of gas turbine blade by using FEM. International Journal of Scientific Research Engineering and Technology. **2**(2):60-5 (2013)

[10] Zhang B, Shirvani K, Taheri M, Beirami K, Wang Y. Effect of TiC and Magnetic Field on Microstructure and Mechanical Properties of IN738 Superalloy Processed by Selective Laser Melting. Journal of Materials Engineering and Performance. **4**:1-6 (2023)

[11] Ashokkumar M, Thirumalaikumarasamy D, Sonar T, Deepak S, Vignesh P, Anbarasu M. An overview of cold spray coating in additive manufacturing, component repairing and other engineering applications. Journal of the Mechanical Behavior of Materials. **31**(1):514-34 (2022)

[12] Gowreesh.R, N.Sreenivasalu Reddy and N.V.Yogananda Murthy. "Convective Heat Transfer Analysis Of An Aero Gas Turbine Blade Using Ansys", International journal of Mechanics of solids. **4** (1),55-62 (2009)

[13] Mohankumar A, Duraisamy T, Sampathkumar D, Sonar T. Electrochemical corrosion performance of friction stir processed cold spray metal matrix composite coatings on AZ31B magnesium alloy under sodium chloride environment. Surface Topography: Metrology and Properties. **10**(3):035010 (2022)

[14] Kauthalkar P, Shikarwar MD, Sharma DP. Analysis of thermal streses distribution pattern on gas turbine blade using ansys. International journal of Engineering Education and technology. **2**(3), (2010)

[15] Chowdhury TS, Mohsin FT, Tonni MM, Mita MN, Ehsan MM. A Critical Review on Gas Turbine Cooling Performance and Failure Analysis of Turbine Blades. International Journal of Thermofluids. **9**:100329 (2023)

[16] Mohankumar A, Duraisamy T, Packkirisamy V, Sampathkumar D. Response Surface Methodology and Mayfly Optimization for Predicting the Properties of Cold-Sprayed AA2024/Al2O3 Coatings on AZ31B Magnesium Alloy. Journal of Materials Engineering and Performance. **2**:1-9 (2023)

[17] Ashokkumar M, Thirumalaikumarasamy D, Sonar T, Vignesh P, Deepak S. Optimization of cold spray coating parameters using RSM for reducing the porosity level of AA2024/Al2O3 coating on AZ31B magnesium alloy. International Journal on Interactive Design and Manufacturing (IJIDeM). **15**:1-5 (2023)

[18] Ashokkumar M, Thirumalaikumarasamy D, Sonar T, Ivanov M, Deepak S, Rajangam P, Barathiraja R. Effect of post-processing treatments on mechanical performance of cold spray coating–an overview. Journal of the Mechanical Behavior of Materials. **32**(1):20220271 (2023)

[19] Mohankumar A, Duraisamy T, Sampathkumar D, Ranganathan S, Balachandran G, Kaliyamoorthy M, Mariappan M, Mulugeta L. Optimization of cold spray process inputs to minimize porosity and maximize hardness of metal matrix composite coatings on AZ31B magnesium alloy. Journal of Nanomaterials. **30** (2022)

[20] Mohankumar A, Duraisamy T, Chidambaramseshadri R, Pattabi T, Ranganathan S, Kaliyamoorthy M, Balachandran G, Sampathkumar D, Rajendran PR. Enhancing the corrosion resistance of low pressure cold sprayed metal matrix composite coatings on AZ31B Mg alloy through friction stir processing. Coatings. **12**(2):135 (2022)

[21] SampathKumar D, Duraisamy T, Pattabi T, Mohankumar A. Static puncture resistance characteristics with various indenter nose shape geometry perforation of shear thickening fluid impregnated polypropylene fabric for soft armour application. Polymers and Polymer Composites. **30**, 09673911211063303 (2022)

[22] Seshaiah S, Sampathkumar D, Mariappan M, Mohankumar A, Balachandran G, Kaliyamoorthy M, Rajendran B, Gopal R. Optimization on material removal rate and surface roughness of stainless steel 304 wire cut EDM by response surface methodology. Advances in Materials Science and Engineering. **29** (2022)

[23] Mathivanan K, Thirumalaikumarasamy D, Ashokkumar M, Deepak S, Mathanbabu M. Optimization and prediction of AZ91D stellite-6 coated magnesium alloy using Box Behnken design and hybrid deep belief network. Journal of Materials Research and Technology. **15**:2953-69 (2021)

[24] AmrutaPasarkar, S.Balaguru, 2022 'Development on Fatigue Testing of Hardfaced Components in Sodium Cooled Fast Reactors', Engineering Failure Analysis, **137**, 106161 (2022)

CFD analysis on the aerodynamic coefficients of an ejection seat system at supersonic speed

Md. Mahbubur Rahman[1*], *Sunil Chandel*[1], and *D. G. Thakur*[1]

[1]Department of Mechanical Engineering, Defence Institute of Advanced Technology (DU), Girinagar, Pune - 411025, India

Abstract. In the present study, a 3-D study of the aerodynamic coefficients of an ejection seat system is performed. The analysis is performed using the Reynolds Averaged Navier-Stokes equations where an unstructured grid of the polyhedral cells is used. The aerodynamic coefficients are calculated using a density-based solver and the standard k-ε turbulence model. The investigation is performed at Mach number (Ma) = 1.25 by varying the angle of attack (α) from -15^0 to 15^0 with the increment of 5^0 where the yaw angle (β) is fixed at 0^0. According to the findings, the magnitude of the axial force coefficient (Cx) increases as α decreases whereas the value of the normal force coefficient (Cz) decreases with the increase of α. Similarly, the value of the pitching moment coefficient (C_m) decreases with the increase of α.

1 Introduction

A military fighter aircraft is made up of several different sorts of equipment, one of which is the ejection seat system. This system is composed of different small components. The main objective of the system is to save the pilot's life when required. With time the ejection seat system has developed a lot. After World War II the modern aircraft was manufactured with the developed ejection seat system that protects the pilot's life from any catastrophic failure.

During the ejection of the pilot along with the ejection seat from the cockpit, the pilot is suddenly exposed to an environment where the free stream velocity of the air can be supersonic. During the ejection procedure, the pressure forces exerted on the pilot's body change as the ejection seat system rotates. The rotation of the ejection seat system is unpredictable during the process. Hence, during the analysis of an ejection seat system, many orientations of the ejection seat are to be analyzed. Initially through experimental methods [1], the ejection seat system was analyzed. However, the development of the CFD which is used in various applications [2-4] of the fluid dynamics makes it possible to analyze the ejection seat system with less time and cost.

Habchi et al. [5] performed a steady-state analysis of a 3-D ejection seat system at Ma = 0.6 and 0.9 by varying α. Initially, they analyzed a 2-D ejection seat system and later on analyzed a 3-D ejection seat system where the effect of the Ma and the effect of the variation of the α and β was analyzed. The effect of a yaw fin along with the ejection seat system was further measured by changing β. Later in 1995, Habchi and Hufford [6] studied an unsteady

* Corresponding author: mahbub.rizvee@gmail.com

state analysis on the ejection seat system at Ma = 0.6. Initially, a classical blunt body (cylinder) was studied for different grids. After that, a 2-D ejection seat system was analyzed before analyzing a 3-D ejection seat system.

A numerical and experimental investigation of the ejection seat was studied by Chen et al. [7] at Ma = 0.6, 0.9, and 1.2. They observed the aerodynamic coefficients by varying α and β. From their study, they concluded that the numerical and the experimental results match pretty well. Zhu et al. [8] observed the detachment of the ejection seat system from the cockpit using an overset chimera grid in CFD-FASTRAN. This analysis was performed after the validation at Ma = 0.6 which was performed by varying α and β. The importance of using a protective device along the ejection seat system was analyzed at Ma = 0.9 by Tian and Chen [9]. An unstructured grid of tetrahedron cells was solved using the Spalart-Allmaras turbulent model to analyze the aerodynamic coefficients in ANSYS-Fluent. They observed that a significant amount of axial force was reduced after the installation of a protective device with the ejection seat system.

In 2012 Zhu et al. [10] analyzed the canopy separation from the aircraft using CFD-FASTRAN. Later on, they performed an unsteady state analysis on the ejection seat system using a standard wall function to observe the detachment of the TY-5 ejection seat from the aircraft at Ma = 1.2. At Ma = 0.9, an unscaled ACES II ejection seat was evaluated to determine the aerodynamic coefficient for various values of α and β by Guan et al. [11]. They validated the numerical approach with the experimental results at Ma = 0.6. Investigation of the aerodynamic coefficients of the ejection seat assuming a steady-state condition was performed by Mahbub et al. [12]. They analyzed the ejection seat system by varying Ma, α, and β to generate a database. This investigation was performed using the standard k-ε turbulence model in ANSYS-Fluent where unstructured grids were utilized.

Very few literatures are available where the ejection seat system analyzed in the supersonic region. Hence, in this study, the ejection seat system is analyzed at Ma = 1.25 for different α to observe the aerodynamic characteristics. Before this study, no investigation was performed at this Ma. This influences the author to study a steady-state analysis on the ejection seat system at Ma = 1.25 for $\alpha = -15^0, -10^0, -5^0, 0^0, 5^0, 10^0$, and 15^0. This investigation concludes that the changing of α has a significant consequence on the aerodynamic coefficients.

2 Methodology

2.1 Computational model and boundary conditions

For the analysis, a 3-D CAD model of the ejection seat system is prepared in SolidWorks. A simplified CAD model is prepared as the real model of the ejection seat system is complicated with sharp edges. The CAD model along with the aerodynamic coefficients and rotational angles is shown in Figure 1(a). The positive and the negative α generated when the model rotates clockwise and counterclockwise with respect to the Y-axis respectively. Air is used as the working fluid for the analysis where the density variation of the air is considered as the Ma of the working fluid is above 0.3.

A cylindrical domain is prepared for this analysis where the far-field boundary condition is set at the inlet of the boundary and the pressure outlet is set at the outlet of the boundary. Symmetric condition is applied for the cylindrical surface and the surface of the ejection seat along with the occupant is assumed as a stationary wall with no-slip condition. Figure 1(b) depicts the boundary conditions used in this study.

 https://doi.org/10.1051/matecconf/202439303002

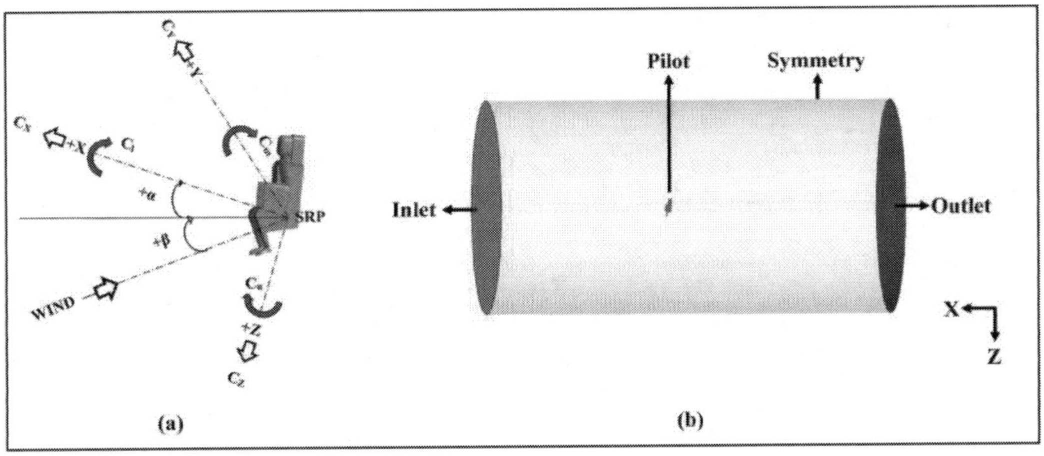

Fig. 1. (a) Coordinate system with the aerodynamic coefficients, **(b)** boundary conditions.

2.2 Nnumerical computations

At the beginning of the grid generation, an unstructured grid with tetrahedron cells is created using ANSYS meshing software. Later on, the tetrahedron cells are converted into polyhedral cells in ANSYS Fluent shown in Figure 2. For this analysis around 15 inflation layer is created around the ejection seat system to resolve the viscous boundary layer.

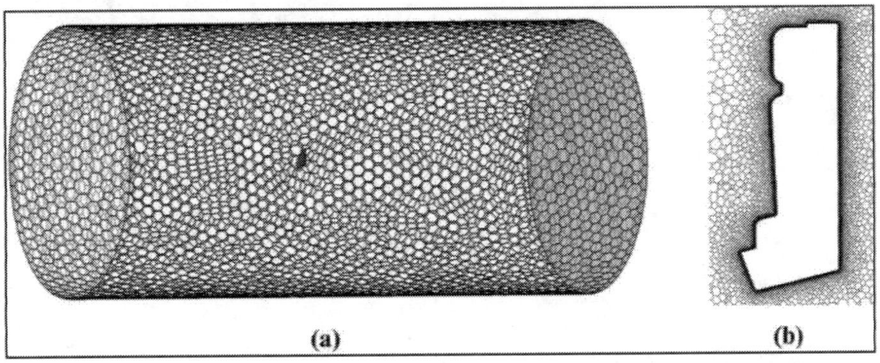

Fig. 2. 3D Mesh model, **(a)** volume mesh, **(b)** close view of the mesh around the model.

A proper mesh independence study is performed for multiple mesh parameters at Ma = 0.6 and $\alpha = 0^0$ shown in Table 1.

Table 1. Grid Independence Study.

Mesh	Number of cells (million)	Cz (Numerical)	Cz (Experimental)
1	2.5	0.169	0.153
2	3.5	0.161	
3	4.0	0.157	
4	4.5	0.157	

From the table, it is observed that Mesh 3 is the ideal mesh that gives close results compared to the experimental result. With the greater number of grids, Mesh 4 generates outcomes which are similar to the Mesh 3. So, for further analysis of the ejection seat system, the parameters of the Mesh 3 are used.

2.3 Validation of the Numerical Approach

At Ma = 0.6, the numerical approaches are validated with the experimental results of White [1] by varying α from -15^0 to 15^0. To improve the accuracy second-order spatial discretization is utilized. The values of all the aerodynamic coefficients are obtained after the residual goes below 10^{-4}. Figure 3 displays the validation of the computed results with the experimental results where the computed and the experimental results match pretty well.

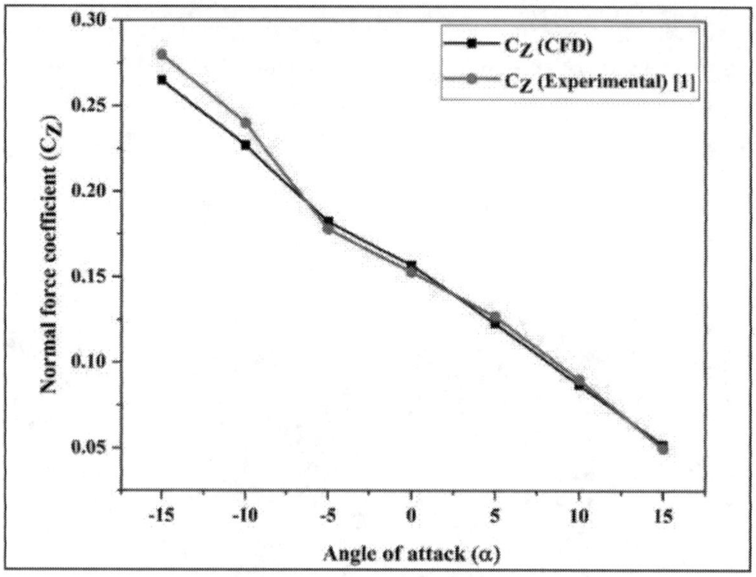

Fig. 3. Validation of the numerical methods.

3 Results and discussion

3.1 Pressure distribution around the ejection seat system

Figure 4 shows the pressure contours on the symmetric plane of the ejection seat system at Ma = 1.25 for α = -15^0, -10^0, -5^0, 0^0, 5^0, 10^0, and 15^0. From the figure it is observed that the maximum pressure region is at the front of the ejection seat system and the low pressure appears at the back of the ejection seat system. The intensity of the pressure changes as the α changes similarly the location of the high-pressure region changes as the α varies from -15^0 to 15^0. The flow separates due to the presence of the sharp edges in the ejection seat system which creates lower pressure at the back of the ejection seat system. An oblique shock appears at the front of the ejection seat system for all the α as the Ma of the free stream is 1.25.

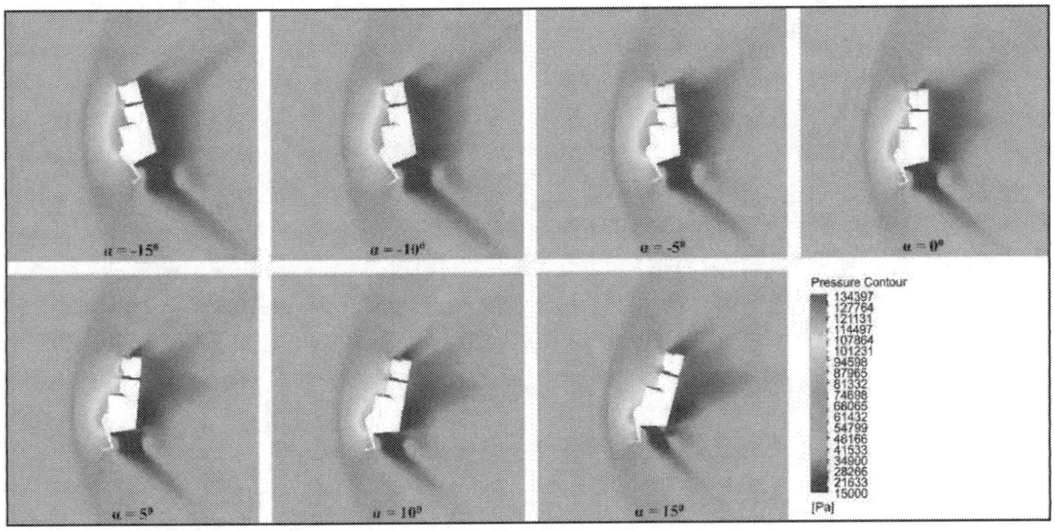

Fig. 4. Pressure distribution on the symmetric plane at Ma = 1.25.

3.2 Aaerodynamic characteristics of the ejection seat system

Pressure contour on the pilot's surface at Ma = 1.25 for α = -15⁰, -10⁰, -5⁰, 0⁰, 5⁰, 10⁰, and 15⁰ are shown in Figure 5. From the figure it is observed that the pressure on the pilot's surface changes with the α. At α = -15⁰, the maximum pressure is found on the surface of the pilot. However, after α = -15⁰, the intensity of the high pressure reduces with the increase of α.

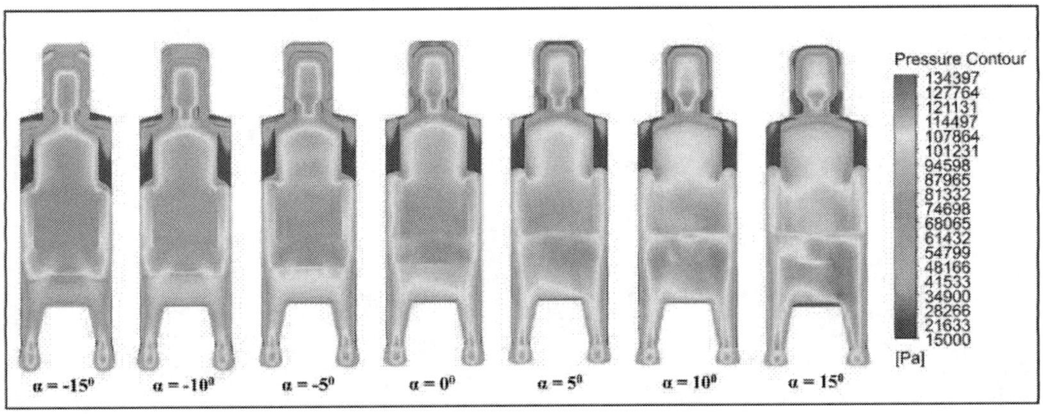

Fig. 5. Pressure contour on the surface of the pilot at Ma = 1.25.

The magnitude of the C_X is increasing with the decrease of α which is shown in Figure 6. Due to the pressure difference between the front and rear surfaces, C_X is generated. The front surface of the pilot is largely exposed to the free stream at α = -15⁰, compared to the other α. Due to the large exposed surface area, the magnitude of the C_X is maximum at this α which can be seen in Figure 5. After α = -15⁰, the exposed surface area is reduced with the increase of α and eventually reduces the magnitude of the C_X. So, at α = -15⁰ and 15⁰, the maximum and minimum magnitude of the C_X is found.

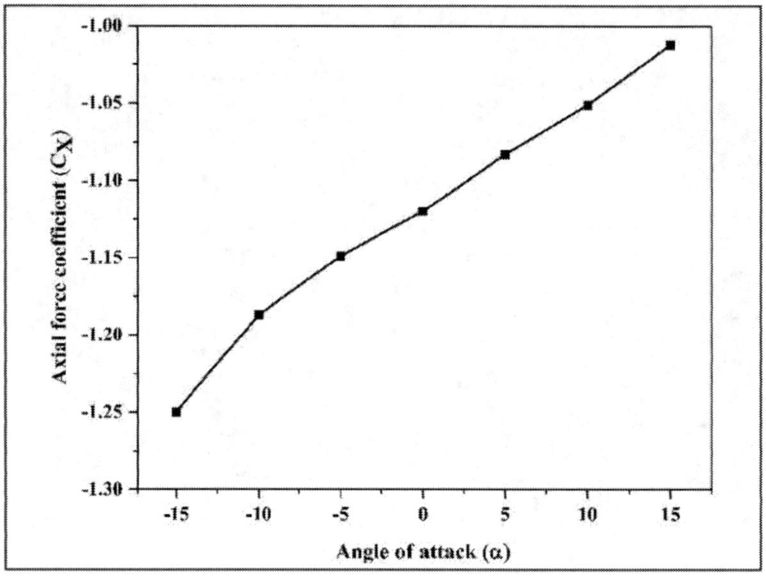

Fig. 6. Variation of C_X with the α at Ma = 1.25.

Figure 7(a) shows the variation of the C_Z with the α at Ma = 1.25. It can be observed that C_Z is inversely proportional to α. C_Z is generated due to the pressure difference between the top and bottom surfaces. With decreasing α, the higher pressure acting on the pilot's surface moves upward which creates a high-pressure difference and eventually increases the value of the C_Z. Hence, at α = -15^0, the maximum value of the C_Z is found and at α = 15^0, the minimum value of the C_Z is observed.

The variation of the C_m with the α at Ma = 1.25 is shown in Figure 7(b). Figure 7(b) depicts that the value of the C_m reduced with the increase of α. At α = -15^0, the maximum value of the C_m is found whereas, at α = 15^0, the minimum value of the C_m is observed.

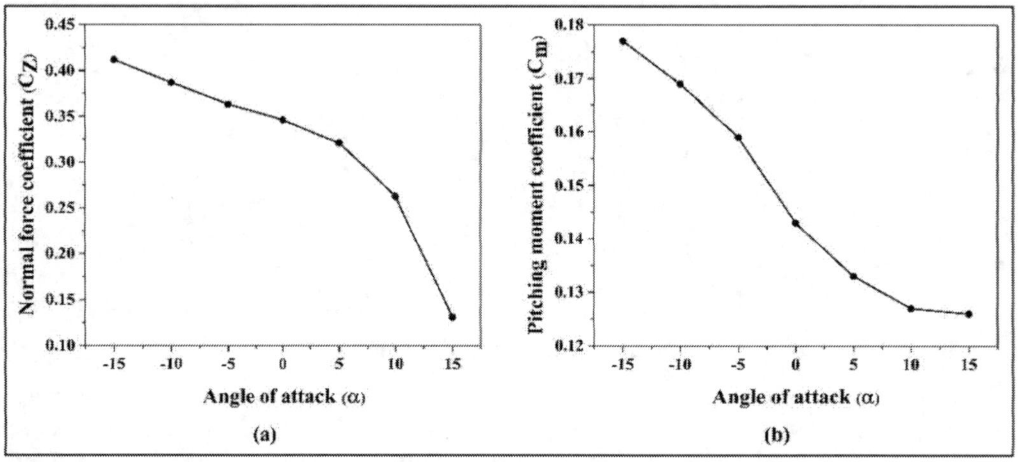

Fig. 7. (a) Variation of C_Z with the α at Ma = 1.25, **(b)** Variation of C_m with the α at Ma = 1.25.

4 Conclusions

In this study, the aerodynamic characteristics of a 3-D ejection seat system are analyzed using an unstructured grid in ANSYS-Fluent at Ma = 1.25 for α = -15^0, -10^0, -5^0, 0^0, 5^0, 10^0, and 15^0 where the β remains at 0^0. The results obtained from this analysis are confirmed and demonstrate good agreement with the experimental data. The findings show that the pressure acting on the surface of the pilot largely depends on the α. Similarly, the aerodynamic coefficients also change with the change of α which is summarized as follows,

• The maximum pressure acting on the surface of the pilot is maximum at $\alpha = -15^0$, as the front surface of the pilot is largely exposed to the free stream and the intensity of the pressure is reduced as the α increases.

• At $\alpha = -15^0$, the maximum magnitude of the C_X is found as the maximum pressure difference is observed and after that, the magnitude of the C_X is reduced as the α increases.

• The minimum value of the C_Z is observed at $\alpha = 15^0$ due to less pressure difference between the top and bottom regions and the value of the C_Z is maximum at $\alpha = -15^0$.

• The value of the C_m is found maximum at $\alpha = -15^0$ and the value of the C_m is found minimum at $\alpha = 15^0$.

• The current study can be utilized while designing and developing an ejection seat system.

Acknowledgements

This research was funded by the AR&DB, DRDO, Ministry of Defence, Govt of India, under Project Number 2006.

References

1. B. J. White, Aeromechanical properties of ejection seat escape systems, Air Force Flight Dynamics Lab Wright-Patterson AFB, OH., Apr. (1974)
2. Z. Wu, Z. Xie, P. Wang, W. Ding, Aerodynamic drag performance analysis of different types of high-speed train pantograph fairing. J. Appl. Sci. Eng. **23**, Issue 3, 509-519 (2020). DOI: 10.6180/jase.202009_23(3).0015
3. M. M. Rahman, A. K. M. Islam, Numerical study of convective heat transfer for turbulent flow in corrugated tube using Al2O3-water nanofluid, in AIP Conference Proceedings **2121**, Issue 1 (2019) DOI: 10.1063/1.5115916
4. S. S. Rajan, M. Santhoshkumar, N. Lakshmanan, N. S. Pillai, M. Paramasivam, CFD analysis and wind tunnel experiment on a typical launch vehicle model. J. Appl. Sci. Eng. **12**, Issue 3, 223-229 (2009). DOI: 10.6180/jase.2009.12.3.01
5. S. D. Habchi, A. J. Przekwas, T. Marquette, P. Ayoub, CFD analysis of ejection seat escape systems, SAE Transactions, 1566-1579 (1992) DOI: 10.4271/921924
6. S. Habchi, G. Hufford, Transient simulation of turbulent flow over blunt bodies (ejection seat configurations), in 13th Applied Aerodynamics Conference, Reston, Virigina: American Institute of Aeronautics and Astronautics, Jun. (1995) DOI: 10.2514/6.1995-1837
7. D. H. Chen, W. H. Wu, J. J. Wang, Y. Huang, Investigation on the aerodynamic performance of an ejection seat. Aeronaut. J. **111**, Issue 1120, 373–380 (2007). DOI: 10.1017/S0001924000004620
8. Y. Zhu, X. Zhao, S. Zhang, Computational studies of aircraft life-support systems, in 49th AIAA Aerospace Sciences Meeting including the New Horizons Forum and Aerospace Exposition, Reston, Virigina: American Institute of Aeronautics and Astronautics, Jan. (2011) DOI: 10.2514/6.2011-1045
9. J. L. Tian, L. Chen, Aerodynamic characteristics numerical simulation of the protective device of ejection seat. Appl. Mech. Mater. **229**, 1811–1814 (2012). DOI: 10.4028/www.scientific.net/AMM.229-231.1811

10. Y. Zhu, H. Guan, X. Zhao, S. Zhang, Computational studies of jettisoned canopy and ejection occupant/seat, in 50th AIAA Aerospace Sciences Meeting Including the New Horizons Forum and Aerospace Exposition (2012) DOI: 10.2514/6.2012-890
11. H. Guan, Y. Zhu, X. Zhao, S. Zhang, Aerodynamic characteristics of ejection seat and occupant, in 51st AIAA Aerospace Sciences Meeting including the New Horizons Forum and Aerospace Exposition, Reston, Virigina: American Institute of Aeronautics and Astronautics, Jan. (2013) DOI: 10.2514/6.2013-386
12. M. M. Rahman, V. Prakash, S. Chandel, D. G. Thakur, R. Čep, N. Khedkar, S. Salunkhe, E. S. Abouel Nasr, Analysis of the aerodynamic characteristics of an ejection seat system using computational fluid dynamics. Front. Mech. Eng. **9**, 1255051 (2023). DOI: 10.3389/fmech.2023.1255051

Numerical investigation on the effect of stroke plane inclination on the aerodynamic performance of dragonfly take-off flight

Shubham Tiwari[1*] *Sejal Madan*[2], and *Sunil Chandel*[1]

[1] Department of Mechanical Engineering, Defence Institute of Advanced Technology (DU), Pune – 411025, India
[2] Department of Information Technology, Guru Nanak Dev Engineering College, Ludhiana, Punjab – 141006, India

Abstract. A numerical investigation is carried out to study the role of inclined stroke plane on the aerodynamic performance of a dragonfly during a take-off flight. A two-dimensional numerical simulation of tandem foils oscillating in-phase along an inclined stroke plane at $Re = 160$ is performed using ANSYS Fluent. The stroke plane angle is varied from $10^o \leq \beta \leq 80^o$ to determine its effect on aerodynamic force coefficients of forefoil and hindfoil. The result shows that the presence of forefoil reduces the hindfoil C_V for low stroke plane angle cases. The cycle-average vertical force coefficient $\overline{C_V}$ of both foils increases with β up to 50° and then decreases. A vortex pair is present in the wake of the foils during each cycle, which induces a downward dipole jet. The dipole jet characteristics such as jet width, location and maximum velocity components are measured for each stroke plane angle. It is observed that the cause of variation in $\overline{C_V}$ and $\overline{C_H}$ with stroke plane angle can be explained with the help of dipole jet characteristics.
Keywords: Dragonfly, forewing, hindwing, take-off, inclined stroke plane

1 Introduction

Flying insects utilize a flapping wing system to generate lift and thrust necessary for a controlled flight. Understanding the underlying principle behind their flight is of fundamental interest for developing Micro Air Vehicles (MAVs). The unsteady aerodynamics associated with insect flight has been studied extensively due to the complex nature of their wing kinematics and their outstanding flying capability.

Dragonflies are unique among the other flying insect as they have remarkable agility and rapid maneuvering abilities. Dragonfly flight has been a focus of research for the past five decades [1-3]. Wang [4] performed a two-dimensional computational study of a dragonfly hovering at Re=157 and studied the effect of wing kinematics by varying the stroke plane angle and mean pitch angle so that the resultant force is vertically upward. It showed that the

* Corresponding author: shubh6tiwari@gmail.com

drag plays a vital role by supporting 75 % of insect weight and a downward dipole jet is generated, which becomes faster and narrower with increasing stroke plane angle.

The three-dimensional shape and kinematics of dragonfly wings result in a highly complex flow structure. Its analysis in 2D can help us understand the significance of 3D flow features at a low Reynolds number. The aerodynamic forces and flow features obtained from the 2D numerical simulation of the flapping wing are consistent with the 3D experimental observations [5]. Deepthi and Vengadesan [6] performed a 2D numerical simulation of an insect wing hovering with different stroke plane angles at $Re = 157$. They studied the characteristics of downward dipole jet generated at different stroke plane angles and its effect on vertical force coefficient. The maximum vertical force coefficient is obtained at the stroke plane angle of 60°, consistent with the experimental observation.

The take-off flight requires a large amount of force to be generated so the insect can become airborne. To achieve this, the insect utilizes different approaches for the take-off, such as flapping with leg jump, clap and fling and slow take-off. Li et al. [7] recorded the voluntary take-off flight of a dragonfly using two high-speed cameras, which showed that the dragonfly has the ability to take off after one wingbeat only. They observed that dragonfly flaps their wings with zero phase difference during take-off and the angle of attack of both wings during the downstroke is significantly larger than the forward flight. The results also showed that the stroke plane angle of forewing and hindwing varied considerably during the take-off procedure. Also, Shanmugam and Sohn [8] found that in-phase stroking with a large angle of attack during downstroke helps generate a higher vertical force coefficient, which is beneficial during take-off flight.

We note from the literature that there have been very limited studies on the dragonfly take-off as well as its stroke plane inclination, an essential mode of flight involving rapid acceleration. Hence, in this study, a two-dimensional numerical simulation is performed for two elliptical airfoils in tandem arrangement oscillating in-phase along an inclined stroke plane at $Re = 160$. The role of stroke plane angle is examined in the aerodynamic performance of dragonflies during take-off flight. Vortex dynamics is studied to understand the role of the downward dipole jet on aerodynamic force generation.

2 Numerical Methodology

2.1 Foil kinematics

An elliptically shaped airfoil is considered with a thickness to chord ratio of 0.1 to represent the cross-section of a dragonfly wing. The foil kinematics used in this study, shown in figure 1, is based on the previous numerical studies and experimental observations [4,7]. The pitch center is located at a half-chord distance from the leading edge of the airfoil. During in-phase oscillation of the foils, the foil spacing of $1.3c$ generates the maximum lift [8,9]. Therefore, the distance between the stroke plane of both airfoils, denoted by L, is $1.3c$.

The flapping kinematics of dragonfly wings is represented by the motion of forefoil and hindfoil undergoing both translational as well as pitching along an inclined stroke plane based on the equations of motion as follows:

For forefoil

$$x_f(t) = \frac{A_o}{2}cos(\beta_f) + \frac{A_o}{2}cos\,(2\pi ft)\,cos(\beta_f) \qquad [1]$$

$$y_f(t) = \frac{A_o}{2}sin(\beta_f) + \frac{A_o}{2}cos(2\pi ft)\,sin(\beta_f) \qquad [2]$$

$$\alpha_f(t) = \alpha_o + \alpha_m cos\,(2\pi ft + \pi/2) \qquad [3]$$

For hindfoil

$$x_h(t) = \frac{A_o}{2}cos(\beta_h) + \frac{A_o}{2}cos\,(2\pi ft + \psi)\,cos(\beta_h) \quad [4]$$

$$y_h(t) = \frac{A_o}{2}sin(\beta_h) + \frac{A_o}{2}cos(2\pi ft + \psi)\,sin(\beta_h) \quad [5]$$

$$\alpha_h(t) = \alpha_o + \alpha_m cos\,(2\pi ft + \pi/2 + \psi) \quad [6]$$

Here, $x(t)$ and $y(t)$ represent the instantaneous displacement of the airfoils along x- and y-axis, respectively, whereas $\alpha(t)$ represents the instantaneous pitch angle relative to stroke plane. The stroke plane angle β is the angle between the stroke plane and the direction of freestream velocity U_∞. A_o represents the flapping amplitude along the stroke plane. The flapping frequency is represented by f and the phase difference between the flapping foils is represented by ψ, which is zero for in-phase oscillation during take-off. α_o is the initial pitch angle and α_m represents the pitch amplitude.

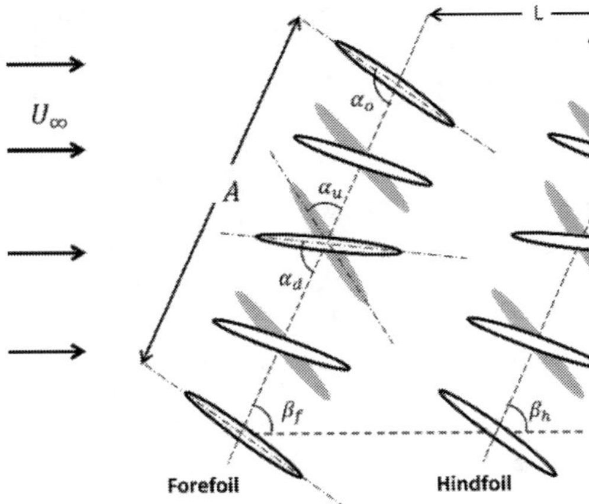

Figure 1: The schematic diagram for forefoil and hindfoil kinematics. White ellipses represent downstroke motion, while grey ellipses denote upstroke motion.

2.2 Computational model

In the present numerical simulation, the flow is considered two-dimensional unsteady and incompressible. The one-equation Spalart-Allmaras (S-A) model is employed to study the flow field evolution. The flow is governed by the continuity equation (eq. 7) and the Navier-Stokes equation (eq. 8).

$$\nabla \cdot u = 0 \quad [7]$$

$$\frac{\partial u}{\partial t} + (u \cdot \nabla)u + \nabla p - \frac{1}{Re}\nabla^2 u = 0 \quad [8]$$

Where u and p represent fluid velocity and pressure, respectively. Equations [7] and [8] are solved using a pressure-based solver in ANSYS Fluent 17.2 software. Pressure-velocity coupling is achieved by using a second-order PISO scheme. The spatial discretization of pressure and momentum is performed using a second-order upwind scheme. The first-order implicit scheme is used for temporal discretization.

The simulation is carried out in a square-shaped computational domain of size $60c \times 60c$, as shown in figure 2. The computational domain is divided into two circular moving zones and an external stationary zone with a non-conformal sliding interface between the moving and stationary zones to allow relative motion. Using this technique, mesh deformation takes place only in the stationary zone. No-slip condition is applied on the surface of both foils. The inlet and outlet are set as velocity inlet and pressure outlet, respectively, whereas the side boundaries are set as symmetry. The motion of both foils and their respective circular zone, as defined by equations [1-6], is achieved by implementing

User-Defined Functions (UDFs) at each time step. The mesh inside the stationary zone is updated using a diffusion-based smoothing method and the remeshing is performed when the skewness of any cell exceeds a user-input value.

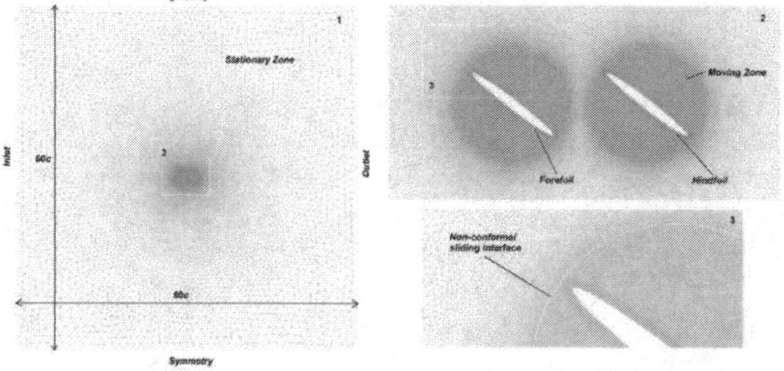

Figure 2: Computational domain with different mesh zones and boundary conditions for present numerical simulation

The Reynolds number is 160 based on freestream velocity and chord length c $(= 0.01\ m)$. The maximum translational velocity of the foil is represented by $U_F = \pi A_o f$. The advanced ratio (J) is the ratio of freestream velocity and maximum translational velocity. The reference velocity is given by $U_R\ (= U_\infty + U_F)$.

The pressure coefficient (C_P) and the aerodynamic force coefficients are calculated using the following equations:

$$C_P = \frac{P - P_\infty}{0.5 \rho {U_R}^2} \quad [9]$$

$$C_V = \frac{F_V}{0.5 \rho {U_R}^2 c} \quad [10]$$

$$C_H = \frac{F_H}{0.5 \rho {U_R}^2 c} \quad [11]$$

Where C_V and C_H are the instantaneous vertical and horizontal force coefficient, respectively. F_V and F_H are the instantaneous vertical and horizontal forces, respectively and ρ is the fluid density. The cycle-average vertical and horizontal force coefficients are represented by $\overline{C_V}$ and $\overline{C_H}$, respectively.

2.3 Grid Independence study & Validation

A grid independence study is performed to ensure that numerical simulation results are independent of the number of mesh elements and time step size. We tested three cases with a different number of mesh elements as well as time step size for a case of in-phase stroke of forefoil and hindfoil with the parameters: $A_o/c = 2.5$, $\beta = 80^o$, $L = 1.3c$, $\alpha_0 = 105^o$, $\alpha_m = 19^o$ (during downstroke), $\alpha_m = 25^o$ (during upstroke), $f = 30\ Hz$, $J = 0.1$ and $Re = 160$. As given in table 1, the maximum difference in the value of $\overline{C_V}$ (total) for cases 2 and 3 is nearly 0.5%. Therefore, the mesh and time step of case 2 is selected for further simulations.

Table 1: Grid and Time-Step Convergence Study

Case	Number of Mesh Elements	Time Step Size	$\overline{C_V}$(Total)
1	170000	T/300	0.7208
2	210000	T/600	0.7317
3	290000	T/900	0.7280

The validation of the present computational model is carried out by reproducing the results of Lua et al. [10]. As shown in figure 3, the result obtained from the present numerical simulation agrees well with the experiment and the numerical results of Lua et al. [10].

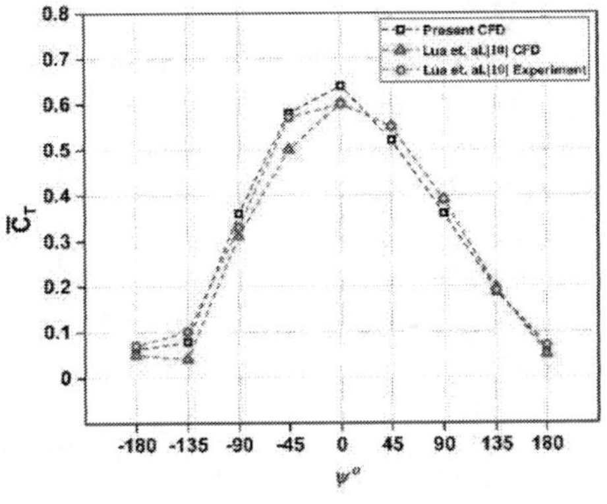

Figure 3: Validation of present computational model.

3 Results and discussion

Cycle-Averaged Force Coefficients

Figure 4(a) shows that the maximum $\overline{C_V}$ is obtained at $\beta = 50^o$ for both the foils, resulting in a peak in total $\overline{C_V}$. The total $\overline{C_V}$ increases with stroke plane angle up to 50°, after that, it decreases. The value of total $\overline{C_V}$ can change heavily by varying the stroke plane angle. The lowest value of total $\overline{C_V}$ (=0.56) is obtained at $\beta = 20^o$ and the highest value of total $\overline{C_V}$ (=1.35) is obtained at $\beta = 50^o$. As shown in figure 4(a), the forefoil $\overline{C_V}$ is greater than hindfoil $\overline{C_V}$ for all cases of β, this shows that the presence of forefoil reduces the performance of hindfoil. Figure 4(b) shows that, the change in forefoil $\overline{C_H}$ is significant as compared to hindfoil by increasing the stroke plane angle. The maximum value of total $\overline{C_H}$ (i.e., maximum drag) is obtained at $\beta = 30^o$ and further increase in the stroke plane angle result in thrust generation. The maximum thrust (negative value represents thrust) is obtained at $\beta = 80^o$. Also, the hindfoil generate thrust for all value of β, whereas the forefoil produces thrust only for $\beta \geq 60^o$.

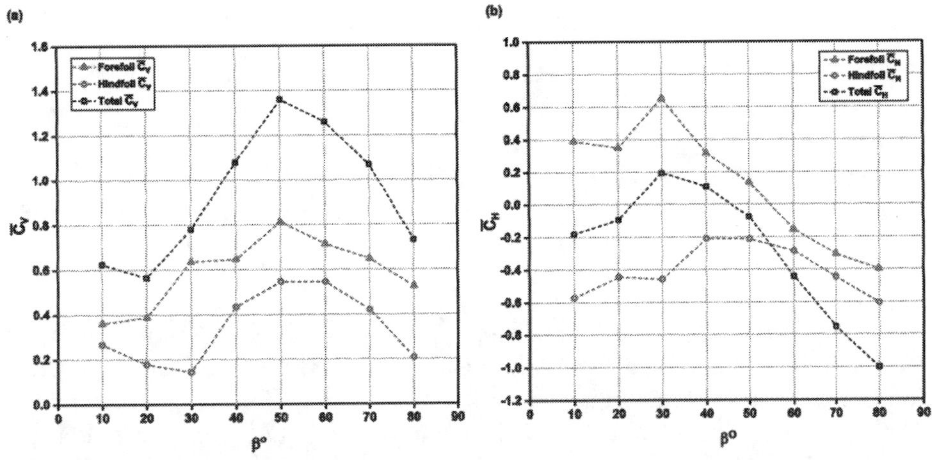

Figure 4: Cycle-averaged vertical ($\overline{C_V}$) and horizontal force ($\overline{C_H}$) coefficient for different stroke plane angle β.

During the upstroke, the hindfoil TEV forms a pair with a CCW vortex generated by the hindfoil trailing edge during the previous downstroke. The strength of the vortex pair and its movement changes for different stroke plane angles. A fluid jet is induced between the vortex pair and is referred to as a dipole jet. The foils experience a reaction force due to the induced dipole jet. The vortex pair generated in each cycle leads to the formation of a vortex street in the wake region. The properties of the dipole jet generated by the vortex pair are studied for different stroke plane angle values.

The properties of the dipole jet are measured at t/T = 0.25 because the maximum value of C_V is obtained near mid-downstroke. The dipole jet for different cases of stroke plane angle is shown with the help of velocity contour in figure 5. The strength of the vortex pair increases with stroke plane angle resulting in a faster dipole jet. In all cases of β, the jet moves downward at a certain angle with the vertical. The inclination angle of the downward jet changes for different stroke plane angles. The reaction force on the foils generated due to the dipole jet can contribute either in thrust or lift depending upon the direction of the dipole jet.

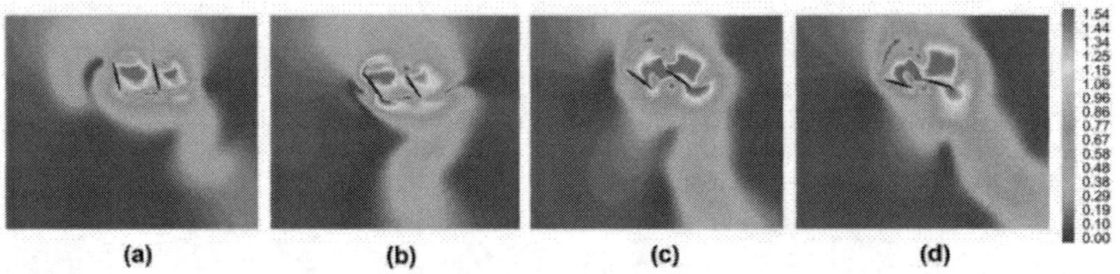

Figure 5: Velocity contours (u/U_R) during mid-downstroke for four different stroke plane angle cases (a) β = 10° (b) β = 30° (c) β = 50° (d) β = 70°

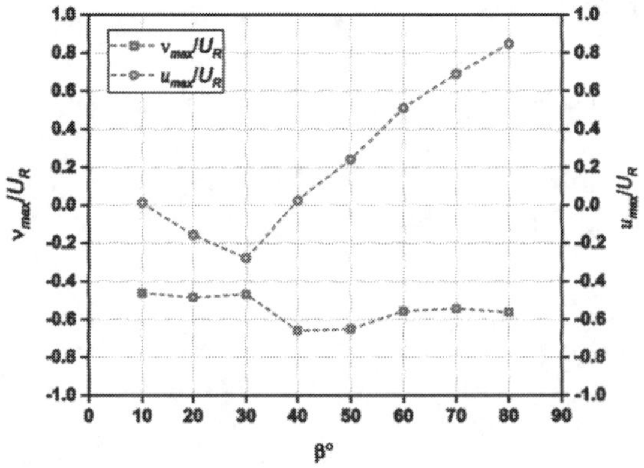

Figure 6: Maximum vertical (v_{max}/U_R) and horizontal (u_{max}/U_R) velocity component of the downward dipole jet during mid-downstroke for different stroke plane angles.

The effect of stroke plane angle on the dipole jet velocity during mid-downstroke is shown in figure 6. The highest value of v_{max}/U_R of the dipole jet is obtained at $\beta = 40^0$. The value of the maximum vertical velocity component increases with β up to $\beta = 40^0$ and then decreases. From figure 6, we can observe that the dipole jet is inclined opposite to the direction of freestream velocity (negative value of u_{max}/U_R) for lower stroke plane angle ($\beta < 40^0$). The highest value of u_{max}/U_R is obtained at $\beta = 80^0$. The value of (u_{max}/U_R) for $\beta = 10^0$ and $\beta = 40^0$ is almost zero, which means the velocity vectors of the dipole jet are nearly vertical for these cases.

The variation of u_{max}/U_R with respect to β, as shown in figure 6 indicates the effect of dipole jet inclination on $\overline{C_H}$. The highest total $\overline{C_H}$ is obtained at $\beta = 30^0$ where the negative peak of u_{max}/U_R is observed. Also, maximum total thrust is obtained at $\beta = 80^0$ where the positive peak of u_{max}/U_R occurs. The value of u_{max}/U_R changes from negative to positive for $30^o < \beta < 80^o$, similarly total $\overline{C_H}$ changes from drag to thrust, as shown in figure 4(b).

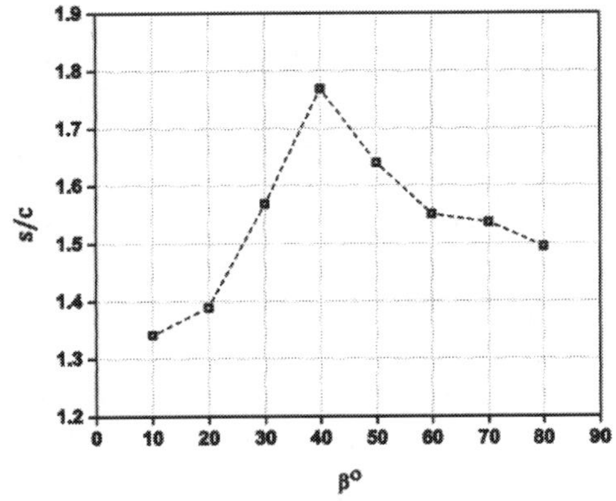

Figure 7: Dipole jet width during mid-downstroke for different stroke plane angles.

The location of vortex pair and dipole jet width (denoted by 's'), which is the distance between centroids of counter-rotating vortices, was measured during mid-downstroke from their respective vorticity contour.

The variation in the width of the dipole jet for various stroke plane angles is shown in figure 7. It clearly shows that the dipole jet gets wider with an increase in β up to 40° and gets narrower with a further increase in stroke plane angle. The effect of increasing the inclination of the stroke plane is such that the foils sweep less horizontal distance for a given stroke amplitude (A_o) at higher β and thus results in a narrower jet. Figure 7 shows that the dipole jet is much narrower at $\beta = 50^0$ as compared to $\beta = 40^0$, whereas there is a very small difference in the value of v_{max}/U_R for $\beta = 40^0$ and $\beta = 50^0$ (as shown in figure 6). Hence, the $\overline{C_V}$ obtained for $\beta = 50^0$ is greater than $\beta = 40^0$.

The analysis of the dipole jet with respect to stroke plane angle clearly shows the effect of the wake generated by the foils on the variation of aerodynamic forces. The faster and narrower jet obtained at $\beta = 50^0$ helps the foils to generate the highest $\overline{C_V}$. Also, the highest jet velocity and the inclination of dipole jet along the direction of freestream velocity for $\beta = 80^0$ case results in the maximum value of thrust.

4 Conclusion

The effect of stroke plane angle on the aerodynamic force coefficients during dragonfly take-off flight is studied by performing a two-dimensional numerical simulation of tandem flapping foils oscillating in-phase at $Re = 160$ and $J = 0.1$. The foils are subjected to sinusoidal pitch and plunge motion along an inclined stroke plane and the stroke plane angle is varied from $\beta = 10^o$ to $\beta = 80^o$. It is observed that the hindfoil performance is reduced by the presence of forefoil for low stroke plane angle ($\beta < 40^o$) cases. The maximum vertical force coefficient $\overline{C_V}$ is obtained at $\beta = 50^o$ whereas maximum thrust and drag is obtained at $\beta = 80^o$ and $\beta = 30^o$ respectively. A downward dipole jet is present in the wake region of the foils for every stroke plane angle. The aerodynamic forces on the foils get affected due to the momentum transfer by the jet. The properties of the dipole jet obtained at each stroke plane angle show the influence of the wake generated by the foils on the variation of aerodynamic forces. These results can help to achieve desired lift during the take-off flight of MAVs.

References

1. Norberg RÅ. Hovering flight of the dragonfly aeschna juncea L., kinematics and aerodynamics. In: *Swimming and flying in nature*. Boston: Springer, 1975, pp:763-781.
2. Choudhry A, Leknys R, Arjomandi M, et al. An insight into the dynamic stall lift characteristics. *Exp Therm Fluid Sci* 2014; 58: 188–208.
3. Dickinson MH, Lehmann FO and Sane SP. Wing rotation and the aerodynamic basis of insect right. *Science* 1999; 284: 1954–1960.
4. Wang ZJ. The role of drag in insect hovering. *J Exp Biol* 2004; 207: 4147–4155.
5. Wang ZJ, Birch JM and Dickinson MH. Unsteady forces and flows in low Reynolds number hovering flight: Two-dimensional computations vs robotic wing experiments. *J Exp Biol* 2004; 207: 449–460.
6. Deepthi S and Vengadesan S. Role of Dipole Jet in Inclined Stroke Plane Kinematics of Insect Flight. *J Bionic Eng* 2020; 17: 161–173.
7. Li Q, Zheng M, Pan T, et al. Experimental and numerical investigation on dragonfly wing and body motion during voluntary take-off. *Sci Rep* 2018; 8: 1011.
8. Shanmugam AR and Sohn CH. Numerical investigation of the aerodynamic performance of dragonfly-like flapping foil in take-off flight. *Proc IMechE Part G: J Aerospace Engineering* 2019; 233: 5801–5815.
9. Gadde SN and Vengadesan S. Lagrangian Coherent Structures in Tandem Flapping Wing Hovering. *J Bionic Eng* 2017; 14: 307–316.
10. Lua KB, Lu H, Zhang XH, et al. Aerodynamics of two-dimensional flapping wings in tandem configuration. *Phys Fluids* 2016; 28: 121901.

Development and Numerical validation of an Aerospike nozzle Contour Design

Sam Daniel Fenny A[1*], *Deepak* Sampathkumar[1,2], *Nishant B* Mayekar[3], and *Sree Adithya* S R[1]

[1]Department of Mechanical and Automation Engineering, Agni College of Technology, Chennai 600130, India
[2]Department of Mechanical Engineering, Government College of Engineering, Bargur 635104, India
[3]Department of Mechanical Engineering, Agnel Institute of Technology and Design, Assagao, Goa, India

Abstract. Aerospike nozzle simplified design and analysis, as well as test results, are presented. New techniques in nozzle design have been used to improve the performance of current rocket engines. The use of an aerospike nozzle is one such technique. It outperforms the existing proven bell nozzle in terms of overall performance. This study analyses the development and flow analysis of an aerospike nozzle at sea level, validating their results by taking into account currently used bell nozzles. The flow over the aerospike pump is investigated in higher depth using computational fluid dynamics under an autonomous fire testing condition. This test showed that the approximate design method used to determine an aerospike nozzle shape can lead to extremely effective nozzles. A model of the current 7.45 kN bell nozzle was selected as a specimen to develop an aerospike Nozzle. Gambit and Fluent software were used to model and examine the flow characteristics of building the aerospike nozzle. The developed aerospike nozzle shows a nearly 2% increase in performance over the existing nozzle.

1 Introduction

To achieve an advanced launching vehicle, one of the most important components is a highly efficient propulsion system. Future aerospace propulsion technology will inevitably focus on researching and developing reusable propulsion systems that are lightweight, affordable, and highly effective. Since the bell-shaped nozzle rocket engine has reached the pinnacle of development, it is challenging to significantly increase its performance in comparison to conventional rocket engines. The aerospike nozzle engine is a good contender to power an advanced launching vehicle in the future because of its small size, lightweight, excellent performance at all altitudes, and improved use of the vehicle base [1]. This kind of nozzle has the ability to continuously adjust for altitude. Since its plume is free to adapt and open to the outside atmosphere, it is thought to function better at off-design altitudes than the usual bell-shaped nozzle. This is because it enables the engine to run at its maximum expansion at all altitudes. This paper initially examines the spike nozzle design and contrasts it with the

* Corresponding author: samdfenny@gmail.com

bell nozzle that has been in use up to this point [2]. The principle behind the aerospike nozzles will then be discussed. Furthermore, this nozzle type's benefits and drawbacks will be examined in relation to the conventional converging-diverging nozzle [3].

1.1 Nozzle concepts, Thrust and Performance

A nozzle is a mechanical device used to regulate the flow characteristics or direction of fluid out of (or into) a pipe or enclosed chamber. A nozzle is often a pipe or tube with a variable cross-sectional area that is used to control or guide fluid flow [5]. The rate of flow, speed, direction, mass, form, and/or pressure of the stream that emerges from nozzles are commonly controlled. The purpose of the rocket nozzle is to efficiently transform the propellant's thermal energy into kinetic energy so that high exhaust velocity can be achieved in the intended direction [7]. The three primary components of a rocket nozzle are the throat, the divergent section, and the convergent section [6].

The ratio of the cross-sectional exit area to the throat cross-sectional area determines a rocket nozzle's performance. We refer to this as the expansion area ratio. Since the geometry of the nozzle usually does not change, this value normally stays constant. During engine operation, the pressure inside the combustion chamber and the mass flow rates also stay relatively constant. Based on the anticipated operating circumstances, the rocket engineer selects the nozzle designs and characteristics. The most basic performance metric for a rocket motor is the thrust it produces. "Thrust, being the force that a motor exerts, is what propels a rocket. Thrust is generated by the expelling of mass (the exhaust) flowing through the nozzle at high velocity. The expression for thrust is given by the following equation (1)."[4]

$$F = \int P \, dA = V_e + (P_e - P_a) \, A_e \tag{1}$$

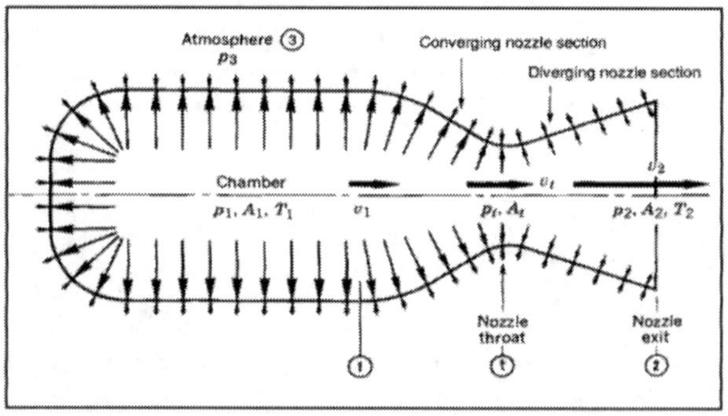

Fig. 1. Thrust and Thrust Coefficient

2 Bell Nozzle Design Methodology

2.1 Parabolic Approximation of Bell Nozzles

Designing a nearly-optimal thrust bell nozzle shape can be accomplished by applying G.VR. Rao's recommended parabolic approximation techniques. Figure 2 (a) depicts the design arrangement of a parabolic approximation bell nozzle [8]. A 1.49 Rt circular arc makes up the nozzle shape directly upstream of the throat T. The diverging section nozzle contour consists of a parabola from point N to exit E and a circular entry section with a radius of 0.379 Rt from the throat T [13].

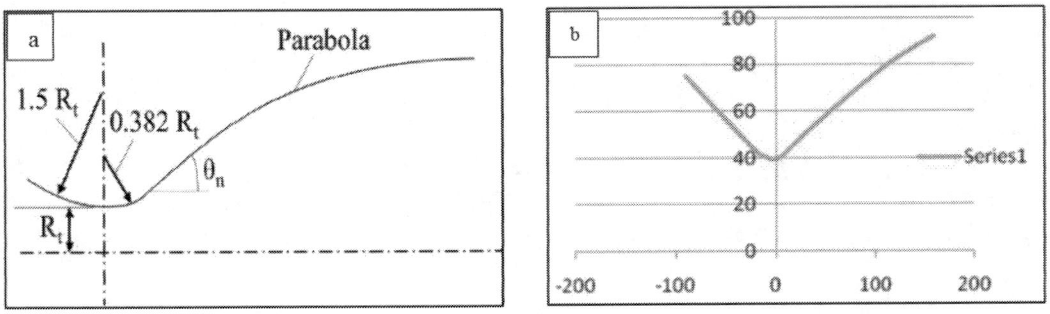

Fig. 2. (a) G.V.R.Rao contour (NASA SP 125) (b) Bell Nozzle Contour

Table 1. Specifications of Bell Nozzle

Parameters	Values
Thrust	F= 7KN
Propellant	Air, Nitrogen
Throat Dia	dt=79mm
Chamber pressure	Po=8.33 bar
Area ratio	A/A*=5.24
Chamber Temperature	To=300k
Gas constant	R=287 KJ/Kg/K
Material	Steel (SS 304)

2.2 Theoretical calculations of Bell Nozzle at sea level

For area ratio A/A*=5.24
πre2/ πrt2 =5.24
re2 = 0.09041
de = 0.1808 m
For Convergent Divergent Nozzle
Lc = 132 mm (given)
Sin 15° = 0.0509 / Z = 0.1966 m Cos 15°= L/Z
LD = Cos 15°× 0.1966
LD =189.9 mm
For Bell Nozzle,
LD = 80% of LD
LD = 150.8 mm
For Sea level Test.
For γ =1.4, Ae/A*=5.2, From Isentropic table, Mach number at exit Me = 3.225

Thrust Calculation:
Thrust F =mp×Ce + (Pe-Pa) Ae
= 9.518×637.826 + (16243.5 – 1.013×105) ×0.02567
F = 3887.42 N

Specific Impulse (I_{sp}):
I_{sp}= C_j/g
= 408.42/9.81
I_{sp} = 41.63 sec Specific fuel Consumption:
= 1/ I_{sp}
=0.024018 1/s

Efficiency Calculation:
The definition for efficiency of bell shaped nozzle isgiven by,

$$\eta = \frac{To4 - Te}{T04 - Tes}$$

$$\eta = \frac{300 - 97.5}{300 \times (1 - 0.32468)}$$

$$= 0.9432$$

$$\eta = 94.32\ \%$$

2.3 Bell Nozzle Flow Characteristics

Calculation of flow over the bell nozzle, which had been designed earlier in this section, under cold flow conditions is covered in this section. The next stage would be to carry out comparable calculations with different flow conditions by adjusting the Mach number and altitude (or ambient pressure). Fluent software was used for pre-processing after Gambit was used to produce the axis symmetry mesh. For the computations, the Spalart-Allmaras turbulence model was employed. By applying total pressure and temperature inside the chamber and estimating the mass flow rate to meet the sonic flow condition at the throat, the input flow condition was produced [9-10]. Figure.3 (a) shows the velocity contour, as the concept of nozzles function; velocity gets increased at its exit as 411 m/s. Also in the nozzle, there will be temperature shock due to altitude variation. Figure. 4(b) shows the pressure contour. For the area ratio of 5.24 and for the constant γ =1.4, the inlet chamber pressure 8.33×10^5 N/m^2 is given as the boundary condition and the outlet pressure is 1.83×10^4 N/m^2. Which is closer to the calculated value. The inlet temperature is 302 K and the exit temperature is reduced to 96.5 K as shown in the Figure 2.2 (b). In the bell nozzle Mach number varies throughout the contour with reference to the velocity of the fluid and velocity of sound. Mach at exit for bell nozzle is 1 as shown in the Mach contour in Figure 4 (a).
You are free to use colour illustrations for the online version of the proceedings, but any print version will be printed in black and white unless special arrangements have been made with the conference organiser. Please check whether this is the case. If the print version will be black and white only, you should check your figure captions carefully and remove any reference to colour in the illustration and text. In addition, some colour figures will degrade or suffer loss of information when converted to black and white, and this should be considered when preparing them.

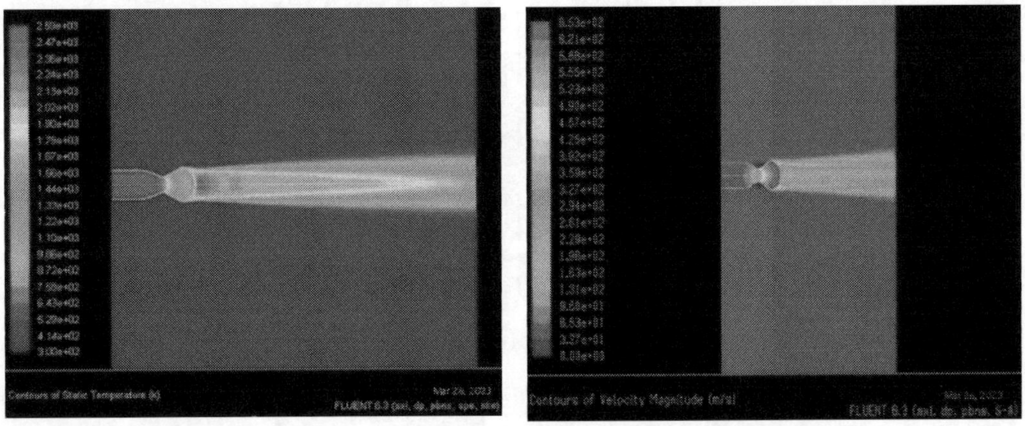

Fig. 3. (a) Velocity contour of Bell Nozzle (b) Temperature contour of Bell Nozzle

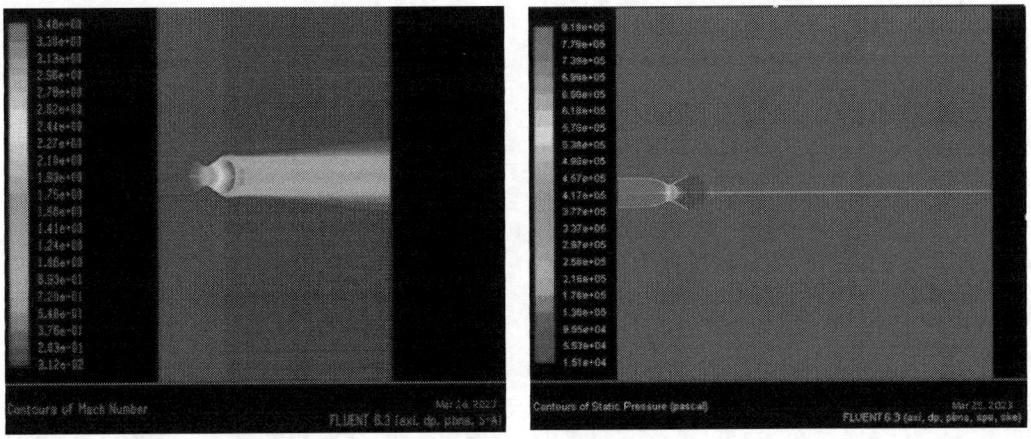

Fig. 4. (a) Mach Contour (b) Pressure contour of Bell Nozzle

3 Aerospike Nozzle Design Methodology

A rocket nozzle design that permits combustion to happen around a spike's (or center plug's) perimeter. The ambient (atmospheric) pressure then shapes and modifies the thrust-producing, hot-exhaust flow. It goes by the names plug nozzle and spike nozzle as well [14]. Using an easy approximation of Rao's method based on calculus variations, it is possible to design the contour of Aerospike Nozzle. The method to be approximated is based on the assumption that a series of hemispherical expansion waves will appear in the cowl lip of the aerospike nozzle. Using this method, the plug nozzle contour for a given expansion ratio, ε, and ratio of specific heats, γ, can be determined. Below is a brief description of the method [15].

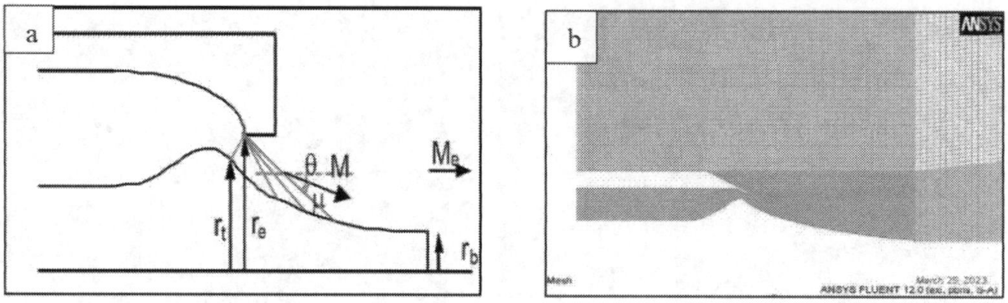

Fig. 5. (a) Plug Contour (b) Meshed Aerospike nozzle contour

For the contour design the Prandtl-Meyer function is used. The thruster angle is provided by θt = v (Me), where Me is the Mach number at the exit and n is the Prandtl-Meyer function, as the flow is assumed to be parallel to the nozzle axis at the exit.

$\nu=\sqrt{((\gamma+1)/(\gamma-1))}\ \tan^{-1}\sqrt{((\gamma+1)/(\gamma-1))\ (M^2-1))}-\tan^{-1}\sqrt{(M^2-1)}$

The throat area is $A_t = \pi\ (r_e^2 - r^2) / \cos\theta_t = F / P_cC_F$

Exit area of the aerospike is

$A_e = \pi\ (r_e^2 - r_b^2)$

The expansion ratio

$\varepsilon = A_e/A_t$ is given, re and rt are ascertainable for a fixed at the aerospike nozzle's radial coordinate is provided by

$X = r_e - r/\tan(\mu+\theta)$

Where $\theta = \theta_t - \nu$ and the Mach angle $\mu = \sin^{-1}(1/M)$.

3.1 Aerospike Nozzle Flow Characteristics

This section deals with the computation of the flow over the truncated aerospike nozzles designed in the previous section of the static test as a first step toward validation of the power on CFD. The next step would be to perform similar calculations at a range of flow conditions by altitude, or ambient pressure, Mach number, and angle of attack. The rocket could then be launched with a number of instruments to collect test data and compare the calculations.

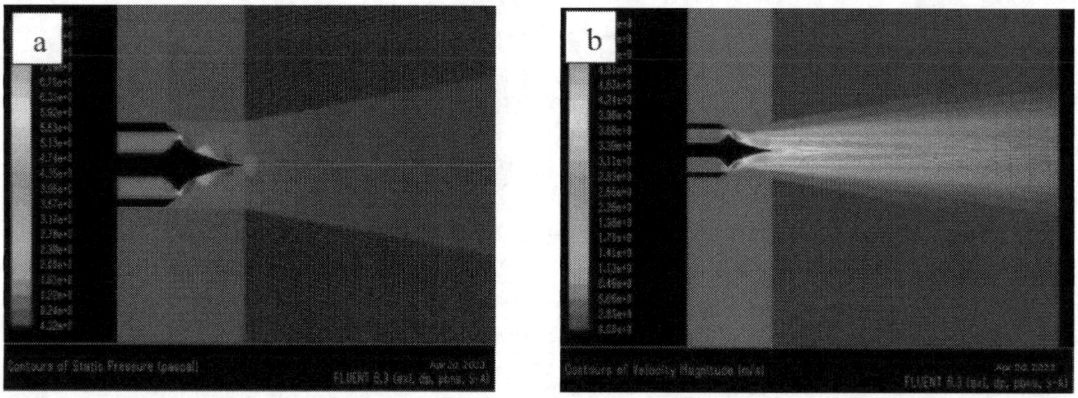

Fig. 6. (a) Pressure Contour of Aerospike Nozzle (b) Velocity contour of Aerospike Nozzle

Figure.6. (a) shows the pressure of the aerospike nozzle the exit pressure is $1.22\text{x}10^4\ \text{N/m}^2$. The conventional one is not capable of receiving shocks, in particular weak ones, it will be able to do so. The velocity at exit is increased as expected, it is 566 m/s. The aerospike engine has, as expected, no divergence losses at design pressure altitudes. In addition, the viscous losses for annular aerospike engines are less than those of an equivalent bell shape nozzle due to its smaller diameter. Mach number at the throat of the plug is 1 and it's shown in the mach contour Fig 7 (a).Temperature Contour is shown in the Fig 7 (b) and the outlet temperature is 148.8 k.

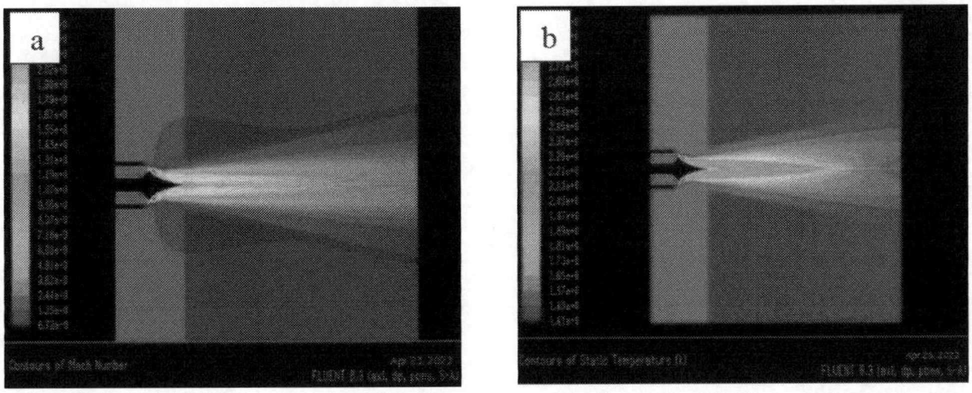

Fig. 7. (a) Mach Number Contour (b) Temperature contour of Aerospike Nozzle

Table 2. Performance Comparison of Bell and Aerospike Nozzle

	Bell Nozzle		**Aerospike Nozzle**
Properties at exit	**Theoretical Result**	**CFD Result**	**CFD Result**
Pressure(P_e)	1.62435 x10^4N/m^2	1.83x10^4 N/m^2	1.22 x10^4N/m^2
Temperature (T_e)	97.35 K	96.5 K	148.8 K
Velocity(C_e)	408.42 m/s	411 m/s	566 m/s
Mach Number(M_e)	3.225	3.2	2.26
Mass flow rate	9.518 kg/s	9.55 kg/s	11.77 kg/s
Efficiency	94.32%		96.26%

4 Conclusion

The design and cold flow analysis of a 7.45 kN thrust engine's bell and aerospike nozzles are presented in this paper. A complete aerospike nozzle contour is suggested, based on the calculus of variation created using the G.V.R.RAO approach. To evaluate the computational capability of current CFD tools to calculate flow over Aerospike nozzles in operating conditions, this flow shall be computed by using FLUENT software. Also, both bell and aerospike nozzle performances have been verified. In comparison to the reference bell nozzle, the aerospike nozzle performs approximately 1.9 percent better in terms of performance when analyzed using CFD for cold flow conditions, taking into account parameters like pressure, velocity, and temperature. Furthermore, the bell nozzle has an efficiency of 94.36% whereas the aerospike nozzle has an overall efficiency of 96.26%. In terms of performance, the bell nozzle is inferior to the aerospike nozzle. Aerospike nozzles may be produced and evaluated based on the comparison of theoretical and CFD values. The current study focuses on two indicators a drop in pressure and an increase in velocity that can be helpful as a basis for designing an aerospike nozzle. However, further work needs to be done to reduce the nozzle's design temperature.

References

1. Wang CH, Liu Y, Qin LZ. Aerospike nozzle contour design and its performance validation. Acta Astronautica,1;**64**(11-12):1264-75, (2009 Jun)
2. Tian H, Guo Z, Hao Z, Hedong L, Li C. Numerical and experimental investigation of throttleable hybrid rocket motor with aerospike nozzle. Aerospace Science and Technology,1;**106**:105983.(2020 Nov)
3. Nair PP, Suryan A, Kim HD. Computational study of performance characteristics for truncated conical aerospike nozzles. Journal of Thermal Science,**26**:483-9, (2017 Dec)
4. Wang G, Chen L, Guan B, Zhang Y, Zhu L. Numerical investigation on thrust characteristics of an annular expansion–deflection nozzle. Physics of Fluids,1;**35**(5), (2023 May)
5. Mishra AK, Goswami V, Sarma P. Aerodynamic thrust characteristics and performance valuations for an aerospike nozzle-review. International Journal of Engineering Applied Sciences and Technology, **5**(11):128-33, (2021)
6. Nagaral V, Narayana G. Study of Aerospike Nozzle and its Thrust Vectoring Characteristics. In25th AIAA International Space Planes and Hypersonic Systems and Technologies Conference, (p. 3101), (2023)
7. Liu XY, Cheng M, Zhang YZ, Wang JP. Design and optimization of aerospike nozzle for rotating detonation engine. Aerospace Science and Technology, 1;**120**:107300, (2022 Jan)
8. Wang Y, Le J. A rotating detonation engine using methane-ethylene mixture and air. Acta Astronautica,1;**188**:25-35 (2021 Nov)
9. Nair PP, Suryan A, Kim HD. Computational study of performance characteristics for truncated conical aerospike nozzles. Journal of Thermal Science, **26**:483-9, (2017 Dec)
10. Dakka S, Dennison O. Numerical analysis of aerospike engine nozzle performance at various truncation lengths. International Journal of Aviation, Aeronautics, and Aerospace,18;**8**(2),(2021 Jun)
11. Balaguru S, Shashi K, Vela M, Chellapandi P. Thermo mechanical analysis of SS304 circular grid plate hard faced with colmonoy. Applied Mechanics and Materials, Nov 30;**229**:710-7, (2012)
12. Reza MS, Arora K. Contour design of aerospike nozzle and comparison of performance. In2017 International Conference on Infocom Technologies and Unmanned Systems (Trends and Future Directions)(ICTUS), 18 (pp. 836-842). IEEE, (2017)
13. Gayathri N, Senthilkumar P, Dineshkumar K, Purusothaman M, Sriharan B. CFD analysis on flow through nozzle of conical type and truncated conical plug type. Materials Today: Proceedings,1;**72**:2387-94, (2023 Jan)
14. Rao GV. Exhaust nozzle contour for optimum thrust. Journal of Jet Propulsion, 28(6):377-82, (1958 Jun)
15. Naik A, Charapalle S, Kalekar Y, Walave G, Balaji K, Bajaj DK. CFD analysis on the effect of Pintle geometry in the performance of expansion Deflection Dual-Bell Nozzle. InAIP Conference Proceedings, (**Vol. 2890**, No. 1). AIP Publishing,(2023)

16. Golliard T, Mihaescu M. Computational Aeroacoustics for a Cold, Non-Ideally Expanded Aerospike Nozzle. Journal of Turbomachinery, 1;**146**(2), (2023 Feb)
17. Balaguru S, Abid M, Gupta M. Investigations on different hardfacing processes for High temperature applications of Ni-Cr-B-Si alloy hardfaced on austenitic stainless steel components. Journal of Materials Research and Technology,1;**9**(5):10062-72, (2020)

MATEC Web of Conferences **393**, 03005 (2024) https://doi.org/10.1051/matecconf/202439303005
STAAAR-2023

Adaptive PID Control for Autopilot Design of Small Fixed Wing UAVs

Tanzil Ahmed Lohani[1], *Abhishek* Dixit[1] and *Pooja* Agrawal[2*]

[1]Department of Aerospace Engineering, Defence Institute of Advanced Technology (DIAT), DU, Pune, India
[2]Assistant Professor, School of Robotics, Defence Institute of Advanced Technology (DIAT), DU, Pune, India

Abstract. The development of a flight control system for the class of small fixed wing Unmanned Aerial Vehicle (UAV) imposes a challenge to the designers to achieve acceptable and stable performance characteristics across the defined flight envelope. This paper focuses on developing a dynamic model of the Aerosonde UAV, incorporating it with an adaptive Proportional, Integral and Derivative (PID) control system for designing both lateral and longitudinal autopilots. The PID gains for each control loop are self-tuned based on an analytical approach, making sure that the desired command value is achieved at each time interval as closely as possible. To check the robust nature of the designed control system in the presence of external wind disturbance, a Dryden wind gust model was developed, and the entire simulation results were carried out in MATLAB®/Simulink environment. The proposed adaptive PID controller proves the effectiveness and robustness of the autopilot control system.

1 Introduction

In the past few decades, Unmanned Aerial Vehicles (UAVs) have gained interest and importance in both the defence and civilian fields. And this is majorly because of their characteristics like accessibility, low cost and no human-life risk involvement. UAVs have predominantly proved their success in high-risk missions ranging from search and rescue operations carried out during natural disasters like earthquakes and floods, in national defence, weather monitoring and so on. In all such scenarios, the presence of a human being in flight was never imagined [1-3]. The execution of a successful UAV flight depends greatly on its autonomous controller termed autopilot. An autopilot's job is to control UAVs flight by continuously tracking the desired flight path accurately and to make sure its design is robust in regard to adverse environmental conditions involving wind gusts. Apart from path tracking autopilot controls the aircraft's attitude, airspeed and altitude [4]. In cases of small UAVs the wind velocity magnitude greatly effects its performance since it matches closely with the actual UAVs speed. As a result of which engineers face a greater

* Corresponding author: poojaa.diat@gmail.com

challenge in designing an efficient autopilot system for UAVs. In recent era usage of modern control theory for designing control system algorithms for UAV autopilots have been successfully established, but these are majorly based on nonlinear control, new and progressive algorithms. Even though such autopilots have shown their success, only a few reports are documented. Due to their complexity, nonlinear behaviour and high computational cost there is less interest in engineers to employ them in physical UAV autopilot systems. An alternative option would be PID (Proportional, Integral and Derivative) autopilots, which are integrated in real-time control systems. Because PID autopilots have a simple structure, easy to implement without sacrificing on acceptable performance characteristics defined for the UAV autopilot. However, PID autopilots need a proper understanding of the mathematical model of the aircraft under consideration and tuning methods to achieve improved performance over the entire flight envelope [5].

The fixed wing UAVs configuration is closely related to the conventional aircraft design for manned operations. The dynamic model of fixed wing UAVs is immensely characterized to be nonlinear and highly coupled. External disturbance like wind gust affects the dynamic model for such an aircraft [4].

This paper aims to design an adaptive PID controller for a fixed wing Aerosonde UAV [6], that has the capability of controlling the aircraft's longitudinal and lateral motion. The controller design process requires firstly a thorough modelling of the Aerosonde aircraft based on kinematics and dynamics equations for rigid-body derived from Newton's second law of motion. The aerodynamic forces (lift and drag) and the aircraft inertia are taken into consideration. The MATLAB®/Simulink environment is used to execute the dynamic model for Aerosonde UAV. Then trimming the Aerosonde dynamic model based on optimization through penalty method is carried out. Next, the adaptive PID based autopilots are developed for longitudinal (pitch and altitude hold) and lateral (roll and yaw attitude hold) cases. The gains of PID control loops are tuned with an analytical method. To check the robust nature of the developed PID controller in the presence of wind disturbance, a Dryden wind gust model is considered.

2 The Aerosonde UAV description and dynamic model

2.1 Aircraft description

The Aerosonde is a small, autonomous, fixed wing UAV designed mainly for weather surveillance. Aerosonde was the first UAV to cross the Atlantic Ocean in 1998. Figure 1, [7] shows the aerosonde UAV model. The general specifications of the UAV are listed in Table 1 [8]. Even though Aerosonde is equipped with a complicated control system, it has high flexibility in its design that makes it suitable to command the aircraft effectively from any location [9].

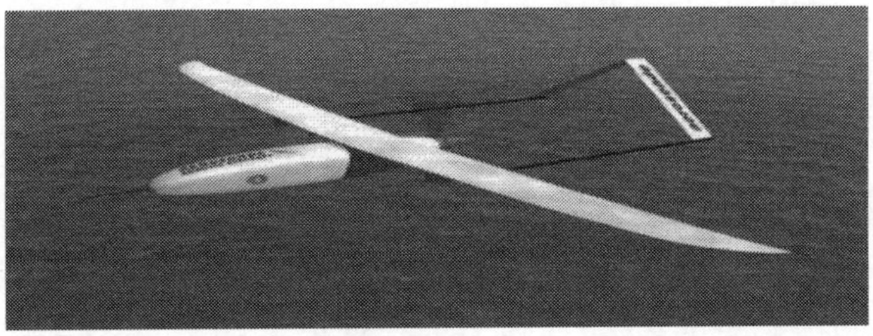

Fig. 1. Aerosonde UAV model [7].

Table 1. Aerosonde UAV specifications.

Specifications	
Mass (m)	13.5 kg
Wing span (b)	2.8956 m
Flight	Fully autonomous
Maximum speed	30-40 m/s
Cruise speed	20-30 m/s
Operational altitude range	~6096 m
Payload	1 kg
Wing chord (c)	0.19 m

2.2 Aerosonde UAV dynamic model

In order to build the 6-DoF UAV model for the aircraft, the UAV is assumed to be a rigid-body, with the entire mass being symmetrically distributed about the aircraft's absolute coordinate system [10], this is based on a non-linear model described by Newton's second law and is represented by 12 dynamic state variables listed in Table 2 using 12 dynamic equations [11].

Table 2. State variables for UAV equations of motion.

Symbol	**Description**
P_n	Inertial north position of UAV along $\mathbf{i}^i$ axis of inertial frame
P_e	Inertial east position of the UAV along $\mathbf{j}^i$ axis of inertial frame
P_d	Inertial down position (negative of altitude) of the UAV along $\mathbf{k}^i$ axis of inertial frame
u	Body frame velocity measured along $\mathbf{i}^b$ axis of body frame
v	Body frame velocity measured along $\mathbf{j}^b$ axis of body frame
w	Body frame velocity measured along $\mathbf{k}^b$ axis of body frame
p	Roll rate measured along $\mathbf{i}^b$ axis of body frame
q	Pitch rate measured along $\mathbf{j}^b$ axis of body frame
r	Yaw rate measured along $\mathbf{k}^b$ axis of body frame
ϕ	Roll angle defined with respect to vehicle frame
θ	Pitch angle defined with respect to vehicle frame
ψ	Yaw angle defined with respect to vehicle frame

Where $\mathbf{i}^b$, $\mathbf{j}^b$ and $\mathbf{k}^b$ are axes that run from the centre of gravity point of the UAV towards the nose, right wing and vertically down respectively. Hence the state variables of UAV are represented in vector form as

$$X = (u, v, w, p, q, r, \phi, \theta, \psi, P_n, P_e, P_d) \tag{1}$$

All the propeller properties, aerodynamic coefficients and moments of inertia values for the aerosonde aircraft which is to be dynamically modelled are required to be substituted in the 12 dynamic equations which are as follows;

Force equation

$$\begin{pmatrix} \dot{u} \\ \dot{v} \\ \dot{w} \end{pmatrix} = \frac{1}{m}\begin{pmatrix} f_x \\ f_y \\ f_z \end{pmatrix} + \begin{pmatrix} vr - qw \\ pw - ur \\ qu - pv \end{pmatrix} \qquad (2)\,(3)\,(4)$$

Moment equation

$$\begin{pmatrix} \dot{p} \\ \dot{q} \\ \dot{r} \end{pmatrix} = \begin{pmatrix} \Gamma_1 pq - \Gamma_2 qr + \Gamma_3 l + \Gamma_4 n \\ \Gamma_5 pr - \Gamma_6 (p^2 - r^2) + \frac{m_m}{I_y} \\ \Gamma_7 pq - \Gamma_1 qr + \Gamma_4 l + \Gamma_8 n \end{pmatrix} \quad (5)\,(6)\,(7)$$

Kinematic equation

$$\begin{pmatrix} \dot{\phi} \\ \dot{\theta} \\ \dot{\psi} \end{pmatrix} = \begin{pmatrix} p + q tan(\theta) s(\phi) + r tan(\theta) c(\phi) \\ q c(\phi) - r s(\theta) \\ q \frac{s(\phi)}{c(\theta)} + r \frac{c(\phi)}{c(\theta)} \end{pmatrix} \quad (8)\,(9)\,(10)$$

Navigation equation

$$\begin{pmatrix} \dot{P}_n \\ \dot{P}_e \\ \dot{P}_d \end{pmatrix} = \begin{pmatrix} u(c(\theta)c(\psi)) + v(s(\phi)s(\theta)c(\psi) - c(\phi)s(\psi)) + w(c(\phi)s(\theta)c(\psi) + s(\phi)s(\psi)) \\ u(c(\theta)s(\psi)) + v(s(\phi)s(\theta)s(\psi) + c(\phi)c(\psi)) + w(c(\phi)s(\theta)s(\psi) - s(\phi)c(\psi)) \\ -u(s(\theta)) + v s(\phi)c(\theta) + w c(\phi)c(\theta) \end{pmatrix} \quad (11)(12)(13)$$

Where f_x, f_y and f_z represent the x, y and z components of forces acting on UAV in the body frame respectively. Similarly, the x, y and z (roll, pitch and yaw) components of torque (moment) acting on UAV in the body frame is represented by l , m_m and n respectively. Here s and c denote the sine and cosine of corresponding angles. Γ_1 to Γ_8 represent inertial parameters.

A dynamic model is then created for the aerosonde UAV in the Simulink workspace. Which is a non-linear 6-DoF plant model. The Figure 2 represents the generic block diagram representation of the dynamic model along with the PID controller connected with it. The control vector U is defined as given in Equation (14).

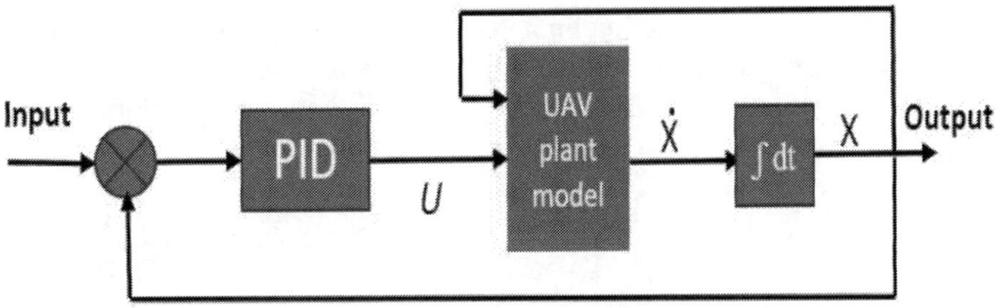

Fig. 2. Dynamic model.

$$U = (\delta_a, \delta_e, \delta_r, \delta_t)^T \quad (14)$$

Here $\delta_a, \delta_e, \delta_r,$ and δ_t represent aileron, elevator, rudder deflections and throttle command respectively.

Thereafter the non-linear 6-DoF plant model is trimmed for steady state and level flight (SSLF) conditions at 50 meters initial altitude. Optimization through Penalty method is used to trim the model [12]. The objective function was to minimize $\dot{X}$ subject to the constraints of SSLF which include Va (airspeed) to be fixed at a constant value of 25 m/s, flight path angle γ, roll angle ϕ, yaw angle ψ and y component of velocity v such that all will achieve zero state. In order to convert from constrained to unconstrained optimization problem penalty method comes in handy, where the objective function $F_0(Q)$ as given by

Equation (15) is now modified to include the SSLF constraints defined earlier along with $\dot{X}$. Here, *fminsearch* function defined in MATLAB is used to find the global minimum of the updated cost function. The optimization process is terminated once all constraints are met and the objective function is zero. The optimization process yielded the trim states X^* and corresponding trim control vectors U^* which are described through Equations (16) and (17).

$$F_0(Q) = Q^T \times H \times Q \tag{15}$$

Where $Q = (\dot{X}, Va\text{-}25, \gamma, \phi, \psi, v)^T$
H = A diagonal matrix of penalty parameters α.

$$X^* = \begin{pmatrix} u^* \\ v^* \\ w^* \\ p^* \\ q^* \\ r^* \\ \phi^* \\ \theta^* \\ \psi^* \end{pmatrix} = \begin{pmatrix} 24.9153 \\ -6.1623e^{-12} \\ 2.0557 \\ -7.8974e^{-12} \\ -5.9597e^{-12} \\ -3.7812e^{-12} \\ -9.9756e^{-12} \\ 0.0823 \\ -1.04e^{-11} \end{pmatrix} \tag{16}$$

$$U^* = \begin{pmatrix} \delta_a{}^* \\ \delta_e{}^* \\ \delta_r{}^* \\ \delta_t{}^* \end{pmatrix} = \begin{pmatrix} -5.0563e^{-13} \\ -0.1093 \\ -8.9993e^{-13} \\ 0.3335 \end{pmatrix} \tag{17}$$

The trimmed non-linear plant model of the UAV is now linearized about this trim point using the linear analysis tool of Simulink. The model is also equipped with a saturation check for the control surface deflection limitations [13]. To validate the SSLF condition the U^* and X^* values are fed into the model as initial conditions and the simulation is executed for 15 seconds time period and the results are shown in Figure 3, the plots prove that the aircraft model maintains itself in the steady state throughout the simulation period.

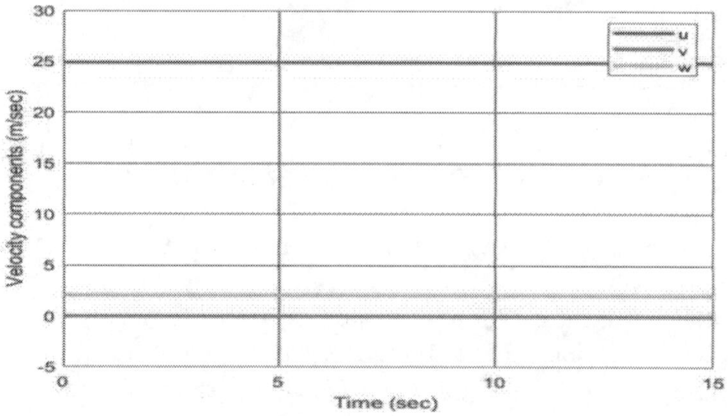

(a) Velocity components

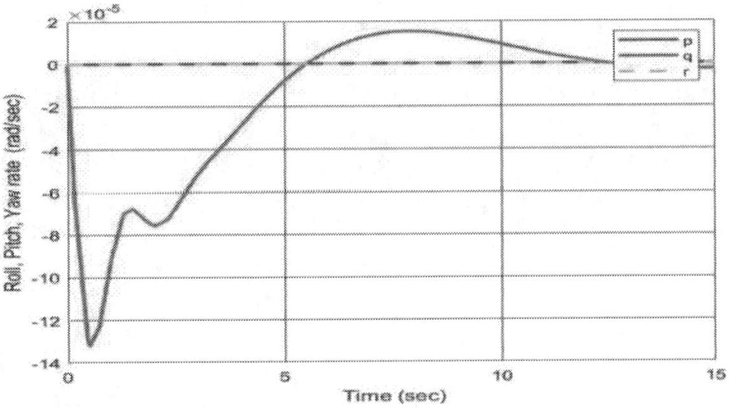

(b) Angular rates

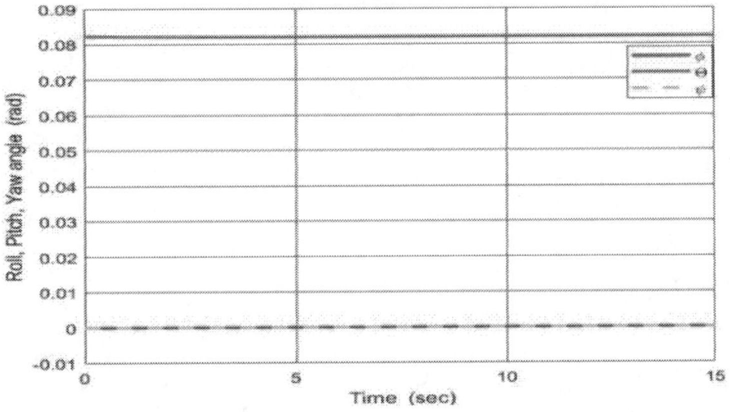

(c) Euler angles

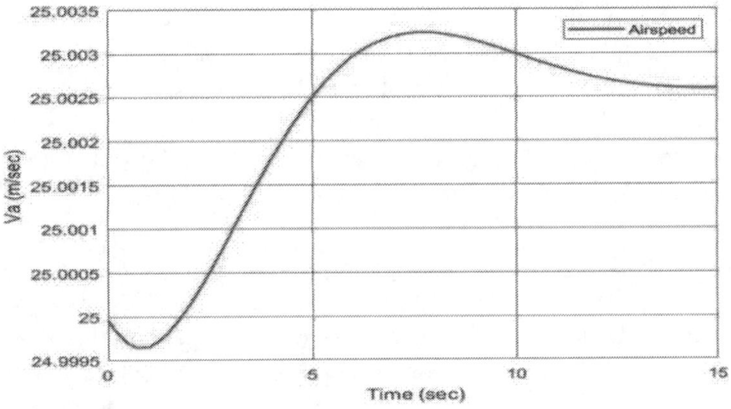

(d) Airspeed

Fig. 3. Trim validation.

3 Controller design

The success of any UAV flight depends on how reliable is the control system designed for carrying out the desired mission. The PID controller is the most widely used and proven controller for UAVs, the fact that PID controllers can easily be implemented in any aircraft system substantiates their extensive usage in today's world of controllers. The UAV control system is bifurcated into two controllers. The primary controller being the "Lateral Controller" and the other "Longitudinal Controller" [14]. The job of the lateral controller is to control the roll angle and heading (yaw) angle of the UAV. The pitch angle, altitude and airspeed are regulated by the lateral controller. Any controller designed achieves the desired performance characteristics of the UAV by continuously reducing the error term (difference between current state and desired state), hence a closed loop control system is necessary [15].

Proportional, Integral and Derivative terms make up a PID controller. The transfer function of the PID controller is given as

$$\mathrm{H(s)} = \frac{K_d.S^2+K_p.S+K_I}{S} \tag{18}$$

Where K_p, K_d and K_I represent proportional, derivative and integral gain constants. In real time PID controllers perform in the presence of external disturbance. Thus often for researchers and engineers, it's a great challenge to appropriately tune the gain constants of the PID controller in the above mentioned situation. In [13], the following techniques for tuning a PID controller are discussed,

- Trial and error method
- Analytical method
- Empirical method

In this work, the PID controller gains are adaptive and their formulation is based on formula based analytical method described in [11].

3.1 Autopilot design

In this work, the successive loop closure technique is utilized to design both lateral and longitudinal autopilot systems for aerosonde UAV [11]. In designing the autopilot with the mentioned technique, an assumption that the longitudinal dynamics are decoupled from the lateral dynamics is considered. This paper aims to design an adaptive PID based lateral autopilot system for controlling roll and yaw commands, similarly a longitudinal autopilot system for controlling pitch and altitude commands. The basic idea behind the successive loop closure technique is to wrap several simple feedback loops in succession around the plant dynamic model created. An important condition for this technique to be implemented is that the inner loop must have a higher bandwidth than the outer loop. Usually, the bandwidth separation between the loops is a factor of around 5 to 10.

3.1.1 Lateral autopilot design

For the lateral autopilot system, the inner loop is the roll attitude hold loop which is used to control roll angle ϕ and roll rate p. The outer loop is used for directional control (yaw angle ψ) of the UAV. The generic block diagram for such a lateral control loop is shown in Figure 4 (a).

The outer PID loop for directional control takes in the commanded yaw angle (ψ^c) as input and an error value is developed between this ψ^c and the current ψ state that is fed back from the plant. Thereafter appropriate PID gain values are obtained by analytical formulas given in [11], which also depend on the current airspeed (V_a). The output of this outer yaw PID loop is ϕ^c (roll command) state.

The inner PID loop for roll control takes in this ϕ^c state as input, this is then added to the ϕ^c actually commanded by the user and an error value is developed between the total ϕ^c and current ϕ state that is fed back from the plant. As mentioned above, the appropriate adaptive PID gain values are obtained by analytical formulas [11]. This inner PID loop has δ_a as output which is fed into the UAV dynamic plant model.

The complete lateral autopilot is modelled in Simulink workspace and connected to plant dynamics. With regard to the yaw control lateral autopilot, a PI controller is developed and for roll control lateral autopilot, a PD controller is developed.

3.1.2 Longitudinal autopilot design

In the case of the longitudinal autopilot system, the inner loop is the pitch attitude hold loop which is used to control pitch angle θ and pitch rate q. The outer altitude hold loop is used to achieve the desired altitude value for the UAV. The generic block diagram for such a longitudinal control loop is shown in Figure 4 (b).

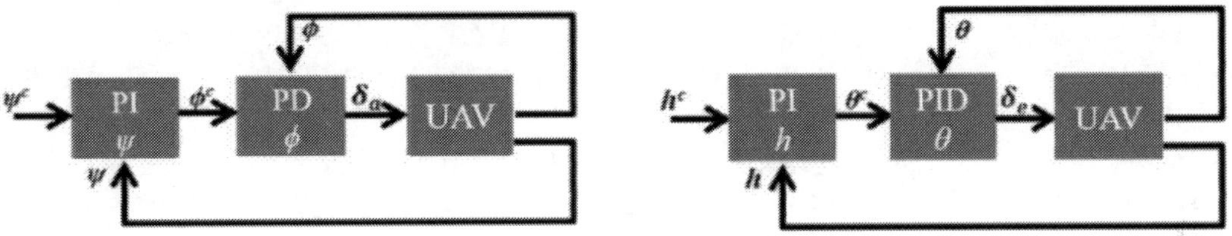

(a) Lateral (b) Longitudinal

Fig. 4. Autopilot system.

The outer PID loop for altitude control takes in the commanded altitude h^c as input. And an error value is developed between h^c and the current altitude h state that is fed back from the plant. The appropriate PID gain values are obtained by analytical formulas given in [11]. The output of this outer altitude PID loop is θ^c (pitch command) state.

The inner PID loop for pitch attitude control takes in this θ^c state as input, along with the θ^c actually commanded by the user and an error value is developed between this θ^c and the current θ state which is fed back from the plant. Then appropriate PID gain values are obtained by analytical formulas as given in [11]. This inner PID loop has δ_e as output which is fed into the UAV dynamic plant model.

Here the altitude control autopilot is based on an adaptive PI controller and for pitch attitude control autopilot an adaptive gain based PID controller is developed.

4 Dryden wind gust model

To check the effectiveness of the proposed adaptive gain based PID control system developed and to analyze the dynamic response of the UAV to external disturbances, the Dryden wind gust model is used in this work. The turbulence intensity σ and spatial wavelengths L are two parameters of the Dryden wind gust model. The Dryden transfer functions defined in [11] are used to build the model. A nominal constant airspeed of 15m/s is assumed for the implementation of Dryden model. The MIL-F-8785C military specification defines the values of these parameters at different altitudes. Suitable parameters for low and medium altitudes with light and moderate turbulence levels are presented in [16] and are shown in Table 3. The plant model is now updated to have additional inputs from the gust model developed.

Table 3. Dryden wind gust parameters.

Gust description	Altitude h (m)	$L_u = L_v$ (m)	L_w (m)	$\sigma_u = \sigma_v$ (m/s)	σ_w (m/s)
Low altitude, light turbulence	50	200	50	1.06	0.7
Low altitude, moderate turbulence	50	200	50	2.12	1.4
Medium altitude, light turbulence	600	533	533	1.5	1.5
Medium altitude, moderate turbulence	600	533	533	3	3

5 Simulation results

The developed autopilots are assessed in the presence of external wind conditions. The roll, yaw, pitch and altitude control loops are given a triangular input command and the corresponding response obtained from the plant is studied. The medium altitude, moderate turbulence class of the Dryden wind gust model is chosen to demonstrate the robustness of the developed adaptive PID autopilot system. The UAV is maintained at the trim condition at zero time instant (initial condition) and at an altitude of 600 meters.

5.1 Lateral autopilot

Figure 5 (a) shows the triangular input response of roll only adaptive control loop. It is observed that the response of ϕ is closely matched with ϕ^c . Figure 5 (b) displays the corresponding $K_d\phi$ and $K_p\phi$ gain variation.

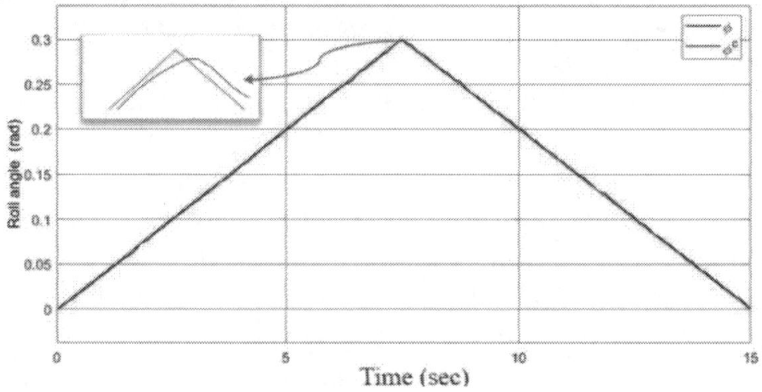

(a) Roll triangular response

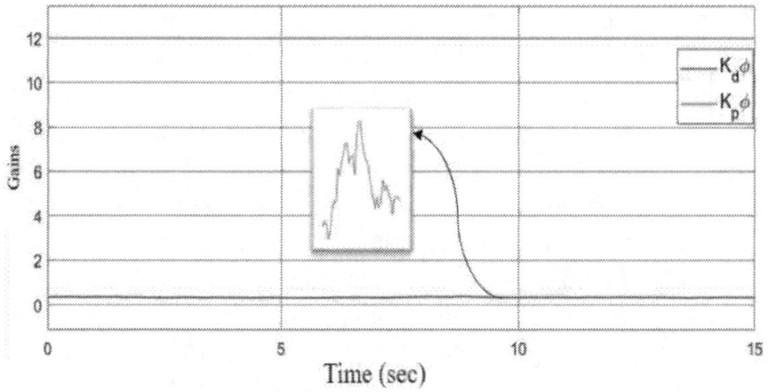

(b) Gain variation

Fig. 5. Roll only adaptive PID.

Figure 6 (a) presents the triangular input response for yaw and roll (complete lateral autopilot) adaptive control loop. It is seen that the response of ψ is closely following ψ^c. Corresponding $K_I\psi$ and $K_p\psi$ gain variations can be observed in Figure 6 (b). The ϕ commanded due to this yaw loop and the roll initially commanded in triangular form earlier (Fig 5), collectively gives the total ϕ^c, Figure 6 (c) depicts the response of the total ϕ^c achieved from the lateral adaptive PID control loop. The corresponding $K_d\phi$ and $K_p\phi$ gain variation is captured in Figure 6 (d).

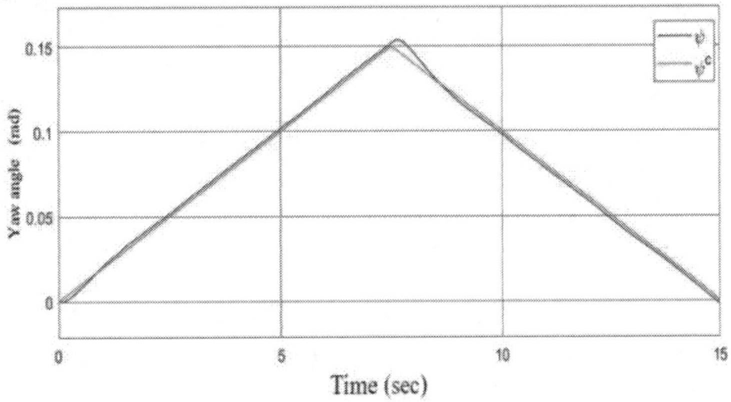

(a) Yaw triangular response

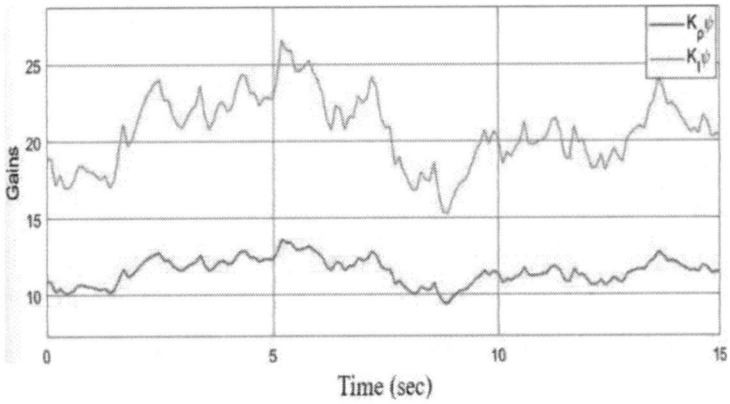

(b) Gain variation

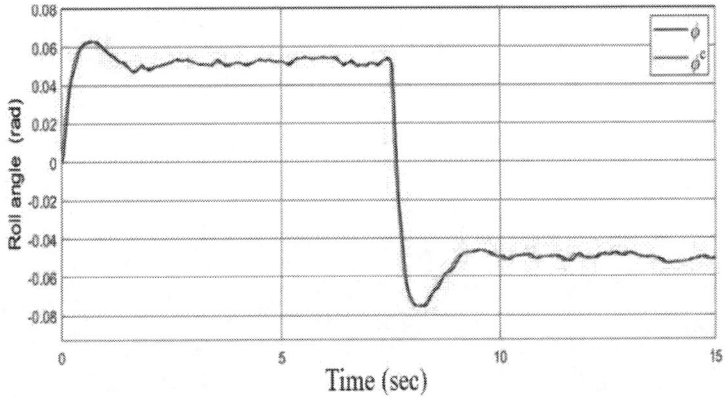

(c) Roll response

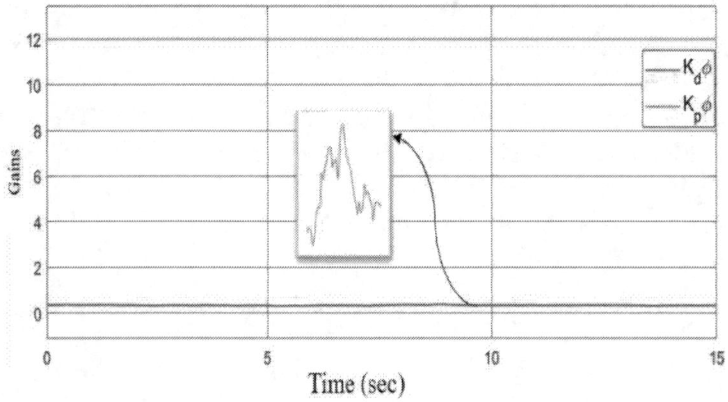

(d) Gain variation

Fig. 6. Complete lateral adaptive PID.

5.2 Longitudinal autopilot

The Pitch only adaptive control loop performance for applied triangular input is displayed through Figure 7 (a). It is observed that the response of θ is closely matched with θ^c. The variation of $K_p\theta$, $K_I\theta$ and $K_d\theta$ gains is visible in Figure 7 (b).

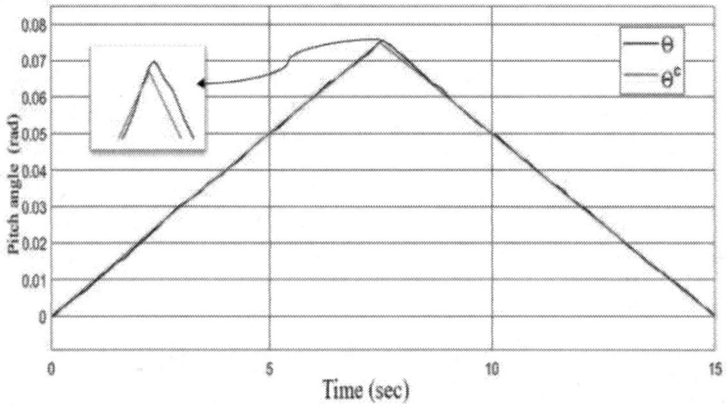

(a) Pitch triangular response

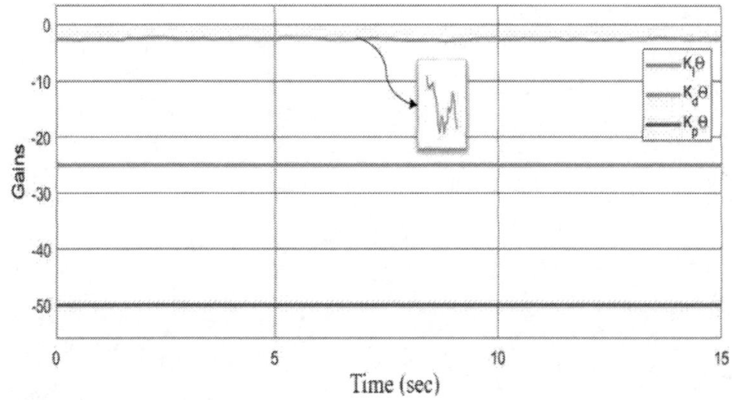

(b) Gain variation

Fig. 7. Pitch only adaptive PID.

At this point, Figure 8 (a) deals with the triangular input response for altitude and pitch (complete longitudinal autopilot) adaptive control loop. It is observed that the response of h is closely tracking h^c. Figure 8 (b) shows the corresponding gains K_ph and K_Ih variation. The change in θ commanded due to this altitude loop and the change in pitch initially commanded in triangular form earlier (Fig 7), collectively provide the total θ^c. Figure 8 (c) shows the response of total θ^c achieved from the longitudinal adaptive PID control loop. Figure 8 (d) depicts the corresponding gains $K_p\theta$, $K_I\theta$ and $K_d\theta$ variation.

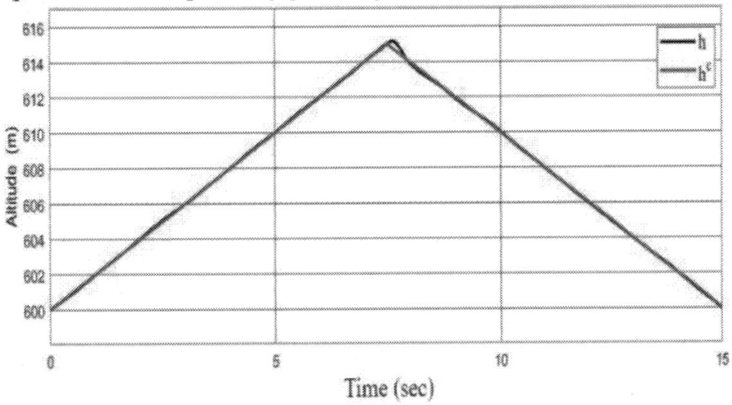

(a) Altitude triangular response

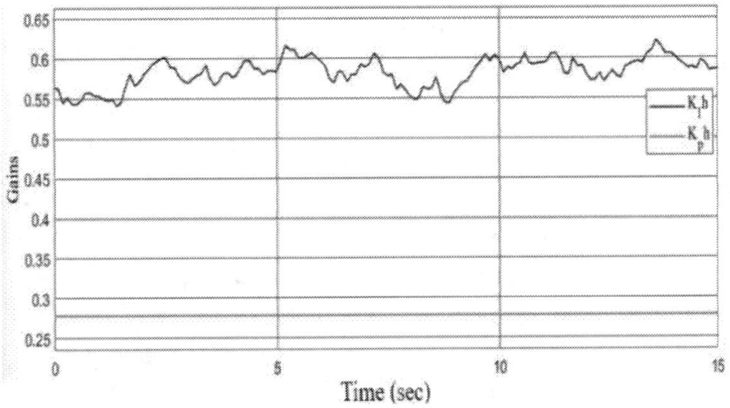

(b) Gain variation

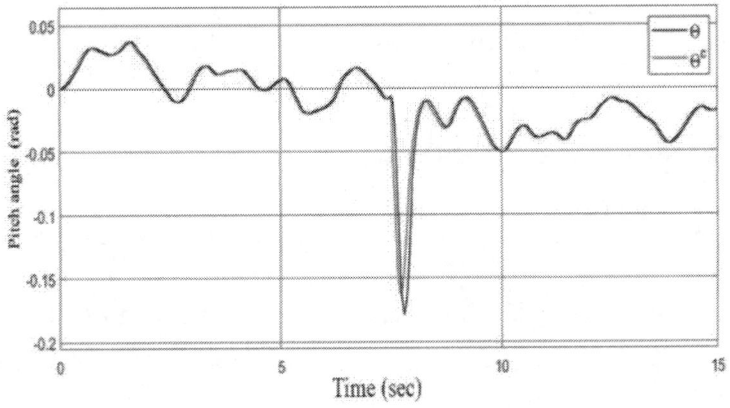

(c) Pitch response

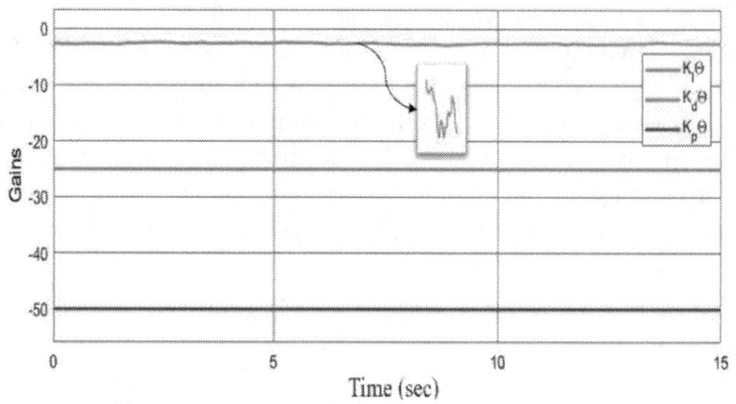

(d) Gain variation

Fig. 8. Complete longitudinal adaptive PID.

6 Conclusion

A Dynamic model of the Aerosonde UAV was developed and the adaptive PID autopilot controller was proposed for controlling both the attitude and altitude of the aircraft, the self-tuning based on adaptive PID gains was possible with the help of an analytical approach. The proposed controller was simulated under the existence of external wind disturbance. The external wind disturbance is based on the Dryden wind gust model. The results successfully validate the robustness of the autopilot system by effectively handling moderate turbulence at medium altitude.

References

1. J. Xu, Q. Guo, L. Xiao, Z. Li, G. Zhang, Autonomous Decision-Making Method for Combat Mission of UAV based on Deep Reinforcement Learning, in 4th (IAEAC) conference, 538-544 (2019)
2. Y. Sato, T. Yamasaki, H. Takano, Y.Baba, Trajectory Guidance and Control for a Small UAV, Int. J. Aeronaut. Space Sci. **7**, 137-144 (2006)
3. Valavanis, P. Kimon, Advances in Unmanned Aerial Vehicles (ISCA, Volume 33) Design and Control of a Miniature Quadrotor, (Springer 2007)
4. H. Castaneda, J. Morales, E. Diaz, Quasi-Continuous Sliding Mode Flight Control for a Fixed-Wing UAV, Congreso Nacional de Control Automatico, Ensenada, Baja California, Mexico, (October 2013)
5. B. Kada, Y. Ghazzawi, Robust PID Controller Design for an UAV Flight Control System, World Congress on Engineering, (2011)
6. https://www.textronsystems.com/products/aerosonde-uas, as visited on 17-November-2023
7. https://commons.wikimedia.org/wiki/Aerosonde, as visited on 17-November-2023
8. M. Niculescu, Lateral Track Control Law for Aerosonde UAV. 39th AIAA conference, USA (2001)
9. A. Sarhan, S. Qin, Robust Adaptive Flight Controller for UAV Systems, 4th ICISCE conference, 1214-1219 (2017)

10. B.Zhu, J. Zhu, Q. Chen, A bio-inspired flight control strategy for a tail-sitter unmanned aerial vehicle, Sci. China Inf. Sci. **63**, 1-10 (2020)
11. R. Beard, T. McLain, Small Unmanned Aircraft: Theory and Practice, (Princeton University Press, Princeton 2012)
12. O. Yeniay, Penalty Function Methods for Constrained Optimization with Genetic Algorithms. Math. Comput. Appl. **10**, 45-56 (2005)
13. A. Elbatal, A. Youssef, M. Elkhatib, Smart aerosonde UAV longitudinal flight control system based on genetic algorithm, Bul. Elect. Eng. Info. 2433-2441 (2021)
14. F.Lin, H. Duan, X. Qu, PID control strategy for UAV flight control system based on improved genetic algorithm optimization, 26th CCDC conference, 92-97 (2014)
15. S. Mathisen, F. Leira, H. Helgesen, K.Gryte, T. Johansen, Autonomous ballistic airdrop of objects from a small fixed-wing unmanned aerial vehicle, Auton. Robots, **44**, 859-875 (2020)
16. J. Langelaan, N. Alley, J. Neidhoefer, Wind Field Estimation for Small Unmanned Aerial Vehicles, J. Guid. Control Dyn. **34**, 1016-1030 (2010)

A comprehensive study on the design and computational analysis of an air swirl burner

Devansh Dhyani[1*] and *Swayam* Phade[1]

[1]VIT Bhopal University, School of Mechanical Engineering, Bhopal, Madhya Pradesh, India

Abstract. This paper presents a comprehensive study on optimizing the design of a vane-type air swirl burner, along with a computational analysis comparing the combustion of diesel and hydrogen fuel within the same burner. For the combustion simulations, a coupled approach was employed involving a Lagrangian approach with the Discrete Phase Model to track the trajectories of fuel particles providing a comprehensive understanding of the particulate dynamics within the system. This research aimed to analyze hydrogen's potential as a sustainable energy carrier in comparison to its conventional counterparts such as diesel fuel. The results showed that the contours of hydrogen fuel revealed a well-defined and intense flame front whereas it was not the same in the case of diesel fuel which had a less intense flame front.

1 Introduction

The critical need for energy-efficient and environmentally conscious combustion technologies have gained prominence amidst the global challenges surrounding traditional energy sources. As the environmental impact of these conventional resources undergoes scrutiny, there is a growing urgency to explore cleaner and more sustainable alternatives. In direct response to this imperative, the current research is carried out, focusing on the design and computational analysis of a vane-type air swirl burner with a fixed swirl number. The vane-type air swirl burner is a technology designed to enhance the combustion process by utilising its vanes to provide a swirling flow to create a well-defined and stable flame pattern. Swirling flows are commonly found in most practical applications, such as gas turbine combustors, furnaces, spraying machines, whirlpools, cyclone separators, and vortex shedding from aircraft wings [1,2]. The swirl intensity for rotating flows is usually characterized by the non-dimensional swirl number $S = G_{\theta},/G_z R$, which represents the ratio of the axial flux of angular momentum, G_θ to the axial flux of axial momentum, G_z times the nozzle radius [3]. In this study, a Lagrangian approach coupled with the Discrete Phase Model (DPM) was employed to investigate the dynamic behaviour of particles within a fluid medium, specifically focusing on the use of liquid hydrogen as fuel [4]. The Lagrangian approach is chosen for its ability to track individual particles throughout their trajectories [5]. It provides a detailed insight into the complex interactions and motions of discrete entities

* Corresponding author: devanshdhyani2022@vitbhopal.ac.in

within the liquid hydrogen flow. By adopting the Lagrangian perspective a better understanding of how particles respond to local flow conditions and navigate through intricate fluid structures was developed which is particularly relevant in the case of liquid hydrogen fuel. The integration of the Discrete Phase Model is used to simulate and analyze the movement of distinct particles, aiding in modelling the discrete nature of the dispersed phase in the liquid hydrogen environment [6]. This combined Lagrangian-DPM framework was chosen to address the specific research questions related to individual particle dynamics and interactions in the context of liquid hydrogen as a fuel, aligning with the intricate nature of the studied phenomena in this unique fuel setting.

There are numerous studies available in the literature that discuss the swirling flow, hydrogen combustion and the Discrete Phase Model. One such study examined isothermal free swirling jets for five different swirl numbers at a fixed Reynolds number [1] Later on the results were further evaluated for local entrainment rate in the initial region of isothermal free swirling jets by Stereo PIV and for coherent structures in the near-field region of an isothermal free swirling jet after vortex breakdown [3,7]. Another study explored the use of hydrogen-containing fuel blends to determine their flashback and blow-off limits in swirl burners using the Lewis and von Elbe critical boundary velocity gradient expressions [8]. The authors of another study examined the effect of hydrogen addition on the combustion behaviours of turbulent unconfined swirling flames. This study used the steady laminar flamelet combustion and the realizable k-$\mathcal{E}$ turbulence models to model the behaviour [9]. Based on the results, the amount of hydrogen in the fuel significantly affects the flame structure. Another study modelled a natural gas diffusion flame, taking into account the effect of the swirl number [10]. It is reported that combustion characteristics change significantly depending on changes in swirl number. A numerical study related to the CH4-H2 mixtures was also conducted, and the results showed that hydrogen addition leads to an increase in flame temperature [11]. An experimental study was conducted on hydrogen-methane blended fuels, and according to measurements, hydrogen-methane blended fuels can be used successfully without any changes to conventional natural gas burners [12]. One study explored the effects of swirl number on the combustion characteristics of hydrogen Another study conducted both experimental and numerical studies of gaseous fuel combustion in a swirl chamber [13]. A study was conducted to discuss the use of the discrete phase model in fluid flow [6]. Authors of another study were verified the discrete phase model with water particle flow in a tundish [5].

Although there are some studies about with and without swirling combustion processes, these studies are generally related to methane or natural gas flames. Investigations related to the combustion of hydrogen-containing fuels in swirl combustors are very restricted in the literature. Therefore, this study focuses on the combustion processes and properties of hydrogen fuel and compares those with the ones of diesel fuel with fixed swirl numbers.

2 Simulation setup

2.1 Apparatus

A vertically oriented vane-type swirl burner was designed with the height of 500 mm and the diameter of 100 mm and is used to carry out the combustion processes of hydrogen and diesel fuel. The fuel is injected through a fuel inlet which has a diameter of 10 mm. The swirl generator is of tangential entry type to promote a strong and controlled swirl, which is beneficial for achieving better mixing of fuel and air, as well as enhancing flame stabilization. The tangential air is introduced through 10 tangential inlets to impart angular momentum upstream from the base of the burner. As a vane-type swirl burner, it is equipped with a system of vanes, where the total number of vanes is 10, each having a length of 50 mm, a width of 4 mm, a height of 26 mm and a vane angle of 139.48 degrees.

2.2 Simulation properties

The software which has been used for the simulation is the Ansys Fluent 2023 R2 Student Version. The solver employed was of the pressure-based type along with the k-omega type of viscosity model. A coupled approach involving the Discrete Phase Model and the Non-Premixed Combustion Model was employed with the Probability Density Function. In the Discrete Phase Model simulation, the particle treatment involves unsteady particle tracking with a particle time step size of 0.0001 seconds over a single time step. The tracking parameters are configured with a maximum number of steps set at 10000 and a length scale of 0.0001 meters. The injection properties specify the use of droplets with a pressure-swirl atomizer, employing 100 particle streams. The injector is positioned 0.04m above the zero point. The point properties for the injected droplets include an initial temperature of 300K, a flow rate of 0.003 kg/s, and a simulation period from 0 to 20 seconds. The injector's inner diameter is 0.001 meters, and the spray is emitted with a half angle of 15 degrees. The total number of iterations is set to 900, reflecting the iterative nature of the solution process employed to achieve convergence and obtain accurate results.

3 Methodology

The first phase of the methodology involved the process of designing Computer-Aided Design (CAD) models for the burner and the vanes using SolidWorks 2022. The dimensions of both components are provided in the Apparatus section (2.1). Utilizing design software, the CAD models serve as the foundational blueprints for the subsequent simulations, ensuring a precise representation of the physical burner within the virtual environment.

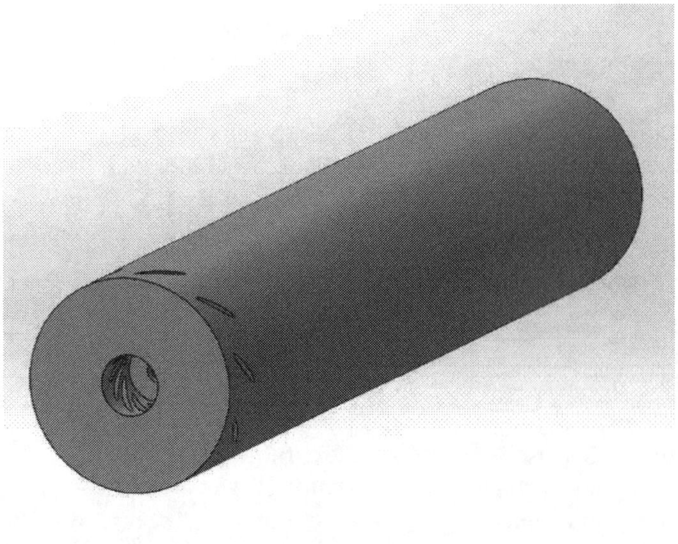

Fig. 1. CAD model of the vane-type air swirl burner

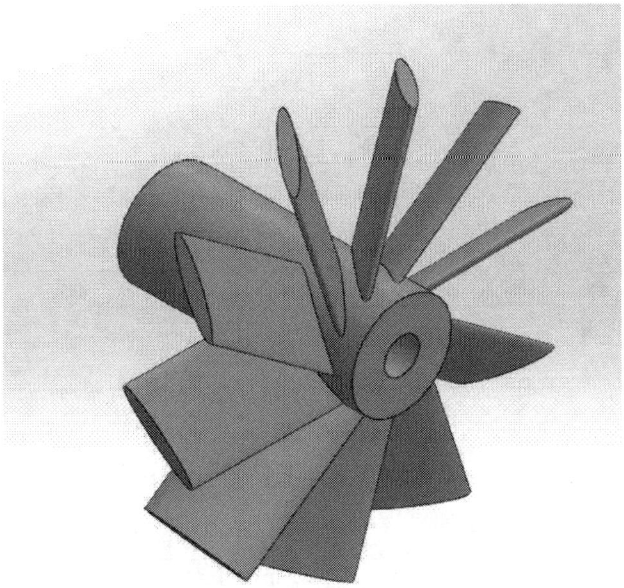

Fig. 2. CAD model of the vanes of the vane-type air swirl burner

Following the CAD design phase, the next step involves meshing the created models using Ansys. Meshing is a crucial process that discretizes the geometric structures into smaller, manageable elements, allowing for the accurate representation of fluid flow and combustion phenomena during simulations.

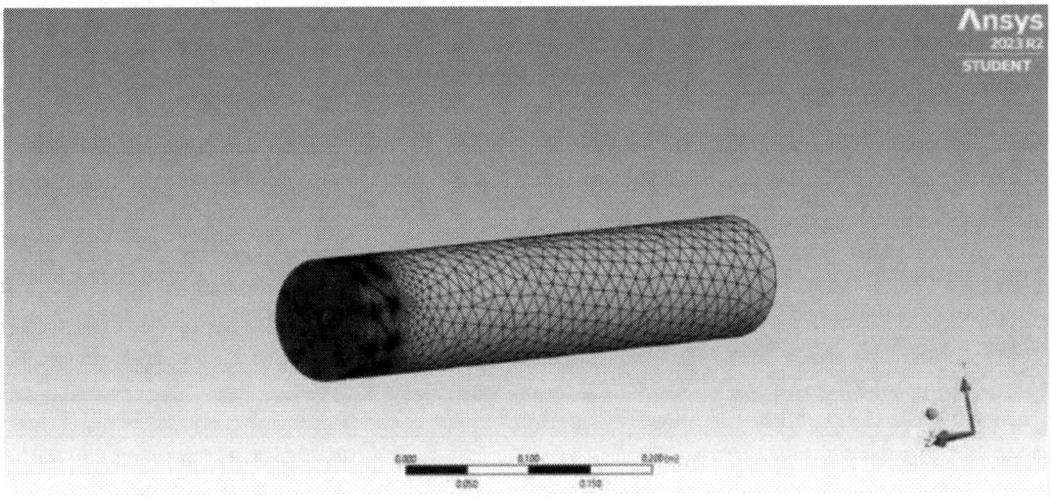

Fig. 3. Meshed model of the burner

Table 1. Properties of the mesh

Display Style	Use Geometry Setting
Physics Preference	CFD
Solver Preference	Fluent
Element Order	Linear
Element Size	9.e-002 m (0.09 m)
Export Format	Standard
Export Preview Surface Mesh	No
Use Adaptive Sizing	No
Growth Rate	Default (1.2)
Max Size	Default (0.18)
Mesh Defeaturing	Yes
Defeature Size	Default (4.5e-004 m)
Capture Curvature	Yes
Curvature Min Size	Default (9.e-004m)
Curvature Normal Angle	Default (18.0 degrees)
Capture Proximity	No
Bounding Box Diagonal	0.51962 m
Average Surface Area	1.138e-002 m^2
Minimum Edge Length	3.1416e-002 m
Check Mesh Quality	Yes
Target Skewness	Default (0.9)
Smoothening	Medium
Mesh Metric	None
Use Automatic Inflation	None
Inflation Option	Smooth Transition
Transition Ratio	0.272
Maximum Layers	5
Growth Rate	1.2
Inflation Algorithm	Pre
View Advanced Options	No
Number of CPUs for Parallel Part Meshing	Program Controlled
Rigid Body Behaviour	Dimensionally Reduced
Triangle Surface Mesher	Program Controlled
Topology Checking	Yes
Pinch Tolerance	Default (8.1e-004)
Generate Pinch on Refresh	No
Nodes	57332
Elements	300692
Show Detailed Statistics	No

4 Results and discusssion

This section deals with the simulation results focusing on the combustion characteristics of hydrogen and diesel fuel within the vane-type air swirl burner. First, the flame stability and the energy release patterns of both fuels will be discussed with the help of the contours of the total temperature and pathlines (coloured by total temperature). Second, the results of the Discrete Phase Model will be presented through its scaled residual graphs.

4.1 Flame stability and energy release patterns

In the case of hydrogen combustion, the total temperature contours reveal a well-defined and intense flame front. The contours (Fig.5.) exhibit elevated temperatures concentrated in the core of the flame, which signifies effective combustion and substantial energy release. The intense heat in this region indicates that hydrogen, a highly reactive and clean-burning fuel, undergoes rapid combustion, resulting in a concentrated and high-temperature flame. This is a characteristic feature of hydrogen's combustion behaviour, known for its ability to produce high-temperature flames [13,14].

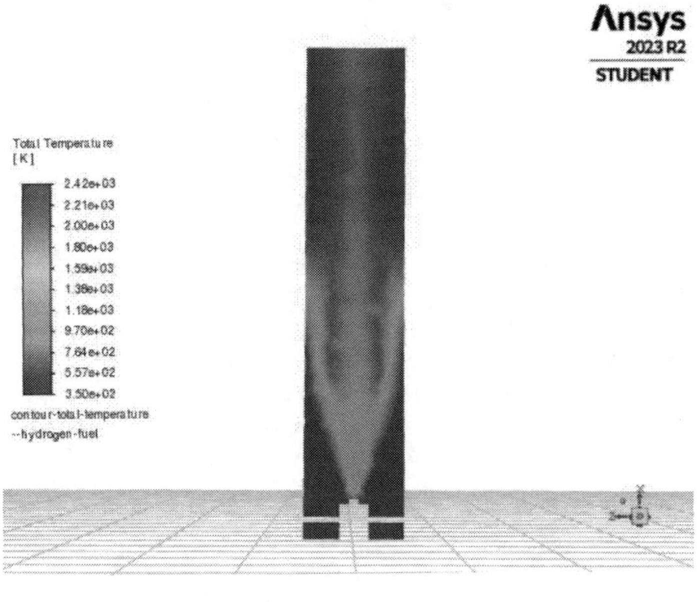

Fig. 5. Contour of Total Temperature (Hydrogen Fuel)

Contrasting the results for hydrogen, the total temperature contours (Fig.6.) for diesel fuel showcase a broader and less intense flame front. The contours depict variations in temperature across the combustion zone, reflecting the different combustion characteristics of diesel. Diesel combustion typically involves a more gradual and distributed release of energy compared to hydrogen. The broader flame front suggests a longer combustion duration and a more dispersed heat release pattern. This behaviour aligns with the characteristics of diesel combustion, where the ignition and combustion processes are influenced by factors such as fuel properties and air-fuel mixing [13,15].

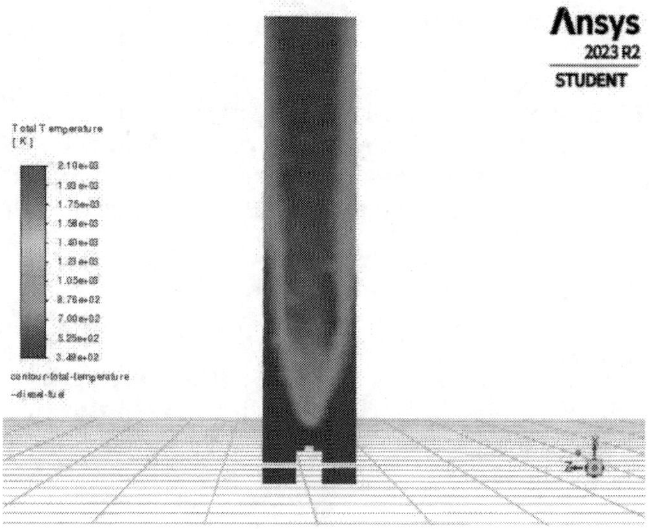

Fig. 6. Contour of Total Temperature (Diesel Fuel)

In the case of hydrogen combustion, the contours of pathline intensity coloured by total temperature reveal a dynamic and well-defined swirling motion within the Air Swirl Burner. The pathlines, colored to represent total temperature variations, showcase a concentrated and intense flame front characterized by high temperatures. The swirling motion contributes to efficient fuel-air mixing, promoting uniform combustion and elevating temperatures in the core of the flame. The result is a visually striking representation of the optimized combustion of hydrogen, with the pathlines vividly illustrating the heat release patterns and the spatial distribution of high-temperature regions. The intense and concentrated pathlines, colored by elevated total temperatures, underscore the effectiveness of hydrogen combustion within the Air Swirl Burner. The visualization reinforces the notion of hydrogen as a high-energy and clean burning fuel, with the pathlines highlighting the controlled and efficient combustion that contributes to the overall performance of the burner.

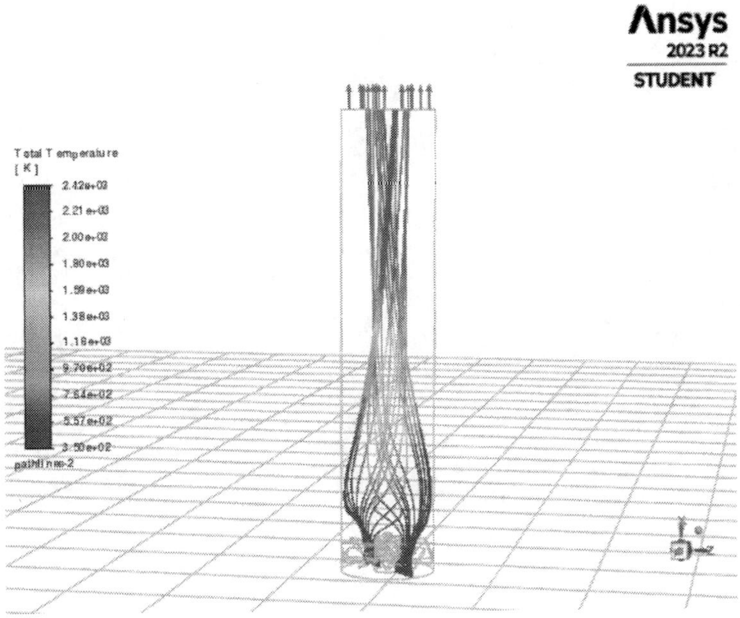

Fig. 7. Contours of pathlines coloured by Total Temperature (Hydrogen)

Contrasting the results for hydrogen, the contours of pathline intensity coloured by total temperature for diesel fuel exhibit a broader and less intense flame front. The pathlines, coloured to represent total temperature variations, convey a distributed and varied temperature profile across the combustion zone. The swirling motion induced by the burner is evident, but the characteristics of the pathlines reflect the differences in combustion behaviour between diesel and hydrogen. The visualization of pathline intensity for diesel fuel provides insights into the airflow dynamics and temperature distribution, showcasing variations in the heat release pattern compared to hydrogen. While diesel combustion still benefits from the swirling motion, the pathlines indicate a more gradual and dispersed distribution of high temperatures, aligning with the expected characteristics of diesel combustion [14].

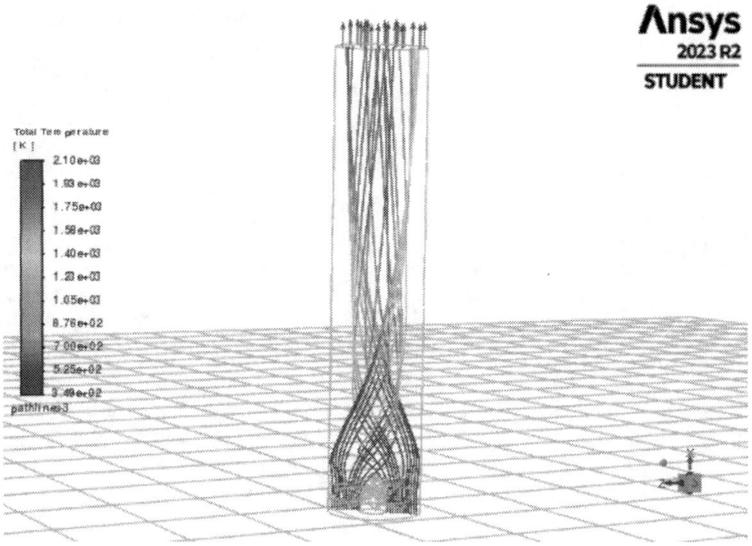

Fig. 8. Contour of Pathlines coloured by Total Temperature (Diesel)

4.2 Scaled Residual Graphs

The scaled residual graphs of the Discrete Phase Model (DPM) provide a quantitative assessment of the convergence and stability of the simulations for both hydrogen and diesel fuels. The scaled residual graph for the Discrete Phase Model during hydrogen combustion exhibits a rapid and consistent decline over the simulation iterations. This trend signifies the convergence of the DPM simulation, indicating that the computational model effectively captures the trajectories, dispersion, and combustion of hydrogen fuel particles within the swirling airflow [16]. The diminishing residuals reflect the robustness of the DPM model in representing the discrete nature of hydrogen particles, providing a high level of accuracy in simulating their interactions with the gaseous phase.

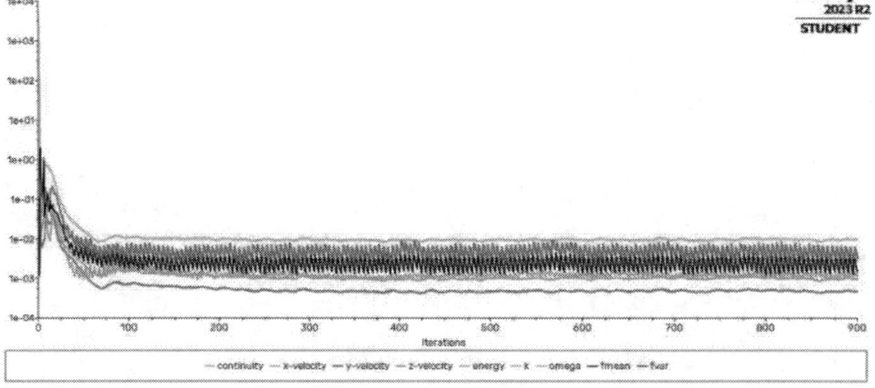

Fig. 9. Scaled Residual Graph (Hydrogen Fuel)

Similarly, the scaled residual graph for the Discrete Phase Model in diesel combustion follows a consistent decline, indicating the convergence of the DPM simulation for diesel fuel particles. The diminishing residuals affirm the computational accuracy of the model in representing the trajectories and combustion of diesel particles within the swirling airflow of the burner. The convergence of the scaled residuals in the DPM graph for diesel fuel underlines the model's effectiveness in simulating the behaviour of discrete fuel particles specific to diesel combustion [16,17].

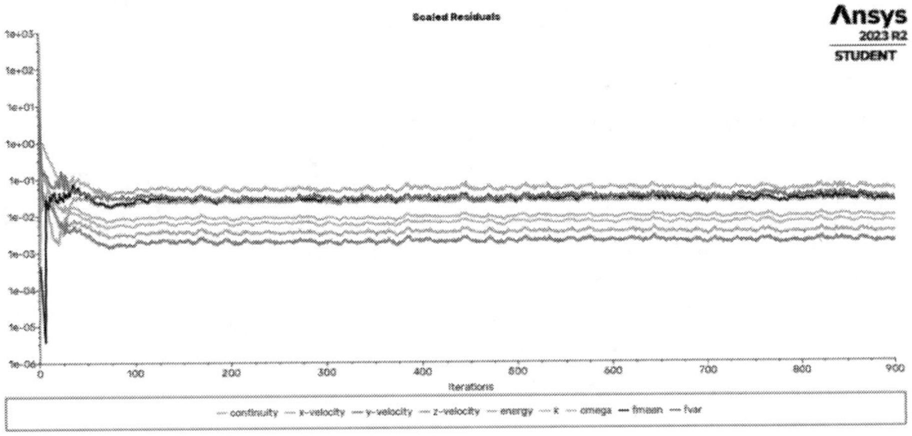

Fig. 10. Scaled Residual Graph (Diesel Fuel)

5 Conclusion

This study addressed the critical need for energy-efficient and environmentally conscious combustion technologies, particularly in the context of hydrogen as a fuel in swirl burners. Employing a Lagrangian approach with the Discrete Phase Model, the research investigated the dynamic behavior of particles in swirling airflow with a fixed swirl number, emphasizing the combustion properties of liquid hydrogen. The methodology involved designing burner and vane models with SolidWorks, meshing with Ansys for precise representation in simulations. Results revealed distinctive features of hydrogen and diesel combustion in the air swirl burner, showcasing the effectiveness of hydrogen combustion with concentrated and intense flame fronts and efficient fuel-air mixing. Diesel combustion, on the other hand, exhibited a broader and less intense flame front. Scaled residual graphs of the Discrete Phase Model validated simulation convergence and stability for both fuels, affirming the accuracy of the computational approach in capturing trajectories, dispersion, and combustion of fuel particles. This research provided valuable insights into the combustion behavior of hydrogen fuel in swirl burners, emphasizing its potential as a clean and efficient alternative, and extended our understanding of combustion processes, contributing to the foundation for further research in energy-efficient and environmentally conscious technologies.

References

1. F. Cozzi, R. Sharma, A. Coghe, F. Arzuffi, *An experimental investigation on isothermal free swirling jet*, in Proceedings of thmhe 38th Meeting of the Italian Section of the Combustion Institute, 20-23 September, Lecce, Italy (2015)

2. R. Sharma, F. Cozzi, A. Coghe, *Phase-averaged characterization of turbulent isothermal free swirling jet after vortex breakdown*, in Proceedings of the 18th International Symposium on the Application of Laser and Imaging Techniques to Fluid Mechanics, 4-7 July Lisbon, Portugal (2016)

3. F. Cozzi, R. Sharma, G. Solero, Exp. Therm. Fluid Sci., **109** (2019). https://doi.org/10.1016/j.expthermflusci.2019.109860

4. N.A. Aziz, M.H. Zawawi, N.M. Zahari, A. Abas, A. Azman, *Simulation of Homogeneous Particle Size in Fluid Flow by Using DPM-DEM*, in the proceddings of ICDSME, Singapore (2019)

5. A. Vakhrushev, M. Wu, *Verification of a Discrete Phase Model with Water-Particle Flow Experiments in a Tundish*, in the proceddings of 5th International Confrence STEELSIM, 10-12 September Technical University of Ostrava, Czech Republic (2013)

6. N. M. Zahari, M. Hafiz, L. M. Sidek, D. Mohamad, Z. Itam, M.Z. Ramli, A. Syamsir, A. Abas, M. Rashid, *Introduction of discrete phase model (DPM) in fluid flow*, in proceddings of AIP Confrence, 29-30 April Ho Chi Minh, Vietnam (2018)

7. F. Cozzi, A. Coghe, R. Sharma, Exp. Therm. Fluid Sci., **94** (2018). https://doi.org/10.1016/j.expthermflusci.2018.01.013

8. N. Syred, M. Abdulsada, A. Griffiths, T. O.'Doherty, P. Bowen, Appl. Energy, **89**, (2012). https://doi.org/10.1016/j.apenergy.2011.01.057

9. B. Rohani, K.M. Saqr, Int. Commun. Heat Mass Transf, **39**, (2012). https://doi.org/10.1016/j.icheatmasstransfer.2012.03.020

10. İ. Yılmaz, J.Energy Resour. Technol, **135**, (2013). https://doi.org/10.1115/1.4024222

11. L. Ziani, A. Caker, K. Chetehouna, A. Malek, B. Mahmah, Int. J. Hydrog. Energy, **38**, (2013). https://doi.org/10.1016/j.ijhydene.2012.11.104

12. I. Yilmaz, M. Ilbas, Int. Commun. Heat Mass Transf, **35**, (2008). https://doi.org/10.1016/j.icheatmasstransfer.2007.06.004

13. M. Ilbas, S. Karyeyen, I Yilmaz, Int. J. Hydrog. Energy, **41**, (2016). https://doi.org/10.1016/j.ijhydene.2015.12.107

14. M.S. Cellek, A. Pınarbaşı, Int. J. Hydrog. Energy, **43** (2018). https://doi.org/10.1016/j.ijhydene.2017.05.107

15. A. Adam, W. He, Y. Fan, D. Han, Int. J. Hydrog. Energy, **49** (2023). https://doi.org/10.1016/j.ijhydene.2023.07.330

16. M. Mayrhofer, M. Koller, P. Seemann, H. Boardbar, R. Prieler, C. Hochenauer, Int. J. Hydrog. Energy, **47** (2022). https://doi.org/10.1016/j.ijhydene.2021.11.236

17. R. Sharma, F. Cozzi, Meccanica, **58** (2023). https://doi.org/10.1007/s11012-023-01695-w

Multipurpose Surveillance Robot using Arduino Mega 2560 and Bluetooth Module HC-05

MD Zobaer Hossain Bhuiyan[1] and **Sabrina Akter Sabina**[1]

[1]Electrical and Computer Engineering, North South University, Dhaka, Bangladesh

Abstract:
Robots are becoming increasingly important for security and surveillance applications. This research paper presents the design and hardware implementation of a low-cost and easy-to-assemble site surveillance robot using an Arduino Mega 2560 microcontroller and a Bluetooth module HC-05. The robot is equipped with various modules and sensors, including a camera and ultrasonic sensors, for capturing images and detecting obstacles, respectively. The Arduino Mega 2560 is used to control the movement of the robot and process sensor data, while the Bluetooth module is used for wireless communication between the robot and an Android device. The robot is capable of patrolling a designated area and sending real-time surveillance footage to an Android device via Wi-Fi. The developed system was tested and evaluated to demonstrate its performance and efficiency. The results showed that the robot is able to patrol a designated area accurately and efficiently and that the real-time surveillance footage is of high quality. The system has the potential to be used in a variety of applications, such as object retrieval, security, and surveillance in residential and commercial areas and spying on certain sites.

Keywords: Remote Monitoring, Security surveillance, Search and rescue, Control system, spying robot

1. Introduction

In recent years, there has been a growing need for efficient and cost-effective surveillance and spying systems. The capability to monitor and detect potential threats in a designated area is crucial for maintaining security and safety. One of the most promising solutions to this problem is the development of robotic systems that are capable of performing a wide range of tasks, including spying, surveillance, and object retrieval. In this research, we present the design and implementation of an area surveillance robot using an Arduino Mega 2560 microcontroller and a Bluetooth module HC-05. The robot is equipped with a camera for capturing images as well as providing live feeds and a motor driver for movement.

The development of multipurpose surveillance robots is important for multiple reasons. First, these robots can provide a more comprehensive and flexible

[1]MD Zobaer Hossain Bhuiyan: zobaer.bhuiyan@northsouth.edu

surveillance solution than traditional surveillance systems. Second, they can be used in a variety of environments, including indoors, outdoors, and in hazardous locations. Third, they can be easily deployed and controlled, making them ideal for use in emergency situations.

The system is designed to be low-cost, portable, and easy to use. The Arduino Mega 2560 microcontroller, which is widely used in robotics and DIY communities, provides a cost-effective solution for controlling robot movement. The Bluetooth module HC-05 enables wireless communication between the robot and the Android device, allowing real-time monitoring of the designated area.

Results of the experimentation show that the robot is able to effectively surveil a designated area and transmit real-time images to a remote device with minimal latency. The robot's ability to move and capture images allows it to cover a larger area than fixed surveillance systems. This design has potential applications in various fields, such as security surveillance, remote monitoring, spying, object retrieval, and search and rescue operations. The goal of this research is to develop a cost-effective, easy-to-use, and efficient surveillance robot that can be used in a variety of applications.

In future work, the authors plan to improve the robot's autonomy and to integrate additional sensors to increase its capabilities. For example, integrating sensors like thermal or LIDAR can help the robot detect potential threats in low-light or obscured environments. Furthermore, the integration of machine vision algorithms can help the robot to analyze the images captured by the camera and detect potential threats in real-time.

2. Related Works

Robotics is very popular in the field of surveillance and rescue. Researchers have developed ways to transmit live video and control robots wirelessly over long distances. The most common method is to use a wireless camera and expensive sensors and mount them on the robot to get the live video and other environmental parameter details. The robot is controlled with an Android Device within the range of a stable Bluetooth connection.

In [1], the authors propose a surveillance robot that can be used to monitor hazardous environments without the need for human intervention. The robot is equipped with different kinds of sensors, including a PIR sensor, ultrasonic sensor, and gas sensor, which allow it to detect movement, obstacles, and hazardous gases. The robot is also equipped with a wireless camera, which allows it to stream live video footage to a remote user. The robot is primarily controlled by an Arduino UNO microcontroller, which is programmed to follow a predefined path or to navigate autonomously. The robot can also be commanded manually by a remote user using a smartphone or PC.The authors tested the robot in a simulated hazardous environment and found that it could successfully detect and monitor movement, obstacles, and hazardous gases. The robot could also stream live video footage to a remote user.The authors conclude that the proposed surveillance robot is a viable solution for monitoring hazardous environments without human intervention.

In [2], the authors propose a small unarmed ground vehicle (SUGV) that can be used for defense and security purposes. The SUGV is designed to be cost-effective and easy to build, making it a viable option for developing countries and small organizations. The SUGV is built on a toy car platform and uses an Arduino Uno microcontroller for control. It is equipped with various sensors, including a PIR sensor, ultrasonic sensor, and gas sensor, which allow it to detect movement, obstacles, and hazardous gasses. The SUGV is also equipped with a wireless camera, which allows it to stream live video footage to a remote user. The SUGV can be controlled manually by a remote user using a smartphone or PC. It can also be programmed to follow a predefined path or to navigate autonomously. The authors tested the SUGV in a simulated hazardous environment and found that it could successfully detect and monitor movement, obstacles, and hazardous gasses. The SUGV could also stream live video footage to a remote user. The authors conclude that the proposed SUGV is a viable solution for defense and security purposes, particularly for developing countries and small organizations.

Authors of [3] have made a spherical spy robot for surveillance operations. The robot is designed to be stealthy and maneuverable, and it is equipped with a variety of sensors, including cameras and microphones. The robot can be controlled remotely using a smartphone or tablet. The robot's spherical shape makes it ideal for surveillance in confined spaces, as it can roll around obstacles and change direction easily. The robot is also equipped with a pendulum-based drive system, which allows it to move over uneven surfaces. The robot is powered by a battery, and it can be programmed to follow a variety of predefined paths or to move autonomously. The robot can also be programmed to avoid obstacles and to return to its starting point when the battery is low. The paper concludes by discussing the applications of the spherical spy robot in surveillance operations. The robot could be used to monitor sensitive areas, such as military bases and government buildings, or to track criminals and terrorists.

Authors of [4] have made an economical and effective robot using an Arduino microcontroller and an Android smartphone. The robot has a robotic claw, various critical sensors and detectors, and a high-resolution camera. It can be controlled remotely using the internet from a laptop or tablet, and the live video feedback is obtained from the wireless camera of the smartphone. The authors argue that modern smartphones are built with hardware that satisfies all the requirements for a surveillance and rescue robot, such as a high-resolution camera, GPS, accelerometer, and internet connectivity. Adopting a smartphone immensely reduces the cost of building a classic surveillance and object retrieval robot. The robot can be used for Surveillance of sensitive areas, military borders, government offices, public places, and homes, Search and rescue operations in disaster zones, Inspection of hazardous or inaccessible areas, and Delivery of medical supplies to remote areas. The authors also discuss the possibility of using solar power to make the robot self-sufficient and capable of operating in any environment. Overall, the journal presents a promising new approach to building surveillance and rescue robots. Using an Android smartphone makes the robot more affordable and accessible to a wider range of users.

In[5], we can see the authors of this journal design an IoT-based Wi-Fi surveillance robot with real-time audio and video streaming. The robot has a camera and microphone to capture audio and video and a Wi-Fi module to transmit the data to a remote device. The robot can be operated using a web browser or mobile app, and the user can view the live video and audio stream from the robot's camera and microphone. The robot is based on an Arduino Uno R3 microcontroller board. The camera and microphone modules are connected to the Arduino board, and the Wi-Fi module is connected to the Arduino board using a serial interface. The Arduino board is responsible for processing the data from the camera and microphone and transmitting it to the remote device over the Wi-Fi network. The robot is powered by a 9V battery. The chassis for the robot is made of acrylic sheets. The motors and wheels are connected to the Arduino board using a motor driver board. The robot has two ultrasonic sensors to detect obstacles. The web browser interface allows the user to control the robot's movement, view the live video and audio stream from the robot's camera and microphone, and take snapshots, and record videos. The mobile app interface provides the same functionality as the web browser interface.

In[6], the authors propose a surveillance robot that can be used to monitor the work performance of humans in hazardous areas. The robot is controlled using an Arduino microcontroller and a Wi-Fi module. It has a camera that can be used to capture live video of the work area. The robot also has sensors that can be used to identify obstacles and humans. The robot can be operated manually or autonomously. The robot was tested in a simulated hazardous environment and was found to be effective in monitoring the work performance of humans. The robot could identify obstacles and humans in its path and avoid them. The robot could also capture live video of the work area and send it to the operator. The authors conclude that the surveillance robot has the potential to be used in a variety of hazardous environments, such as construction sites, chemical plants, and nuclear power plants. The robot can be used to improve safety and productivity in these environments.

The authors of [7], have proposed the design and development of an autonomous multi-purpose surveillance and rescue robot based on the Internet of Things (IoT). The robot is designed to perform a range of functions such as surveillance, rescue, and bomb detection. It has several sensors, including a camera, infrared sensor, gas sensor, and metal detector. The robot also has a microcontroller and a Wi-Fi module, allowing it to be operated remotely and send data to a central server. The proposed robot has the potential to be a useful tool in a wide range of applications. It can be used to detect bombs, monitor risky regions, and rescue individuals from disaster zones. Because it can work without human assistance, the robot's autonomous capabilities make it well-suited for certain jobs.

In [8], the authors propose a border monitoring robot system. An Arduino microcontroller, a camera, and a mechanical arm comprise the system. The robot can traverse over rocky and watery terrain and send live video to a receiver screen. The authors hope that in the future, this type of robot can save lives and replace soldiers.

The authors of [9], have proposed the design and development of a Bluetooth-controlled robot using Arduino. The authors used HC-05 Bluetooth Module, and L298N Motor Driver Module for this project. Android is the operating system of smart phones, and it is capable of developing effective remote control programs. At the same time, this program communicates with the robot via Bluetooth. It has been demonstrated that significant two-way communication between the Android phone and the robot is possible, allowing a non-expert to interact with and change the system's functioning. Authors functionalized the project for a better future.

3. Methodology

3.1 System Architecture

The Arduino Mega 2560 microcontroller is the central processing unit of the robot. It is responsible for controlling the robot's movement, processing sensor data, and communicating with the Android device via Bluetooth. The HC-05 Bluetooth module is used for wireless communication between the robot and the remote control device, which, in this case, is an Android phone. It allows the user to control the robot. The camera captures images and live feeds of the surrounding area. An ultrasonic sensor is used to encounter obstacles in the robot's path. The servo motor controls the movement of the robot's claw. It is also responsible for grabbing any object by the robot to remove it from its path or take it back to its navigator. The L298N motor driver controls the robot's movement by controlling the DC motors. The battery pack provides power to the robot. The circuit diagram of the whole system is described in **Fig 1**.

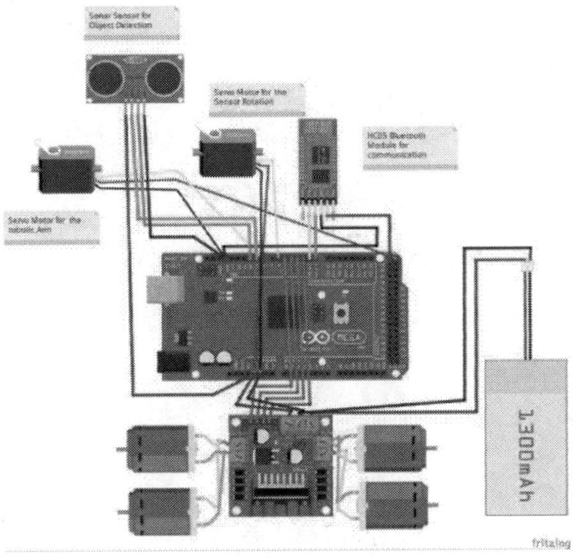

Fig. 1. Circuit Diagram of the Robot

The robot is controlled using the Arduino Bluetooth Controller app[4] for Android. The app allows the user to send logical strings to the robot to move forward or backward, turn left or right, and grab objects. If an obstacle is detected, the robot will change its direction to avoid it. The robot is primarily controlled manually using the app. The user can use the app to move the robot to a specific location or to grab an object and return it back to the starting point.

Fig. 2. Workflow Steps

Algorithm Description: The algorithm works by first initializing all of the necessary pins and variables. Then, it enters a while loop that runs as long as the distance to the obstacle is less than or equal to 10. Inside the while loop, the algorithm first checks if any data is available on the serial port. If there is, it reads the state from the serial port. If the state is equal to 'S', the robot stops. If the state equals 'B', the robot moves backward. If the state is equal to 'F', the robot moves forward. If the state equals 'R', the robot turns right. If the state equals 'L', the robot turns left. If the state equals 'W', the robot moves the servo motor to 90 degrees, making the claw open to grab any object. If the state equals 'E', the robot moves the servo motor to 0 degrees, closing the claw to grab the object successfully.

The algorithm was designed to be simple and efficient at the same time so that anyone can understand the robot's code. It is also very flexible, as the user can change the robot's state by sending commands to the serial port.

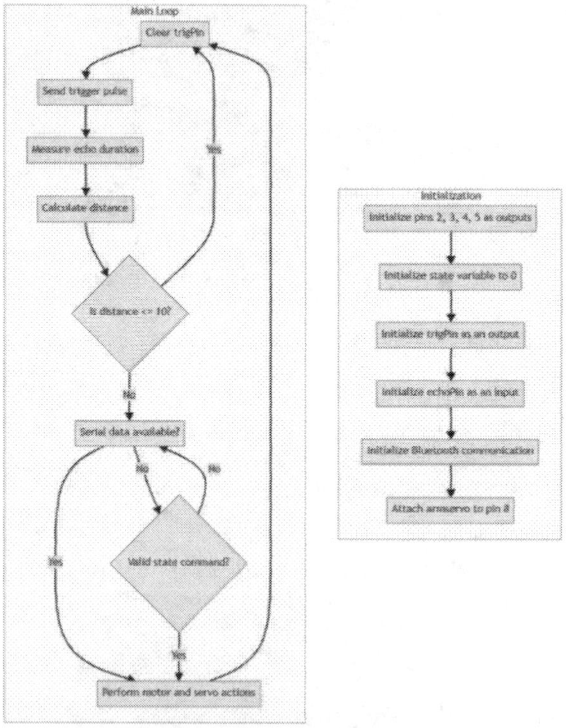

Fig. 3. Flowchart Diagram of the Algorithm

3.2 Results

1. Remote Control via Android Device: The developed surveillance robot demonstrated remarkable versatility by being easily controllable using an Android device. Through the integration of the Bluetooth module HC-05, the robot can be remotely operated, making it accessible for users to navigate and monitor specific areas with ease. This feature enhances the robot's adaptability in various surveillance scenarios, as users can exercise real-time control through a user-friendly interface.

2. Live Video Transmission: One of the key accomplishments of the research is the robot's ability to transmit live video footage in real-time. Equipped with a camera module, the robot captures high-quality images and sends them wirelessly to an Android device through the combined capabilities of the Arduino Mega 2560 microcontroller and the Bluetooth module. This live video feed enhances the monitoring and surveillance capabilities of the robot, enabling users to gain immediate insights into the designated area under surveillance.

3. Object Retrieval Capability: Besides its surveillance functions, the robot showcased a remarkable feature – the capability to grab and bring objects back. This functionality extends its utility beyond monitoring and can be particularly valuable in search and rescue missions or other scenarios where the retrieval of objects is essential. The integration of sensors and the control system enabled the robot to effectively interact with its environment, fulfilling a range of potential applications beyond traditional security and surveillance.

These results underscore the efficacy and versatility of the developed system, making it a valuable tool for applications such as security and surveillance in residential and commercial areas, search and rescue operations, and other tasks that require a remotely controlled robot with the ability to capture live video footage and interact with its surroundings. Additionally, the robot's compatibility with Android devices enhances its user-friendliness and accessibility in real-world scenarios.

Fig. 4. Final Robot

4. Hardware Specification

4.1. *Arduino Mega Microcontroller* [10]

The central component driving the functionality of our surveillance robot is the Arduino Mega 2560 microcontroller. This versatile microcontroller boasts a clock speed of 16 MHz and a generous 256 KB of flash memory, providing ample space for intricate program logic. With a plethora of digital and analog pins, including UART, SPI, and I2C interfaces, the Arduino Mega 2560 facilitates seamless integration with various sensors and communication modules. Its compatibility with the HC-05 Bluetooth module enables wireless control and real-time video transmission. The robust processing power of the ATmega2560 architecture, coupled with extensive community support within the maker community, makes the Arduino

Mega 2560 an ideal choice for our project, ensuring efficient development and troubleshooting. In summary, the Arduino Mega 2560 serves as the cornerstone, offering the computational capability and versatility required for the diverse functionalities essential to our surveillance robot.

4.2. *HC-05 Bluetooth Module* [11]

A well-liked module labeled the HC-05 has given our project two-way (full-duplex) wireless connectivity. This module can be used to connect with any Bluetooth-enabled device, including a phone or laptop, as well as two microcontrollers like an Arduino. The choice of the HC-05 Bluetooth module in our surveillance robot project is pivotal for several reasons. The HC-05 provides a seamless wireless communication interface, enabling efficient and user-friendly remote control of the robot via Android devices. With its easy integration with the Arduino Mega 2560, the HC-05 facilitates real-time data transmission, allowing users to navigate and monitor specific areas effortlessly. Its compatibility with the Arduino platform simplifies the implementation of the communication protocol, ensuring a reliable link between the robot and the user's Android device. The module can easily be interfaced with any microcontroller that supports USART because it communicates with USART at a 9600 baud rate. Using the command mode, we can also set the module's default values. However, it cannot transfer files like photos, videos, or audio.

4.3. *L298N motor driver*

A L298 Motor Driver IC, a 78M05 Voltage Regulator, resistors, capacitor, a Power LED, and a 5V jumper are all included in the L298N Motor Driver module. Only after the jumper is inserted will the 78M05 voltage regulator be activated. The internal circuitry will be powered by the voltage regulator when the power source is less than or equal to 12V, and the 5V pin can be utilized as an output pin to power the microcontroller. Motors A and B's ENA and ENB pins regulate speed, respectively, while their IN1&IN2, IN3&IN4, and IN3&IN4 pins control the direction of the rotation. [12]

4.4. *DC Motors*

The motors are battery-powered, geared DC motors. Gears allow a vehicle to move slower while producing more torque. It is referred to as gear reduction. Used motors have a spin at 800 RPM. The setup component aids in boosting torque while lowering motor speed.

4.5. *Rubber Wheels*

For the robot's proper movement, we incorporated rubber wheels with the DC motors and shaft. Which also allows a strong grip.

4.6. *Battery*

These days, many consumer electronics products rely on lithium polymer (sometimes known as "LiPo") batteries. They have become the most widely used option for those searching for long run times and high power in the radio control market over the past few years. LiPo batteries have several advantages, but users must determine whether those advantages exceed the disadvantages. For this project, we utilized an 11.1V lipo battery. However, we tested 3.7V lithium-ion batteries in a battery case to test the system shown in Fig. 4. We have decided to go with a lithium polymer battery since it was more efficient for the system.

Fig. 5. System testing with Lithium-Ion batteries

4.7. *Jumper Wires*

The microcontroller and other stated devices, the prototype, or the internal circuit are connected to other related instruments via jumper cables.

4.8. *Wifi Camera*

One of the key parts of this project is the wifi camera. With this component, we broadcast live video so that the particular coordinates can be inspected.

4.9. *Servo Motor and Sonar Sensor*

To enable the robot to stop independently, the sonar sensor automatically detects impediments. After pausing, the servo motor causes the sonar sensor to move and look for open spaces so that the machine can travel in an obstacle-free route.

4.10. *Metal Claw*

We used a metallic claw that was driven by a servo motor to grab an object that was close to the robot and remove it. This object was not seen by the sonar sensor but only by the camera.

Fig. 6. Complete System Demonstration

5. Software Specification

5.1. Arduino IDE

The programming language Java was used to create the cross-platform Arduino Integrated Development Environment (IDE), which is available for Windows, macOS, and Linux. On the Arduino board, it is used to write and load programs. The GNU General Public License, version 2, publishes the IDE's source code. The Arduino IDE has specific code organization guidelines to support the C and C++ languages (Souza et al., 2017). The wiring project's software library, which offers numerous input and output methods, is provided by the Arduino IDE (Reche et al., 2015).

5.2. Fritzing Circuit Designing Software

An open-source hardware program called Fritzing empowers anyone to use electronics as a creative medium. In the spirit of Processing and Arduino, it provides a software tool, a community website, and services. Establishing a creative ecosystem, it enables users to document their prototypes, share them with others, teach electronics in a classroom, and layout and create excellent PCBs.

5.3. Arduino Bluetooth Controller App[13]

To transmit data to the Bluetooth module, we used the above-mentioned Android app. Then, using the data we had received from HC05 to the microcontroller, we operated the machine's logic.

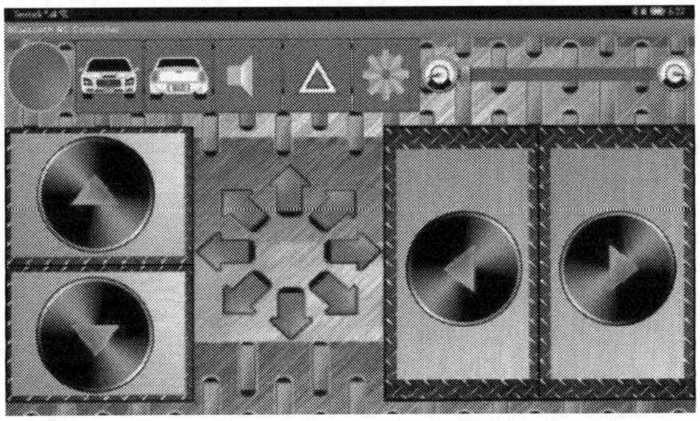

Fig. 7. Controlling app interface [13]

5.4. LookCam App[14]

To get the live broadcast from the Wi-Fi camera, we used the LookCam mobile app.

6. DISCUSSION

The research yielded several significant results for the designed and implemented site surveillance robot utilizing an Arduino Mega 2560 microcontroller and a Bluetooth module HC-05:

6.1 Applications for deploying this robot

The Robot is right now in a prototype phase. However, after a versatile viable product, the robot can be operated in the following fields:

1. Security and Surveillance in Residential Areas:

The robot can be deployed for monitoring and patrolling residential neighborhoods, enhancing overall security by providing real-time video surveillance and remote navigation capabilities.

2. Commercial Area Monitoring:

The fully functional robot can be utilized to patrol and monitor commercial spaces, such as shopping malls, parking lots, and business complexes, to ensure the safety and security of the premises.

3. Search and Rescue Operations:

The robot's object retrieval capability can be leveraged in search and rescue missions. The ability to navigate challenging terrains and retrieve objects makes it a valuable tool in locating and assisting individuals in distress.

4. Industrial Surveillance and Inspection:

Implementation of the robot for industrial applications, conducting surveillance and inspections in environments where human access might be restricted or hazardous. This includes monitoring equipment, detecting anomalies, and ensuring workplace safety.

5. Emergency Response:

The robot can be integrated into emergency response teams to assess disaster-stricken areas. Its remote control, live video transmission, and object retrieval capabilities can assist in assessing damage and retrieving critical items in emergency situations.

6. Radiation Detection for bomb finding:

With the integration of the GM Counter Module with the system, the robot will be able to detect high radiation and find any hidden bomb in a security application.

6.2 LIMITATIONS

The research paper presents a low-cost and easy-to-assemble site surveillance robot with several noteworthy features; however, it is important to acknowledge its limitations. Firstly, the robot's reliance on a Bluetooth module (HC-05) for wireless communication constraints its operational range. The Bluetooth range, though suitable for certain applications, may limit the robot's ability to monitor areas that require extended distances or traverse large premises. Secondly, the absence of an Internet of Things (IoT) framework means that the footage transmission is restricted to the Wi-Fi range, thereby restricting its coverage area to locations with an accessible Wi-Fi network. Lastly, the control of the robot is also bound by Bluetooth range, which may hinder its ability to carry out surveillance tasks in areas that demand remote monitoring beyond Bluetooth's limited reach. Recognizing these limitations, the proposed system can still prove valuable for various security and surveillance applications within its operational boundaries.

6.3 FUTURE SCOPE

The future scope of this research paper holds promising opportunities for further enhancement and expansion of the presented site surveillance robot system.

6.3.1. Removing Limitations: The first area of focus for future research could be the removal of existing limitations in the system. While the current design is cost-effective and easy to assemble, there may be scope for improving its robustness, battery life, and adaptability to different environments. Research can explore advanced materials and components that could make the robot more durable, weather-resistant, and adaptable to various terrains.

6.3.2 Implementing IoT for Controlling: Integrating the Internet of Things (IoT) technology into the system can take its capabilities to the next level. By connecting the surveillance robot to a cloud-based platform, users can have remote and real-time control over the robot's movements and sensor data. This would enable operators to control the robot from anywhere worldwide, enhancing its reach and usability. Additionally, incorporating machine learning algorithms and AI for autonomous decision-making can optimize the robot's performance.

6.3.3 Implementing IoT for Live Footage Transmission: Building upon the IoT integration, research can focus on implementing IoT for live footage transmission. By enabling the robot to stream high-quality surveillance footage directly to a cloud server, authorized users can access the video feed securely from their devices. This feature could enhance the system's efficiency and accessibility, making it an even more valuable tool for security and surveillance applications.

6.3.4 Multi-Robot Collaboration: Investigating the feasibility of deploying multiple robots that can collaborate and coordinate their actions can be an exciting avenue for future research. This could include swarm robotics, where robots work together to cover larger areas efficiently, enhancing their overall surveillance capabilities.

The future scope of this research paper offers a wide array of possibilities for improving and expanding the capabilities of the site surveillance robot, making it a valuable tool in various security and surveillance applications. By addressing the above points, researchers can contribute to advancing this technology and its broader adoption in the field.

ACKNOWLEDGMENT

The authors acknowledge the use of generative AIs like ChatGPT and Google Bard. We have inputted various prompts to improve the grammar and sentence structure. As well as adding and subtracting some of the descriptions in this research paper.

REFERENCES

[1] Akilan, T. *et al.* (2020) 'Surveillance Robot in hazardous place using IOT Technology', *2020 2nd International Conference on Advances in Computing, Communication Control and Networking (ICACCCN)* [Preprint]. doi:10.1109/icacccn51052.2020.9362813.

[2] Aniruddha Prabhu, B.P. and Hebbal, S. (2017) 'Small unarmed robot for defense and security:a cost-effective approach using Arduino Uno', *2017 2nd International Conference On Emerging Computation and Information Technologies (ICECIT)* [Preprint]. doi:10.1109/icecit.2017.8453325.

[3] Rangapur, I., Prasad, B.K.S. and Suresh, R. (2020) 'Design and development of spherical spy robot for surveillance operation', *Procedia Computer Science*, 171, pp. 1212–1220. doi:10.1016/j.procs.2020.04.130.

[4] Shah, M.S. and Borole, P.B. (2016) 'Surveillance and rescue robot using Android smartphone and the internet', *2016 International Conference on Communication and Signal Processing (ICCSP)* [Preprint]. doi:10.1109/iccsp.2016.7754413.

[5] Singh, D. and Nandgaonkar, A. (2018) 'IOT-based Wi-Fi surveillance robot with real-time audio and video streaming', Advances in Intelligent Systems and Computing, pp. 639–647. doi:10.1007/978-981-13-1513-8_65.

[6] Sirasanagandla, S Pachipulusu, M. and Jayaraman, R. (2020) 'Development of surveillance robots to monitor the work performance in Hazardous Area', *2020 International Conference on Communication and Signal Processing (ICCSP)* [Preprint]. doi:10.1109/iccsp48568.2020.9182126.

[7] Dudeja, A., Yadav, A., Parashar, A., Kabtiyal, A. and Arora, S., IoT Based Autonomous Multi-Purpose Surveillance and Rescue Robot.

[8] Imran, M.D., Zafar, A., Khan, A. and Wankhade, M.M., MULTI-PURPOSE SURVEILLANCE ROBOT.

[9] Sridhar, G., Potharaju Raju, P. and Sudha, Y., Bluetooth Controlled Robot using ARDUINO.

[10] Arduino Mega 2560 REV3 (no date) Arduino Official Store. Available at: https://store.arduino.cc/products/arduino-mega-2560-rev3 (Accessed: 26 October 2023).

[11] HC-05 - Bluetooth module (no date) Components101. Available at: https://components101.com/wireless/hc-05-bluetooth-module (Accessed: 26 October 2023).

[12] L298N motor driver module (no date) Components101. Available at: https://components101.com/modules/l293n-motor-driver-module (Accessed: 26 October 2023).

[13] Arduino Bluetooth Controller - apps on Google Play (no date) Google. Available at: https://play.google.com/store/apps/details?id=com.giumig.apps.bluetoothserialmonitor&hl=en&gl=US (Accessed: 26 October 2023).

[14] Lookcam - apps on Google Play (no date) Google. Available at: https://play.google.com/store/apps/details?id=com.view.ppcs&hl=en&gl=US (Accessed: 26 October 2023).

[15] Reddy, M.M.L.N., Akhil, C.H., Goud, G.A., Venkatesh, Y. and Chandana, N., WEB CONTROLLED MULTI-PURPOSE SECURITY ROBOT USING ARDUINO.

[16] Team, T.A. (no date) *Uno R3, Arduino Documentation*. Available at: http://arduino.cc/en/Main/arduinoBoardUno (Accessed: 26 October 2023).

[17] *Home* (no date) *Arduino*. Available at: http://arduino.cc/en/Reference/HomePage (Accessed: 26 October 2023).

MATEC Web of Conferences **393**, 04002 (2024) https://doi.org/10.1051/matecconf/202439304002
STAAAR-2023

Object Detection for Self-Driving Car in Complex Traffic Scenarios

Biplab Das[1*], *Pooja* Agrawal[2]

[1, 2]School of Robotics, Defence Institute of Advanced Technology (DIAT), DU, Pune, India

Abstract. The application of convolutional neural networks (CNNs) in particular has greatly enhanced the object detection capabilities of self-driving cars, because of recent advancements in artificial intelligence (AI). However, striking a balance in vehicular settings between high precision and fast processing continues to be a persistent challenge. Developing nations such as India, possessing the second-largest global population, introduce unique intricacies to road scenarios. Numerous challenges arise on Indian roads, such as unique vehicle kinds and a variety of traffic patterns, such as auto-rickshaws, which are only seen in India. This study presents the outcomes of evaluating the YOLOv8 models, which have demonstrated superior performance in Indian traffic conditions when compared to other existing YOLO models. The examination utilized the dataset, compiled from data collected in the cities of Bangalore and Hyderabad, as well as their surrounding areas. The investigation's findings demonstrate how well the YOLOv8 models work to address the unique problems that Indian road conditions present. This study advances the development of autonomous vehicles designed for intricate traffic situations such as those found on Indian Roads.

1 Introduction

The incorporation of artificial intelligence (AI) technology is causing the automotive sector to evolve quickly [1,2]. Through a yearly average growth rate of 36.15%, worldwide demand for AI in cars is predicted to increase significantly, with a projected USD 6.6 billion in sales of AI technology for cars by 2025 [3,4]. Detecting objects is a very important part of self-driving cars to locate the object as well as to classify the object in front of it and its surroundings so that based on it cars can decide to drive efficiently. In recent years there has been huge growth in computer vision for that reason object detection for autonomous cars has also improved. Yolo models showed great performance compared to other detection models [5] As we all know, speed is crucial for real-time applications. Yolo models, being single-stage detection systems, outperformed double-stage detection systems like R-CNN and Faster R-CNN in terms of locating objects and classification speed [6]. Many kinds of research happened on object detection for self-driving cars but most of them were done on the dataset which does not consist of that much complex data. It is extremely difficult to assure safety and dependability in extreme corner instances for self-driving cars using the

* Corresponding author: biplabdas369ee@email.org

IDD dataset [7], which contains very complex and diverse road scene scenarios. The dataset used here consists of complex traffic scenarios of roads in Hyderabad, Bangalore cities, and their outskirts in India. The objects in the dataset have used car, person, bicycle, bus, traffic sign, traffic light, truck, autorickshaw, animal, motorcycle, and rider. By training the most recent Yolo algorithms on an unconstrained complicated environment dataset and comparing their accuracy. This paper makes significant advances in the field of object localization and classification for self-driving cars.

Object localization and classification are very important parts of vision-based self-driving cars. In recent years various deep learning methods come up with higher accuracy and speed to detect objects in traffic road scenarios. Here we discussed some of the recent papers that focused on object detection for self-driving cars.

1.1 Related Works

The authors of [8] suggested an improved Yolov4 model to identify ten different types of objects and an improved model to forecast whether or not people would cross the street. For object detection, they have used the BDD100k dataset and for pose estimation of pedestrians, they have collected and validated data by themselves. By optimizing they have reduced the complexity of the model of yolov4 to achieve faster speed and accuracy.

In [9] they have improved the capability of the YOLOv5 object detector for detecting tiny objects, specifically in the context of autonomous racing. The researchers propose architectural modifications to the YOLOv5 model, including changes to the backbone, neck, and other elements. The authors suggested further research and testing with different datasets to validate and refine the proposed techniques. Overall, the research advances the development of autonomous vehicle vision systems by offering insights into enhancing tiny object recognition in the YOLOv5 model.

A novel hybrid model for object detection and tracking in autonomous cars is presented in this paper [10]. It combines a tracking model based on Kalman filters with a two-stage object detector based on Faster R-CNN. In-depth tests on the KITTI dataset demonstrated faster and more accurate results than previous approaches, which is essential for real-time object recognition and vehicle safety. Extensive experiments on the KITTI dataset showcased superior accuracy and speed compared to existing methods, crucial for real-time object detection and vehicle safety. It's important to remember, though, that the intricacy of the design can need a large amount of processing power.

The authors of [11] emphasized the importance of exclusively testing self-driving cars on Indian roads, which require a specialized dataset due to the unique and unstructured complexities presented in India. Consequently, they developed this dataset. Within this dataset, the authors employed Faster R-CNN for object detection and conducted an evaluation. Various metrics, such as classifier accuracy, loss RPN classifier, loss RPN regression, loss detector classifier, and loss detector regression, were used to evaluate the model's performance after it had been trained on 1200 images.

In this study enhancement of object detection based on the yolov8 model for self-driving cars has been proposed under the complex traffic environment. Here also comparison with other YOLO models has been done. Fig. 1 shows the architecture of YOLOv8.

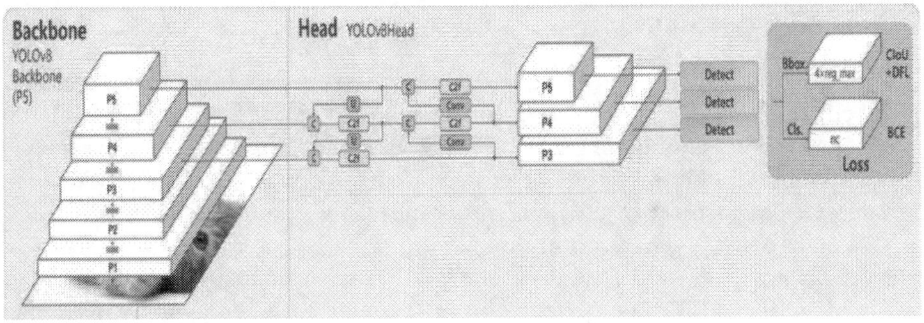

Fig. 1. Architecture of YOLOv8 [12].

2 Methodology

The below subsections discussed the data collection and preprocessing as well as the strategies for training, validating, and testing are examined alongside their corresponding assessment standards.

2.1 Data Preparation

The unstructured and diverse dataset used here is collected from open source, a total of 9300 annotated images were collected with annotation names 0-animal, 1-autorickshaw, 2-bicycle, 3-bus, 4-car, 5-motorcycle, 6-person, 7-rider, 8-traffic light, 9-traffic sign, 10-truck. The collection includes several road scene views of Indian cities, as well as their surrounding areas. These images feature a combination of towns and villages with varying lighting conditions. Fig. 2 shows some images from the dataset. YOLO model prefers some particular input image sizes like 640x640 and 1280x1280 and image annotation in YOLO format. For that reason before training the whole dataset was preprocessed with the online tool available Roboflow (https://app.roboflow.com) for auto-orientation, resizing the images to 640x640 so that training time will be less and converting the annotations from XML to TXT format, also known as YOLO format. Additionally, images are resized to 1280x1280 to assess model performance. The dataset from the Roboflow website has been split between 70, 20, and 10 percent ratios for training, validating, and testing.

Fig. 2. Some unstructured road scene images from the dataset.

2.2 Testing and Evaluation

All the training and evaluation were done with the Python programming language. The deep learning-based YOLOv8 model was used for training on the unstructured complex traffic dataset through transfer learning. The YOLOv8 model works faster and more accurately than existing YOLO models for detecting objects, and this study assisted in the detection of objects in an extensive variety of traffic circumstances, such as on Indian roads. This work uses the YOLOv8 model given by Ultralytics [13] for training, saving models, and validation and testing. There are a total of 6 versions of the YOLOv8 model presented those base models that could be used like yolov8n.pt for nano object, yolov8s.pt used for small object detection, yolov8m.pt, yolov8l.pt, and yolov8x.pt for medium, large, and extra-large object detection purposes. Those base models were used for training, validation, and testing, around 6510 data was used for training, 1860 data was used for validation, and 630 data for testing. Fig. 3 shows the workflow of this paper.

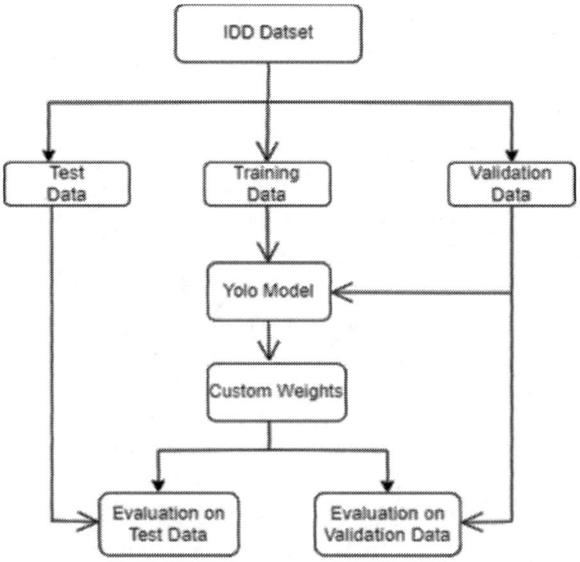

Fig. 3. Object Detection Model.

Determining the extent to which bounding boxes overlap about ground truth data indicates successful recognition is a crucial step in evaluating object detection algorithms. IOUs are utilized for this, and mAP0.5 is the accuracy when IOU=50, meaning that the detection is deemed successful if there is more than 50% overlap. The bounding box must be detected with greater accuracy. All the evaluations were done based on mAP0.5, mAP0.5-0.95, Precision-Recall curve, and other Object Detection Metrics. The formulas for IOU, Recall, and Precision are presented below.

$$IOU = (Area\ of\ Intersection)/(Area\ of\ Union)\ \ldots\ldots \quad (1)$$

$$Precision = (Correct\ Predictions)/(Total\ Predictions)\ldots\ldots \quad (2)$$

$$Recall = (Correct\ Predictions)/(Total\ Ground\ Truth)\ldots\ldots \quad (3)$$

3 Experimental Results

In this section, the evaluation of various object localization and classification models is presented, which are used for detecting complex traffic scenarios. These object detection models were trained, validated, and tested on a complex dataset of 9700 labeled images.

Table 1. Performance of trained Yolo model.

Model	Image size used for training	mAp0.5	mAP0.5-0.95
Yolo5l	640 x 640	0.58	0.374
Yolov8m	640 x 640	0.579	0.371
Yolov8l	640 x 640	0.597	0.388
Yolov8x	640 x 640	0.603	0.390
Yolov8x	1280 x 1280	0.688	0.446

The dataset contains various types of labeled images, such as person, traffic lights, motorcycles, bicycles, traffic signs, animals, riders, buses, cars, trucks. The mean average precision of different models is presented in Table 1. Furthermore, Fig. 4 displays the precision-recall curves of trained models. YOLOv8x (1280x1280), YOLOv8x, YOLOv8l, and YOLOv8m trained models are displayed in (a), (b), (c), and (d). The balance between recall and accuracy for various thresholds is displayed on the precision-recall curve. High accuracy is correlated with a low false positive rate, while high recall is correlated with a low false negative rate. An area under the curve with a high value indicates high recall and precision. When comparing the Precision-Recall curve of the YOLOv8x (1280x1280) trained model to other models, it has shown good results. Fig. 5 shows the validation results of trained models. YOLOv8x (1280x1280), and YOLOv8x (640x640), trained models are displayed in (a) and (b). Fig.5.a) The YOLOv8x(1280x1280) trained model has shown good accuracy. Fig. 6 displays the sample input images (1280x1280) used for testing. Additionally, Fig. 7 displays the objects that the Yolov8x model detected. Consequently, the YOLOv8x model demonstrates suboptimal performance in these underrepresented classes, despite delivering strong results for other categories. When the input image resolution was increased to 1280x1280, the YOLOv8x model performed approximately 14% better than 640x640 resolution images as seen in Table 1.

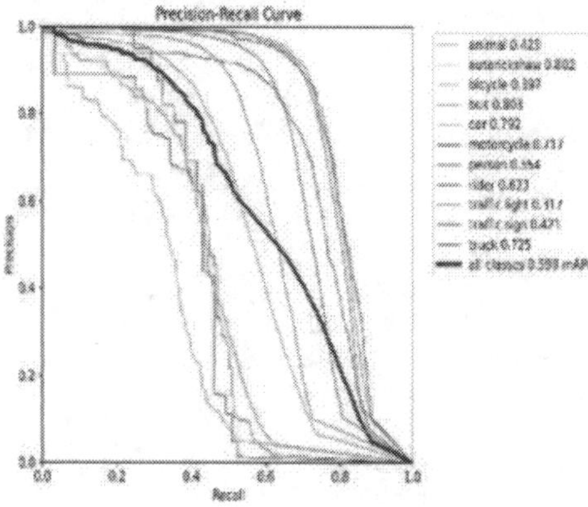

(a)

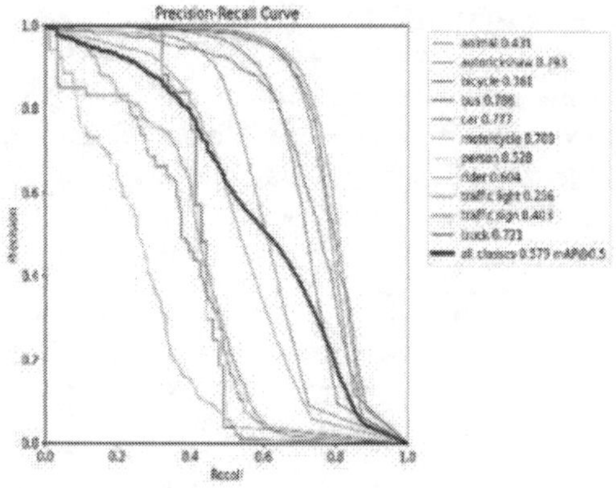

(b)

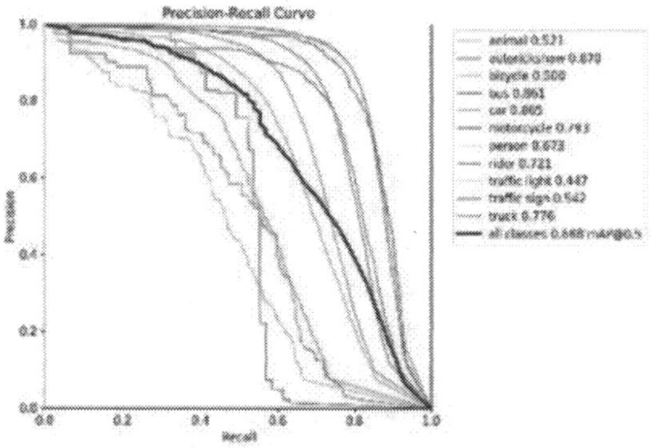

(c)

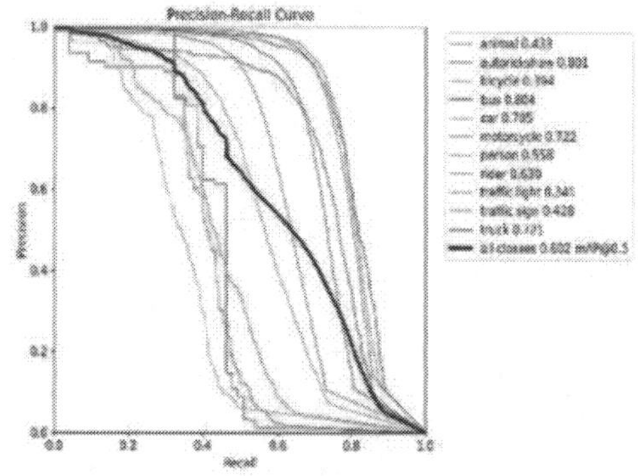

(d)

Fig. 4. Precision-Recall curve of different Yolov8's models after validation. a) Yolov8x (Image size 1280x1280) b) Yolov8x(Image size 640x640) c) Yolov8l(Image size 640x640) d) YOLOv8m(Image size 640x640).

Class	Images	Instances	Box(P	R	mAP50	mAP50-95)
all	1859	39975	0.785	0.619	0.688	0.446
animal	1859	65	0.713	0.523	0.521	0.355
autorickshaw	1859	3049	0.874	0.798	0.87	0.644
bicycle	1859	152	0.619	0.467	0.5	0.307
bus	1859	1167	0.823	0.8	0.861	0.667
car	1859	8389	0.887	0.791	0.865	0.626
motorcycle	1859	8334	0.877	0.694	0.793	0.489
person	1859	6947	0.799	0.571	0.673	0.384
rider	1859	8599	0.856	0.615	0.721	0.428
traffic light	1859	583	0.703	0.336	0.447	0.187
traffic sign	1859	1060	0.683	0.486	0.542	0.317
truck	1859	1718	0.799	0.729	0.776	0.557

(a)

Class	Images	Instances	Box(P	R	mAP50	mAP50-95)
all	1859	39974	0.758	0.541	0.604	0.393
animal	1859	65	0.487	0.4	0.438	0.291
autorickshaw	1859	3049	0.877	0.74	0.813	0.592
bicycle	1859	152	0.532	0.386	0.377	0.23
bus	1859	1167	0.814	0.741	0.806	0.604
car	1859	8389	0.859	0.721	0.797	0.57
motorcycle	1859	8334	0.867	0.667	0.722	0.44
person	1859	6947	0.761	0.479	0.562	0.313
rider	1859	8598	0.835	0.554	0.628	0.371
traffic light	1859	583	0.745	0.214	0.318	0.135
traffic sign	1859	1060	0.652	0.4	0.451	0.255
truck	1859	1718	0.782	0.684	0.732	0.523

(b)

Fig. 5. Validation accuracy of different Yolov8 models. a) Yolov8x(Image size 1280x1280) b) Yolov8x(Image size 640x640).

Fig. 6. Sample images for testing.

Fig. 7. Detected images of the YOLOv8x model.

4 Conclusion

Object detection is a critical component of self-driving cars, but it can be difficult in a complex, varied traffic environment. This paper suggests using the YOLOv8 model to improve object detection in unstructured complex traffic scenarios. YOLOv8x (1280x1280) trained model demonstrated superior performance over other models. By training the model on a larger dataset with nearly identical and higher instances of each class, model performance can be improved. despite delivering strong results for other categories. When the input image resolution was increased to 1280x1280, the YOLOv8x model performed approximately 14% better than 640x640 resolution images as seen in Table 1.

References

1. H. Mankodiya, D. Jadav, R. Gupta, S. Tanwar, W.C. Hong, R. Sharma, "OD-XAI: Explainable AI-Based Semantic Object Detection for Autonomous Vehicles," Appl. Sci. 12, 5310 (2022).
2. S.A. Khan, H. Lim, "Novel Fuzzy Logic Scheme for Push-Based Critical Data Broadcast Mitigation in VNDN," Sensors 22, 8078 (2022).
3. A. N. Bhavana, M.M. Kodabagi, "Exploring the Current State of Road Lane Detection: A Comprehensive Survey," Int. J. Hum.-Comput. Interact. 2, 40-46 (2023).
4. S.A. Khan, H. Lim, "Push-Based Forwarding Scheme Using Fuzzy Logic to Mitigate the Broadcasting Storm Effect in VNDN," in Proceedings of the Artificial Intelligence and Mobile Services–AIMS 2022: 11th International Conference, Held as Part of the Services Conference Federation, SCF , Honolulu, HI, USA, December 10–14, (2022), pp. 3–17. Springer, Berlin/Heidelberg, Germany.

5. J. Redmon, S. Divvala, R. Girshick, and A. Farhadi, "You Only Look Once: Unified, Real-Time Object Detection," in 2016 IEEE Conference on Computer Vision and Pattern Recognition (CVPR, 2016), pp. 779-788.
6. S.A. Khan, H.J. Lee, H. Lim, "Enhancing Object Detection in Self-Driving Cars Using a Hybrid Approach," Electronics 12, 2768 (2023).
7. G. Varma, A. Subramanian, A. Namboodiri, M. Chandraker, and C.V. Jawahar, "IDD: A Dataset for Exploring Problems of Autonomous Navigation in Unconstrained Environments," in IEEE Winter Conference on Applications of Computer Vision (WACV, 2019).
8. Y. Li, H. Wang, L.M. Dang, D. Han, H. Moon, T. Nguyen, "A Deep Learning-Based Hybrid Framework for Object Detection and Recognition in Autonomous Driving," IEEE Access, vol. 8, (2020).
9. A. Benjumea, I. Teeti, F. Cuzzolin, and A. Bradley, "YOLO-Z: Improving Small Object Detection in YOLOv5 for Autonomous Vehicles," (2021).
10. Y. Kortli, S. Gabsi, Y. Lew Yan Voon, J. Lew, M. Jridi, M. Maher, M. Marzougui, and M. Atri, "Deep Embedded Hybrid CNN-LSTM Network for Lane Detection on NVIDIA Jetson Xavier NX," Knowledge-Based Systems, (2022).
11. G.N.V.V. Satya Sai Srinath Namburi, Athul Zac Joseph, S. Umamaheswaran, Ch. Lakshmi Priyanka, Malavika Nair M, and Praveen Sankaran, "NITCAD - Developing an Object Detection, Classification, and Stereo Vision Dataset for Autonomous Navigation in Indian Roads," Procedia Computer Science, (2020).
12. J. Solawetz and F. Francesco, "What is YOLOv8? The Ultimate Guide," (2023). Available online: https://blog.roboflow.com/whats-new-in-yolov8/.
13. G. Jocher and AyushExel, "YOLO by Ultralytics," (2023). Available online: https://docs.ultralytics.com/.

MATEC Web of Conferences **393**, 04003 (2024) https://doi.org/10.1051/matecconf/202439304003
STAAAR-2023

A Robotic Exoskeleton Assistive Device: Design and Kinematic Analysis of a Hand Exoskeleton for Assistance

Arpeet Dhal[1*], *Durgeshwar* Majhi[1], *Ashish Kumar* Rout[2] and *Soumya* S.[1]

[1]School of Robotics, Defence Institute of Advanced Technology (DU), Pune, India
[2]Computational Mechanics, Indian Institute of Technology, Ropar, India

Abstract. Addressing the challenges faced by individuals with motor disabilities, which impedes ease in movement, coordination, or sensation, a low-weight hand exoskeleton was developed considering accessibility, low cost, adaptability, and mass manufacturing to overcome the impact of hand functioning disorders on people's physical and social well-being. In this paper, a 12-DOF Hand exoskeleton consisting of a 6-DOF Arm and 6-DOF finger exoskeleton was designed and later its kinematic analysis was performed. By taking into consideration the constraints provided by the exoskeleton to human hand motion, the Forward kinematics, using Denavit-Hartenberg (D-H) parameters were conducted to determine its workspace.

1 Introduction

The evolution of exoskeleton robotics has garnered significant attention in recent times. While the concept of it dates to the late 1800s, the pioneering success of the Hardiman exoskeleton [1] in the 1960s marked a milestone in the field. Initially designed for military purposes to augment strength and performance, the Hardiman utilized a master-slave configuration with hydraulic actuators.

Subsequent advancements, such as the upper-limb exoskeleton introduced by Kazerooni et al [2], emphasized physical human-robot interaction (pHRI), paving the way for direct mechanical power transmission without a master-slave system. Additionally, the University of Tsukuba's development of the Hybrid Assistive Limb (HAL) showcased the potential for assisting impaired individuals, both in daily life and industrial settings [3]. ADL was designed for consumer use, but later versions were also created for industrial use.

Over the past two decades, there has been a remarkable surge in the utilization of upper-limb exoskeletons, garnering considerable attention from both the biological and technical communities. This technology is increasingly being recognized as a promising solution for individuals facing physical challenges or disabilities. [4] Ekos Vest and FORTIS [5] are two systems that have been designed to boost the wearer's performance and strength.

* Corresponding author: dhalarpeet100@gmail.com

Coming on the design of finger exoskeleton, IOTA [6] is a two degree of freedom thumb exoskeleton that can actuate the carpometacarpal (CMC) and metacarpophalangeal (MCP) joints through the actuation of flexible cables. Additionally, a novel Haptive Hand exoskeleton [7] that allows exerting-controlled forces on the fingertip of the index and thumb, along with several design solutions, was developed that gave some basic insights into the current development of finger exoskeleton.

In the exploration of the exoskeleton field, a notable gap in research, particularly in the realm of lightweight, portable exoskeletons tailored for rehabilitation, was observed. Motivated by this observation, efforts were initiated to contribute to the advancement of exoskeleton technology, with a specific focus on enhancing both active and passive rehabilitation, as well as addressing industrial applications. The journey commenced with the creation of a preliminary arm model, delving into the details of measurements, degrees of freedom, and input joints. Recognizing the complexity of finger design, with spatial constraints and a multitude of movements, developing a unique approach. The objective was to streamline efficiency by controlling a finger's complete range of movements using a single motor, thereby optimizing space utilization. Subsequently, design efforts extended to the development of a lightweight forearm and bicep arm exoskeleton, engineered for the purpose of effortlessly lifting everyday weights. The incorporation of a back support link played a pivotal role, strategically transferring the major load to the waist and thereby reducing the net weight borne by the individual. Navigating through the design intricacies, meticulously addressed joint dynamics aligned with human physiology. This culmination of design and engineering efforts resulted in an exoskeleton poised as an extension of active and passive rehabilitation, with potential applications in the industrial landscape.

The following sections delve deeper into the methodology, results, and discussions surrounding the exoskeleton design, aiming to fill a crucial gap in the evolving landscape of exoskeleton research.

2 Design of Hand Exoskeleton

The primary objective of this exoskeleton encompasses a wide range of applications, serving as assistive devices in the fields of medical rehabilitation and industries, along with defence and military applications. The foremost aim is to provide revolutionary solutions for differently abled people and those with paralysis, offering motion assistance that enhances their quality of life and promotes greater independence. Moreover, this technology has the potential to improve mobility and agility in industrial workspaces and has specific relevance for military applications. Taking these into consideration, the primary goal was to provide a lighter and more robust solution by addressing space constraints. Achieving 12 Degrees of Freedom (DOFs), consisting of 10 active DOFs and 2 passive DOFs, enhances comfort and flexibility in motion.

2.1 Finger Assembly

The design of the finger component was done considering space constraints. The limited space between each finger posed a challenge for effective individual motor control of each joint. Considering this, a design approach was adopted that enabled the coordinated movement of multiple joints of a finger using a single actuator for power transmission. This was achieved by the implementation of a combination of two four-bar link mechanisms, which facilitated the relative motion within the finger joints.

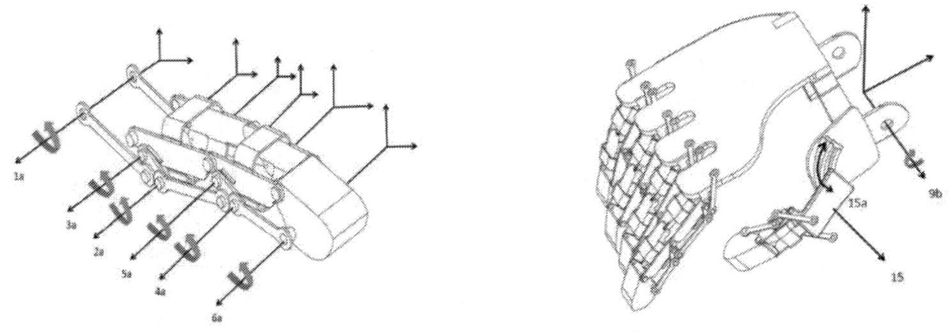

(a) (b)

Fig. 1: (a) Design of Finger assembly showing an under-actuated model being controlled by a single actuator giving rotation about 1a axis, (b) Design of full Finger exoskeleton showing extra passive motion given to thumb and rotation of hand at the wrist joint.

The finger assembly consists of eight links. The whole mechanism is eccentrically mounted upon the base link, which is connected to an actuator, resulting in the input rotation about the **1a** axis. This input rotation causes a general motion of link**1**, which was further attached to link **2**. As the link **2** was making a turning pair with fixed link **8**, the input from link **1** resulted in rotation of link **2** about **2a** axis passing through the hinge point. The combination of the links **2**, **3**, **4**, **8** acts as first four bar mechanisms and links **4**, **5**, **6**, **7** acts as second four bar mechanism. In first mechanism, **2** and **4** are fixed upon **8**, resulting only rotation of **2**, **4** and general motion of link **3**. Link **8** has been fixed upon the finger. Thus, the resultant general motion of link **3** is the motion provided to the finger. The rotation motion of link 4 acts as an input for the second four-bar mechanism and results in rotation and general motion of link **5** and **6** respectively. Here **4** and **6** are fixed upon **7**. The general motion of the link **5** is the actual motion given to the finger. Thus overall, the link **1** gives output as general motion of link **3** and **5**.

Here link **2** rotates about axis **2a**, Link **3** rotates about axis **3a**, Link **4** rotates about axis **4a**, Link **5** rotates about axis **5a**, and Link **6** rotates about axis **6a**. In fig**1**, the thumb **15**, is governed by one four-bar mechanism and it is also provided with one slotted mechanism on a curved path along curved path **15a**, which resembles the exact movement of a finger during enclosing and opening. By successfully coordinating all the fingers' desired motion is achieved.

2.2 Arm Assembly

In humans, the wrist has 3 DOFs, one is flexion/extension, then supination/pronation, and third one is radial deviation /ulnar deviation. Active control of flexion/extension and supination/pronation has been implemented. The flexion/extension movement was facilitated with respect to the forearm. A rope and pulley mechanism were provided to achieve twisting movement, which was attached to the forearm. The length of the rope was regulated by a motor attached to the forearm. This configuration helps in bidirectional motion control of the forearm, resulting in the twisting motion of the wrist. The forearm and upper arm were designed with a cage-like structure, so as to provide comprehensive support to the arm. The forearm was given flexion/extension movement about the elbow joint of the upper arm.

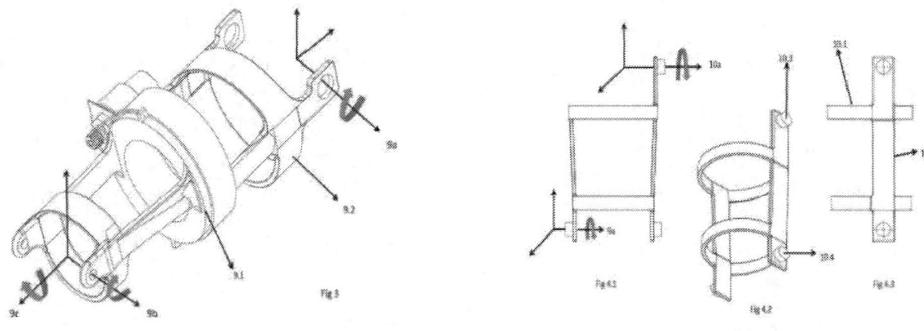

(a) (b)

Fig. 2: (a) Design of forearm exoskeleton with bidirectional motion control, (b) Design of bicep exoskeleton

The upper arm acts as a connector for the forearm and shoulder. Its design encloses the upper portion of the limb, ensuring it doesn't impose any restrictions on shoulder movement. This whole arm assembly is mounted onto the "shoulder blade", which is meticulously designed as an assembly of three parts, taking into consideration of the natural range of motion of the human shoulder. These three components of the shoulder blade include the shoulder curve, shoulder rotor, and the shoulder rotor casing. These parts are made to provide a similar range of motion and flexibility as provided by the shoulder joint of the Scapula, Clavicle, and Humerus.

(a) (b)

Fig. 3: (a) Design of shoulder curve exoskeleton, (b) Design of shoulder rotor exoskeleton.

The Shoulder curve acts as the connecting link between the upper arm and the shoulder rotor. It facilitates the flexion/extension movement of the upper arm. This shoulder curve is mounted on the extreme part of the shoulder rotor, which results in internal/external rotation. The shoulder made a turning and free sliding pair with shoulder rotor casing. Together, these three elements of the shoulder blade serve as a load-transferring component, transferring the load from the arm exoskeleton to the back support. This results in an even distribution of the resulting load upon the body accommodating the need for three rotations along distinct axes at the shoulder.

https://doi.org/10.1051/matecconf/202439304003

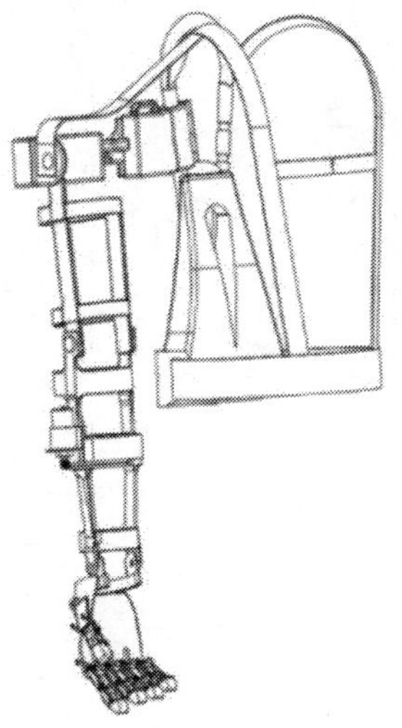

Fig. 4: Design of 12 DOF exoskeleton model

3 Forward Kinematics

The kinematic analysis of the robotic hand was done using Denavit-Hartenberg (D-H) parameters to find the rotation and position vectors of the end-effector. The analysis was bifurcated into two distinct components: the arm assembly and the finger assembly. This paper predominantly focuses on the study of forward kinematics of arm assembly, considering finger assembly as the end-effector. The aim was to achieve an understanding of the arm assembly so that the range of motions and reachability of the exoskeleton could be known. Looking ahead, the plan involves delving into a kinematic analysis of the finger assembly in subsequent phases and studying different grasping strategies to assess the adaptability of the exoskeleton, resulting in improvement in versatility and applicability.

3.1 Finger Assembly

Using D-H parameter notation, four parameters are needed to describe how one frame is connected to its previous frame. These parameters are:

- θ : joint angle between the links
- d : distance of previous joint center to next in z-direction
- α : twist angle between joint axes
- a : distance of previous joint center to next center in x-direction

Table 1. D-H Parameters.

θ	d	α	a
θ_1	0	$-(\pi/2)$	l_1
θ_2	l_2	$-(\pi/2)$	l_3
θ_3	0	0	l_4
θ_4	0	$-(\pi/2)$	l_5
θ_5	l_6	$\pi/2$	0
θ_6	0	0	0

Next, using D-H parameters, a transformation matrix is written for each joint. Here we have considered,

$$c_1 = \cos(\theta_1),\ c_2 = \cos(\theta_2),\ c_3 = \cos(\theta_3),\ c_4 = \cos(\theta_4),\ c_5 = \cos(\theta_5),\ c_6 = \cos(\theta_6);$$

$$s_1 = \sin(\theta_1),\ s_2 = \sin(\theta_2),\ s_3 = \sin(\theta_3),\ s_4 = \sin(\theta_4),\ s_5 = \sin(\theta_5),\ s_6 = \sin(\theta_6);$$

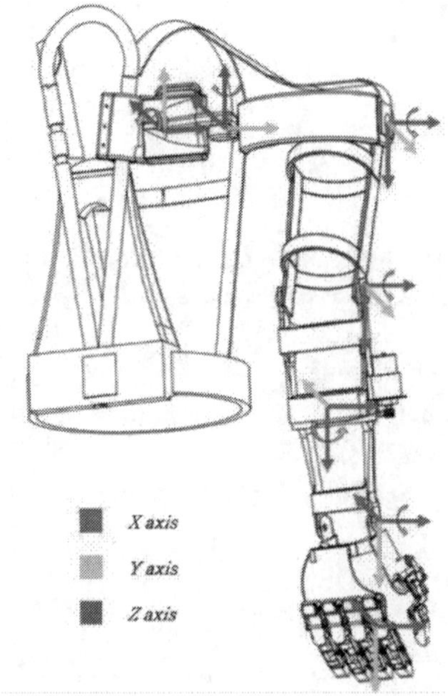

The transformation matrix for each joint is as follows:

$$[T_1^0]=\begin{bmatrix} c_2 & 0 & -s_2 & l_3 * c_2 \\ s_2 & 0 & c_2 & l_3 * s_2 \\ 0 & -1 & 0 & 0 \\ 0 & 0 & 0 & 1 \end{bmatrix} \qquad [T_2^1]=\begin{bmatrix} c_1 & 0 & -s_1 & l_1 * c_1 \\ s_1 & 0 & c_1 & l_1 * s_1 \\ 0 & -1 & 0 & l_2 \\ 0 & 0 & 0 & 1 \end{bmatrix}$$

$$[T_3^2]=\begin{bmatrix} c_3 & -s_3 & 0 & l_4 * c_3 \\ s_3 & c_3 & 0 & l_4 * s_3 \\ 0 & 0 & 1 & 0 \\ 0 & 0 & 0 & 1 \end{bmatrix} \qquad [T_4^3]=\begin{bmatrix} c_4 & 0 & -s_4 & l_5 * c_4 \\ s_4 & 0 & c_4 & l_5 * s_4 \\ 0 & -1 & 0 & 0 \\ 0 & 0 & 0 & 1 \end{bmatrix}$$

$$[T_5^4]=\begin{bmatrix} c_5 & 0 & s_5 & 0 \\ s_5 & 0 & -c_5 & 0 \\ 0 & 0 & 1 & l_6 \\ 0 & 0 & 0 & 1 \end{bmatrix} \qquad [T_6^5]=\begin{bmatrix} c_6 & -s_6 & 0 & 0 \\ s_6 & c_6 & 0 & 0 \\ 0 & 0 & 0 & 0 \\ 0 & 0 & 0 & 1 \end{bmatrix}$$

The transformation matrix of the end-effector was calculated using the transformation matrices of individual joints.

$$T_0^6 = T_1 * T_2 * T_3 * T_4 * T_5 * T_6$$

$$[T_5^4]=\begin{bmatrix} a_{11} & a_{12} & a_{13} & a_{14} \\ a_{21} & a_{22} & a_{23} & a_{24} \\ a_{31} & a_{32} & a_{33} & a_{34} \\ a_{41} & a_{42} & a_{43} & a_{44} \end{bmatrix}$$

$$a_{11} = s_6 * \left(c_4 * (c_3 * s_1 - c_1 * c_2 * s_3) - s_4 * (s_1 * s_3 + c_1 * c_2 * c_3)\right) + c_6 * \left(c_5 * \left(c_4 * (s_1 * s_3 + c_1 * c_2 * c_3) + s_4 * (c_3 * s_1 - c_1 * c_2 * s_3)\right) + c_1 * s_2 * s_5\right);$$

$$a_{12} = c_6 * \left(c_4 * (c_3 * s_1 - c_1 * c_2 * s_3) - s_4 * (s_1 * s_3 + c_1 * c_2 * c_3)\right) - s_6 * \left(c_5 * \left(c_4 * (s_1 * s_3 + c_1 * c_2 * c_3) + s_4 * (c_3 * s_1 - c_1 * c_2 * s_3)\right) + c_1 * s_2 * s_5\right);$$

$$a_{13} = s_5 * \left(c_4 * (s_1 * s_3 + c_1 * c_2 * c_3) + s_4 * (c_3 * s_1 - c_1 * c_2 * s_3)\right) - c_1 * c_5 * s_2;$$

$$a_{14} = l_6 * \left(c_4 * (c_3 * s_1 - c_1 * c_2 * s_3) - s_4 * (s_1 * s_3 + c_1 * c_2 * c_3)\right) + l_1 * c_1 - l_2 * s_1 + l_5 * c_4 * (s_1 * s_3 + c_1 * c_2 * c_3) + l_5 * s_4 * (c_3 * s_1 - c_1 * c_2 * s_3) + l_3 * c_1 * c_2 + l4 * s_1 * s_3 + l_4 * c_1 * c_2 * c_3;$$

$$a_{21} = -s_6 * \left(c_4 * (c_1 * c_3 + c_2 * s_1 * s_3) - s_4 * (c_1 * s_3 - c_2 * c_3 * s_1)\right) - c_6 * \left(c_5 * \left(c_4 * (c_1 * s_3 - c_2 * c_3 * s_1) + s_4 * (c_1 * c_3 + c_2 * s_1 * s_3)\right) - s_1 * s_2 * s_5\right);$$

$$a_{22} = s_6 * \left(c_5 * \left(c_4 * (c_1 * s_3 - c_2 * c_3 * s_1) + s_4 * (c_1 * c_3 + c_2 * s_1 * s_3)\right) - s_1 * s_2 * s_5\right) - c_6 * \left(c_4 * (c_1 * c_3 + c_2 * s_1 * s_3) - s_4 * (c_1 * s_3 - c_2 * c_3 * s_1)\right);$$

$$a_{23} = -s_5 * \left(c_4 * (c_1 * s_3 - c_2 * c_3 * s_1) + s_4 * (c_1 * c_3 + c_2 * s_1 * s_3)\right) - c_5 * s_1 * s_2;$$

$$a_{24} = l_2 * c_1 - l_6 * \left(c_4 * (c_1 * c_3 + c_2 * s_1 * s_3) - s_4 * (c_1 * s_3 - c_2 * c_3 * s_1)\right) + l_1 * s_1 - l_5 * c_4 * (c_1 * s_3 - c_2 * c_3 * s_1) - l_5 * s_4 * (c_1 * c_3 + c_2 * s_1 * s_3) + l_3 * c_2 * s_1 - l_4 * c_1 * s_3 + l_4 * c_2 * c_3 * s_1;$$

$$a_{31} = s_6 * \left(c_2 * s_5 + c_5 * (s_2 * s_3 * s_4 - c_3 * c_4 * s_2)\right) + s_6 * (c_3 * s_2 * s_4 + c_4 * s_2 * s_3);$$

$$a_{32} = c_6 * (c_3 * s_2 * s_4 + c_4 * s_2 * s_3) - s_6 * \left(c_2 * s_5 + c_5 * (s_2 * s_3 * s_4 - c_3 * c_4 * s_2)\right);$$

$$a_{33} = s_5 * (s_2 * s_3 * s_4 - c_3 * c_4 * s_2) - c_2 * c_5;$$

$$a_{34} = l_6 * (c_3 * s_2 * s_4 + c_4 * s_2 * s_3) - l_3 * s_2 - l_4 * c_3 * s_2 - l_5 * c_3 * c_4 * s_2 + l_5 * s_2 * s_3 * s_4;$$

$$a_{41} = 0; \qquad a_{42} = 0; \qquad a_{43} = 0; \qquad a_{44} = 0;$$

The rotational vector (R) and position vector (P) of the end-effector were obtained from this transformation matrix.

$$\mathrm{R} = \begin{bmatrix} a_{11} & a_{12} & a_{13} \\ a_{21} & a_{22} & a_{23} \\ a_{31} & a_{32} & a_{33} \end{bmatrix} \qquad \mathrm{P} = \begin{bmatrix} a_{14} \\ a_{24} \\ a_{34} \end{bmatrix}$$

3.2 Workspace Analysis

A MATLAB code was implemented to determine all the reachable points of the end effector from the forward kinematics of the robotics system. The workspace of the 6-DoF arm assembly was calculated using the matrix 'P' in MATLAB.

For this, the link length of each joint and joint angle have been defined. The link lengths for the arm assembly were done subject-specific and those are as follows:

$$l_1 = 80.2, l_2 = 211.5, l_3 = 76.3, l_4 = 232.57, l_5 = 154.4, l_6 = 127.69$$

The θ range for each assembly joint was done by taking into consideration human motion as follows:

- $\theta_1 = (-\frac{\pi}{9})\ to\ (\frac{7*\pi}{18})$
- $\theta_2 = (-\frac{\pi}{9})\ to\ (\frac{\pi}{2})$
- $\theta_3 = (-\frac{\pi}{36})\ to\ (\frac{2*\pi}{3})$
- $\theta_4 = (-\frac{\pi}{36})\ to\ (\frac{23*\pi}{36})$
- $\theta_5 = (-\frac{17}{36})\ to\ (\frac{17*\pi}{36})$
- $\theta_6 = (-\frac{5*\pi}{36})\ to\ \left(\frac{\pi}{9}\right)$

Utilizing the link lengths and range of angles in MATLAB code, the workspace for the arm assembly was determined.

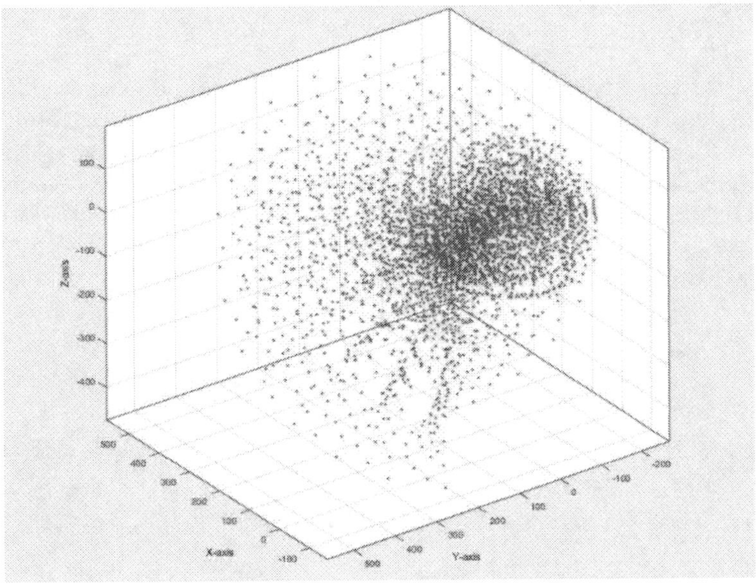

Fig. 6: Constrained workspace of the 6 DOF exoskeleton model scattered plotting in MATLAB

4 Discussions

The fabrication of the exoskeleton and its ability to facilitate a wide range of arm movements with minimum restriction represents a significant achievement and shows practicality in its assistance. The versatility of the exoskeleton is evident in its ability to fit a wide range of individuals, irrespective of their physical dimensions. This adaptability is primarily attributed to modifications made in the shoulder blade assembly and back support. Testing has demonstrated the exoskeleton's ability to provide support when the arm is positioned above the horizontal level. This addresses a common challenge for individuals with weak joints, especially when lifting objects above shoulder height.

All the tests have consistently shown that the exoskeleton prototype effectively provides the necessary range of motion for both the arm and fingers. The specially designed fingers significantly improve the process of grasping objects, promoting better rehabilitation. Notably, the prototype is lightweight and features ample space between the fingers, ensuring user-friendliness. A single motor efficiently controls the multiple finger joints, resulting in a strong grip while minimizing space requirements. The arms offer support at higher angles and enable lifting objects with ease. These joints enable a complete range of motion and proficiently distribute the load. The exoskeleton has proven to be beneficial for individuals with weaker joints and also for industrial workers.

The potential for this exoskeleton to be used in industrial settings, particularly for load handling, holds promise. To facilitate its industrial use, careful consideration should be given to manufacturing structural components according to the expected load-bearing requirements, as well as configuring motors appropriately. The exoskeleton has successfully achieved its goal of assisting patients with finger mobility disorders.

5 Conclusion

The exoskeleton effectively provides a full range of motion for both arms and fingers along with a lightweight, user-friendly environment resulting in improved rehabilitation. Hence, the exoskeleton's design and testing have demonstrated promising results regarding mobility, adaptability, and practical application. Its potential to address challenges in both medical and industrial settings suggests a bright future for this technology. Additional research and refinement, alongside professional input, will further enhance its clinical and real-world utility. A thorough evaluation of clinical benefits will ensure a connection with the unique needs of patients and proactively identify potential complications, which will help in the potential growth of the exoskeleton assistance.

References

1. PF Groshaw. Hardiman i arm test. hardiman i prototype. Rep. S. S-70-1019, (1969).
2. Homayoon Kazerooni. Human-robot interaction via the transfer of power and information signals. IEEE Transactions on systems, Man, and Cybernetics, 20(2):450–463, (1990).
3. Hiroaki Kawamoto, Suwoong Lee, Shigehiro Kanbe, and Yoshiyuki Sankai. Power assist method for hal-3 using emg-based feedback controller. In SMC'03 Conference Proceedings. 2003 IEEE International Conference on Systems, Man and Cybernetics. Conference Theme-System Security and Assurance (Cat. No. 03CH37483), volume 2, pages 1648–1653. IEEE, (2003).
4. Md Rasedul Islam, Christopher Spiewak, Mohammad Habibur Rahman, and Raouf Fareh. A brief review on robotic exoskeletons for upper extremity rehabilitation to find the gap between research porotype and commercial type. Adv. Robot. Autom, 6(2):10–4172, (2017).
5. Sunwook Kim, Maury A Nussbaum, Mohammad Iman Mokhlespour Esfahani, Mohammad Mehdi Alemi, Bochen Jia, and Ehsan Rashedi. Assessing the influence of a passive, upper extremity exoskeletal vest for tasks requiring arm elevation: Part ii–"unexpected" effects on shoulder motion, balance, and spine loading. Applied ergonomics, 70:323–330, (2018).
6. Patrick Aubin, Kelsey Petersen, Hani Sallum, Conor Walsh, Annette Correia, and Leia Stirling. A pediatric robotic thumb exoskeleton for at-home rehabilitation: the isolated orthosis for thumb actuation (iota). International Journal of Intelligent Computing and Cybernetics, 7(3):233–252, (2014).
7. Marco Fontana, Andrea Dettori, Fabio Salsedo, and Massimo Bergamasco. Mechanical design of a novel hand exoskeleton for accurate force displaying. In 2009 IEEE international conference on robotics and automation, pages 1704–1709. IEEE, (2009).

Design of Hybrid Sprinkler: The IOT-Powered Robot for Watering Plants

Dr. *Jayasheel* Kumar KA[1], *Mannaan* Akhtar[2]*, C. *Mohammed* Talha[3], and *Gokul* VG[4]
[1]Senior Assistant Professor, Department of Automobile Engineering New Horizon College of Engineering, India
[2]Students, Department of Automobile Engineering New Horizon College of Engineering, India

Abstract. Innovative methods for watering plants have been developed as a result of the growing demand for efficient and automated systems in agriculture. The design of a hybrid sprinkler, an IoT-powered robot created specifically for effective and precise plant irrigation, is presented in this study. The suggested system enables autonomous operation and intelligent control by fusing the benefits of conventional sprinklers with those of IoT technology. To continuously monitor the environmental conditions, the hybrid sprinkler integrates a variety of sensors, including soil moisture sensors, temperature sensors, and humidity sensors. A central control unit receives this real-time data and uses sophisticated algorithms to establish the best watering schedule for each plant depending on its individual needs. The hybrid sprinkler's design prioritizes sustainability and energy economy. The robot has a rechargeable battery system, and to increase battery life, it uses energy-saving features like sleep modes and sophisticated power management. In a variety of agricultural and horticultural applications, this system assures optimal watering practises, conserves resources, and fosters healthy plant growth by fusing IoT technology, cutting-edge sensors, precision spraying mechanisms, and remote control capabilities.

Keywords: Hybrid, Sprinkler, IoT , Sensors , Agriculture.

1 Introduction

The world is trending towards long-term fixes for many issues. Water scarcity is one of the major problems that we are now dealing with. We must devise creative, environmentally responsible methods to address this problem. One such idea is the creation of a sprinkler robot driven by solar energy and battery powered based on an IoT strategy for watering plants. The goal of this project is to create a robot that can effectively and autonomously irrigate plants using hybrid powered. The robot will water the plants appropriately based on the amount of soil moisture it detects using IoT technology. The way we water our plants and conserve water could be completely altered as a result of this initiative. the design of a hybrid sprinkler, a cutting-edge IoT-powered robot that transforms plant watering procedures, is

Corresponding author: *Mannaan2002@gmail.com

presented in this study. The hybrid sprinkler makes watering more effective and targeted by fusing the advantages of conventional sprinklers with those of IoT technology. The system can monitor and react to real-time environmental data by utilizing a variety of sensors and clever algorithms, assuring ideal watering schedules and water usage. The hybrid sprinkler has a strong emphasis on sustainability and energy economy. In order to increase battery life, the design incorporates a rechargeable battery system and makes use of energy-saving features including sleep modes and intelligent power management. The technology lowers operating expenses and encourages environmentally responsible behaviour by limiting energy use. This system offers a productive, accurate, and environmentally friendly method of watering plants by utilizing Internet of Things (IoT) capabilities, cutting-edge sensors, precision spraying mechanisms, and remote-control features. The individual parts and capabilities of the hybrid sprinkler will be examined in more detail in the sections that follow, illustrating how they might revolutionize horticultural and agricultural practices.

2 Methodology

2.1 Literature study:

- Make review on focusing on how to make it simple and relevance to the project title.

2.2 Conceptual design:

- Sketch up the basic design of the robot
- State the dimension for all parts.
- Draw the scaled sketching model and prepare blueprint by manual sketching.

2.3 Computer Aided Designing (CAD) model:

- Model each parts using blueprint and assemble them in position

2.4 Materials selection:

- Select the appropriate material based on model design and strength criteria.
- Select those which are Light, easy to joining and easy to manufacture and cost effective.

2.5 Report making:

- Finally make a report that includes all the things done in project starting from introduction till the modelling and material selection.

3 Design of Sprinkler

3.1 Infrastructure Design:

The LDR sensors, which absorb light and distinguish between the colors according on how much of it they take in, require that the plant's path be black on a white background. An RFID tag has been attached to each plant. The RFID EM-18 Reader Module has a range of approximately 4 inches. The autonomous robot must be 4 inches or less away from the plant. It is required that the RFID tag face the RFID Reader Module. The wireless Xbee modules in use are set up to communicate with one another and form a network. Due to the maximum weight that DC motors can drive, the system has a restricted capacity for water transport, which is 3 litre.

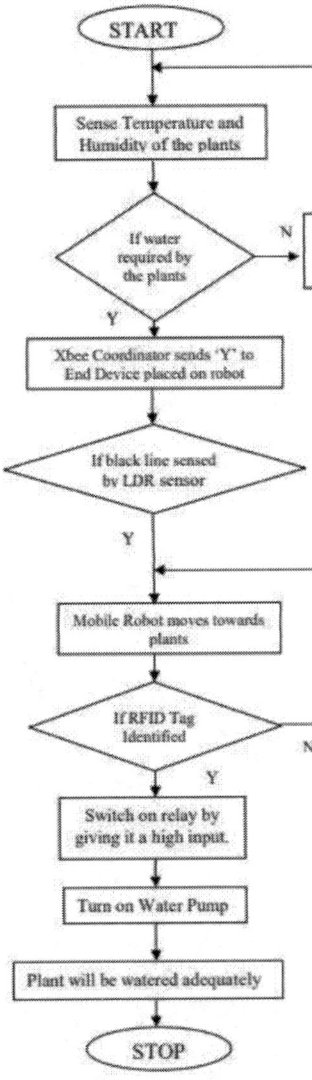

Fig. 1. Infrastructure of sprinkler

3.2 Circuit Design:

3.2.1 Circuit design of transmitter module:

- YL-69 Soil Moisture Sensor: The YL-69 soil moisture sensor is inserted into the soil around the plants that need to be watered in order to measure the soil's moisture content values and determine how much water the plants actually need. It consists of a two-legged lead that is inserted into the ground or other location where the amount of water has to be measured. Two header pins on it connect to an amplifier/A-D circuit, which is connected to the Arduino through a separate circuit.
- 1602 Liquid Crystal Display (LCD): The LCD provides real-time data on the soil's moisture content at a distance from the irrigation robot (receiver module). To make connecting to the Arduino simpler, it interfaces using an I2C module. The module reduces the LCD's 16 pin interface to 4 pins.
- Xbee (Series 1) Module (Transmitter Configuration): The mobile robot is equipped with a wireless connection device called an Xbee that is used to relay information about the moisture content of the soil. It covers a linear area of around 100 meters (300 feet) long.
- Power Supply: A 9 V battery attached to the power port of the Arduino provides power to the transmitter module. The Arduino provides power to other components.

3.2.1 Circuit design of receiver module:

- L293D and DC motors: The irrigation robot was propelled by two 12 V geared motors. There was only one L293D IC in charge of them.
- IR Sensor: This IR sensor uses a TCRT5000 to measure distance and color.
- Relay and Water Pump: Due to the voltage and current the 12 V pump draws, the Arduino Uno is unable to control it directly. Instead, a 12 V single channel relay module (SPDT) was used to switch the pump between its ON and OFF states.

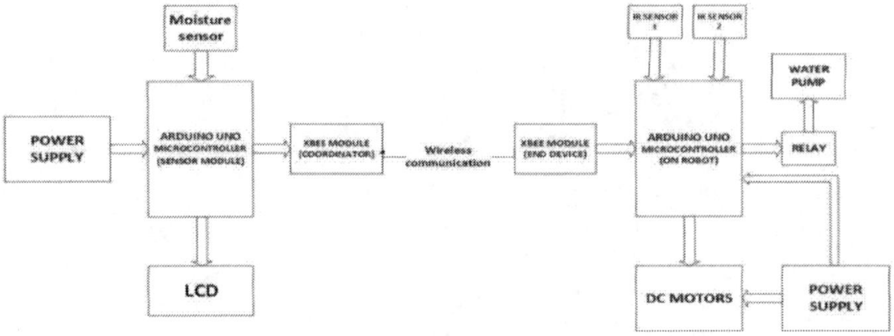

Fig. 2. Block diagram of system

3.3 Computer aided design model:

Initially a rough sketch is drawn and dimensions are decided and then the 3 dimensional model is created using the CATIA software, A tank is provided at the back for storage of water and pump and other auxiliary systems are to be mounted within the robot body, it travels with the help of 4 wheels driven by motor.

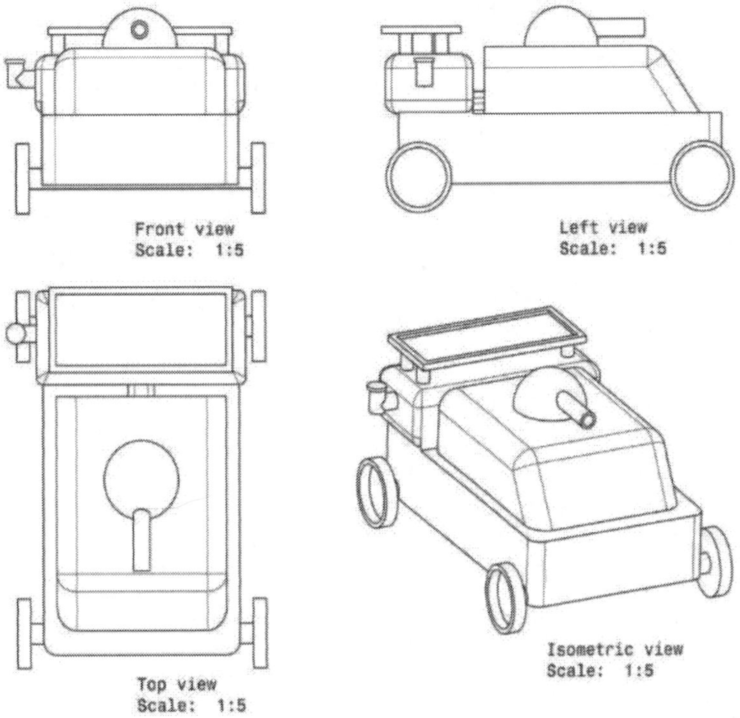

Fig. 3. Views of sprinkler

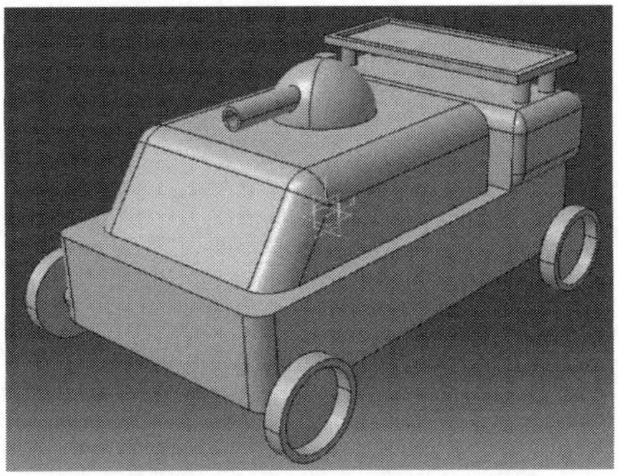

Fig. 4. Isometric front view

Fig. 5. Isometric rear view

3.4 challenges faced during the design or implementation phases:

The development of dependable software algorithms to enable intelligent functionalities and the smooth integration of hardware components like sensors and spraying mechanisms are the fundamental technical challenges. Although crucial, ensuring the consistency and accuracy of sensor data is vulnerable to things like environmental interference and sensor deterioration. Similar to mission-critical, reliable internet access is similarly susceptible to interruptions that could obstruct real-time control and monitoring. An efficiency challenge is to optimise power management to balance energy conservation with sensor sampling rates. Software designers must take careful UX factors into account when creating a user interface that is feature-rich and intuitive while encapsulating system complexity.

Moreover, adaptive algorithms are required to enable adaptability across various conditions and plant species to suit specific watering requirements. Even if more technological investments are needed for upgraded capabilities, cost control is essential for accessibility. To ensure reliability, thorough testing and validation across simulated scenarios are necessary yet time-consuming. The optimisation of the watering schedule may also be impacted by the dependence on other datasets, like as weather predictions. The complex technical rigour is further enhanced by other aspects such as regulatory compliance, maintaining performance through component maintenance, and scaling without sacrificing efficiency. An intelligent watering system that is robust must ultimately consider hardware, software, and environmental elements holistically.

4 Advantages of proposed sprinkler robot:

- Precise and Targeted Irrigation
- Improved Water Efficiency
- Adaptability to Environmental Conditions
- Intelligent Decision-Making
- Remote Monitoring and Control
- Energy Efficiency
- Customization and Adaptation
- Data-Driven Insights
- Reduction in Manual Effort
- Sustainable Plant Irrigation

5 Conclusion:

- The system aims to enable precise, efficient, and adaptable plant irrigation by integrating sensors, algorithms, spraying mechanisms and remote monitoring/control capabilities.
- It makes use of RFID tags on each plant for identification and LDR sensors to determine the plant's route. This makes it possible to irrigate particular plants only.
- The RFID reader's limited range is around 4 inches. For the robot to come close enough to read the tags, it must navigate precisely in this situation.
- Xbee wireless modules allow components to communicate with one another and create a network for data transfer.

- The DC motors' weight limits impose limitations on the water carrying capacity of three litres. This could need to be refilled quite often.
- Important information for figuring out watering requirements and schedules is provided by the soil moisture sensor that is embedded in the soil.
- The robot is powered by 12V DC motors, while the transmitter module is powered by a 9V battery.
- Potential challenges to consider: upfront costs; technical complexity; sensor reliability issues; connectivity disruptions; ongoing maintenance requirements. When well-implemented and managed, the system can promote sustainable agriculture through optimized water usage and plant health monitoring. Further evaluation of tradeoffs is warranted. Overall, it is a promising innovation for efficient, tech-driven irrigation if adoption barriers can be addressed.

References

1. B.C. Wolverton, Anne Johnson, and Keith Bounds, "Interior Landscape Plants for Indoor Air Pollution Abatement: Final Report", National Aeronautics and Space Administration (NASATM-101768) Science and Technology Laboratory, Stennis Space Center, (1989).
2. E.J. Van Henten, J. Hemming, B.A.J. Van Tuijl, J.G. Kornet, J. Meuleman, J. Bontsema and E.A. Van Os; "An Autonomous Robot for Harvesting Cucumbers in Greenhouses"; Autonomous Robots; Volume **13** Issue 3, November (2002).
3. Kevin Sikorski, "Thesis- A Robotic PlantCare System", University of Washington, Intel Research, (2003).
4. Ayumi Kawakami, Koji Tsukada, Keisuke Kambara and Itiro Siio, "PotPet: Pet-like Flowerpot Robot", Tangible and Embedded Interaction (2011), Pages **263-264** ACM New York, NY, USA, (2011).
5. T.C.Manjunath, Ph.D. (IIT Bombay) & Fellow IETE, Ashok Kusagur , Shruthi Sanjay,Saritha Sindushree, C. Ardil, "Design, Development & Implementation of a Temperature Sensor using Zigbee Concepts", International Journal of Electrical and Computer Engineering 3:12 (2008).
6. Christos Tatsiopoulos and Aphrodite Ktena, "A Smart ZIGBEE Based Wireless Sensor Meter System", IEEE Electrical Engineering (2009), ISBN: 9781424445301, (2009)
7. Ragheid Atta, Tahar Boutraa and Abdellah Akhkha, "Smart Irrigation System for Wheat in Saudi Arabia Using Wireless Sensors Network Technology", International Journal of Water Resources and Arid Environments 1(6): **478-482**, (2011), ISSN 2079-7079, © PSIPW, (2011)
8. Manual by Martin Hebel and George Bricker, "Getting Started with XBee RF Modules: A Tutorial for BASIC Stamp and Propeller Microcontrollers", Version **1.0**, ISBN 9781928982562.
9. Manual by Martin Hebel and George Bricker, "Getting Started with XBee RF Modules: A Tutorial for BASIC Stamp and Propeller Microcontrollers", Version **1.0**, ISBN 9781928982562.
10. Connection of RFID and Arduino board", http://www.instructables.com/id/Arduino-and RFID-from-seeedstudio.

Novel design of VTOL Unmanned Aerial Vehicles for transportation of medical supplies

Ashish Tripathi[1*] and *Shiv Manjaree Gopaliya*[1]

[1]School of Mechanical Engineering, VIT Bhopal University, Indore Bhopal Highway, Kothrikalan Sehore 466114, Madhya Pradesh, India

Abstract. Vertical Take-Off and Landing (VTOL) based Unmanned Aerial Vehicles (UAV) allow quick, safe, and effective transit of resources from one place to another. Its primary objective is to complete its mission in a short time, thus, making it more useful during emergency cases such as road accidents. A novel design for VTOP UAVs has been proposed in this paper. A mobile application has been created to send and receive commands. The cost estimation and analysis have been carried out for the development and deployment of this device in a real-life scenario. VTOL Unmanned Aerial Vehicles are a novel design for the transportation of medical supplies that aim to significantly increase the flight range and endurance compared to conventional aircraft rotorcraft configurations. The overall conceptual design, modelling and computational simulation of this design are presented.

Keywords: novel design, vertical take-off and landing, unnamed aerial Vehicles, accidents, mobile application

1 Introduction

VTOL UAVs represent a significant breakthrough in the field of aviation. Their unique design and capabilities have made them a popular choice for a range of applications, from military operations to civilian use [1]. The ability to take off and land vertically without the need for a runway makes these aircraft ideal for use in urban areas or remote locations where traditional aircraft may not be able to operate. Additionally, their maneuverability and versatility have made them an attractive option for a range of industries, including transportation, logistics, and healthcare. One area where VTOL UAVs have already shown tremendous promise is in the delivery of medical supplies and emergency aid to remote or underserved areas [2]. In many parts of the world, access to healthcare is limited, and delivering essential medicines and supplies can be a major challenge. By using VTOL UAVs, healthcare providers can rapidly deliver vital supplies to these areas, saving lives and improving health outcomes. In addition to their use in healthcare, VTOL UAVs have a range of other potential applications. For example, they can be used for disaster relief efforts, delivering food, water, and other supplies to people in need. They can also be used for environmental monitoring, allowing scientists to collect data on air and water quality, weather patterns, and other

* Corresponding author: ashish.tripathi340@gmail.com

important metrics. One of the most exciting aspects of VTOL UAVs is the potential for innovation and collaboration across different fields and industries. By bringing together experts in bio, electronics, avionics, and IT, we can unlock new possibilities and develop innovative solutions to some of the most pressing challenges facing our society [3]. This could include developing new sensors and monitoring technologies, improving battery life and energy efficiency, and creating more intuitive and user-friendly interfaces for controlling and operating these aircraft. As technology continues to evolve and new applications are discovered, these aircraft will play an increasingly important role in shaping the future of aviation and society as a whole. Whether delivering life-saving medical supplies to remote areas or collecting vital data on the environment, VTOL UAVs offer a powerful and transformative technology that has the potential to drive progress and innovation for years to come. Inadequate quality of healthcare services can result in more fatalities compared to a lack of access to healthcare facilities. This fact is exemplified in the case of India, where in 2016, the number of deaths caused by poor healthcare quality was 1.6 million, which is almost twice the number of deaths resulting from the non-utilization of healthcare services, which was 838,000 people [4]. This means that the quality of healthcare provided to patients is an essential factor in determining the success of healthcare services. If healthcare services are available but of substandard quality, they can still lead to devastating consequences, including unnecessary illnesses, injuries, and even death. In India's case, the high number of deaths resulting from poor quality of care can be attributed to several factors, including inadequate training and resources for healthcare providers, a lack of standardized guidelines and protocols for treatment, and insufficient regulation and oversight of healthcare facilities. These issues must be addressed to ensure that patients receive quality healthcare services and that their lives are not put at risk. The present investigation discusses current approaches that have been published in the literature. Additionally, a novel approach to the design of VTOL (Unmanned Aerial Vehicles) is suggested for the transportation of medical supplies in various areas.

2 Literature Review

Many approaches with different methodologies have been discussed in the literature. A paper explains how to use drone technology to deliver medical supplies during extended field campaigns [5]. This investigation used an unmanned, rotary-wing drone to replicate the delivery of a 4.5 kg cargo of medical supplies, such as bandages, tourniquets, analgesics, and blood products. A fictitious victim was placed in a far-off location. Grid coordinates were used to preprogramme the aircraft, which was autopiloted beyond visible range while high-altitude Shadow drones recorded real-time data (altitude, flying time, and route). A collaborative process to create a highly automated drone service for the delivery of pharmaceuticals at San Raffaele Hospital, Milan, Italy [6] . This study shown how the introduction of a drone delivery service, which lessens the stress of logistic movements on foot, might assist the healthcare logistics industry. An independent flight management system control approach for unmanned Vertical Take-off and Landing (VTOL) aircraft was demonstrated in another work. Appropriate actions can be made in emergency states based on the VTOL aircraft location algorithm that was created [7]. A paper emphasises that all military crafts and armaments must participate in joint research and development to enhance our capacity to capitalise on emerging technology [8]. This will probably become more and more of the case as new innovative treatments and instruments are added to the prehospital care setting, enhancing the value of UAVs. Additionally, drone-based medical delivery models provide a novel way to solve persistent problems with fairness and access to healthcare, and they are especially pertinent in the context of COVID-19 and stopping the spread of pandemics [9]. Numerous configurations for UAV design are currently available in the literature [10]. In this paper, a novel configuration that resembles the quadplane+tilt rotor one is selected, Figures and tables, as originals of good quality and well contrasted, are to be in their final form, ready for reproduction, pasted in the appropriate place in the text. Try to ensure that the size of the text in your figures is approximately the same size as the when converted to black and white, and this should be considered when preparing them.

3 Methodology

In this paper, a 3D model for the proposed UAV has been created using Solidworks software. Each part of the UAV was designed separately, starting with the wing part and working towards the fuselage. The coordinates of RG15 and NACA0012 were imported to create the wing, the horizontal stabilizer, and the vertical stabilizer, respectively. NACA0012 is a basic conventional symmetric airfoil and its use as an elevator in the design was relatively convenient. In the final proposed design, the rudder was eliminated as yaw motion was achieved using the yaw moment of the quad configuration by adjusting the RPM of the diagonal pair of propellers.
The step-by-step approach to designing the 3D model of the proposed UAV can be seen in Fig.1. The 3D design of the finally assembled VTOL UAV can be seen in Fig.2.

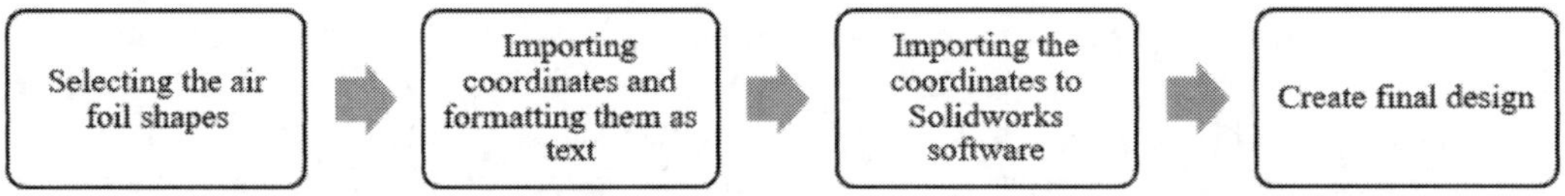

Fig.1. 3D model development in Solidworks software

Fig.2. Views of 3D model of VTOL UAV

3.1. Dropping Mechanism

The major objective of proposing a novel design for VTOL UAVs is for providing early medical help to road accidents [11]. In other words, the drop-off of medicine boxes on site is a crucial task. In this paper, a contactless medicine delivery system has been proposed. Dropping mechanisms are an essential component of drones used in various applications such as aerial photography, search and rescue, and surveying. There are various types of dropping mechanisms available for drones viz. servomotor dropping, magnetic release, pneumatic release, spring-loaded release, gravity release, and parachute release, respectively. The selection of the right type of dropping mechanism depends on several factors such as the size and weight of the payload, the level of control required, and any safety or regulatory requirements that may apply. A servo lock mechanism has been used to lock and hold the medicine box with the help of UAV bottom arms. The good accuracy and repeatability factors of the servo motor mechanism make it a good choice for the work presented in this paper. Fig.3 (a) shows the use of the four bottom arms of the UAV and Fig.3 (b) shows the slots to handle the box. A servo motor is attached with a supportable handle (shown in a yellow circle). Each servo motor has to handle a mass of 1 kg. After the servo handles hold successfully to the medicine box, it tightly fits the bottom and then is uplifted from the ground.

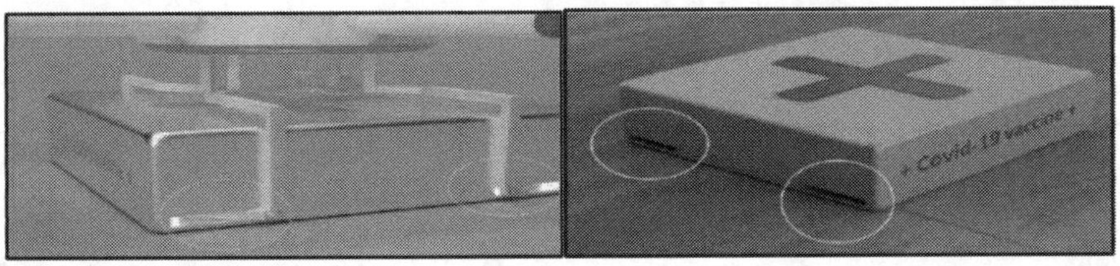

Fig.3. Views of 3D model of VTOL UAV

3.2. Emergency Medical Kit

'Medkit', an emergency medical kit, to be uplifted and provided at the target location by VTOL UAV has been designed and proposed in this paper. To address the payload capacity as well as to protect medicines, Medkit has been designed using lightweight materials viz. carbon fiber-reinforced composites (CRFC) mixed with thermoplastic materials (aluminum polystyrene). The selected materials act as a thermal case and protect medicines/vaccines under extreme working conditions like heat or pressure.
Four accidental cases viz. roadside mishap, heart attack, burns, and electric shocks have been considered, as shown in Figure. The proposed Medkit is modular, highly flexible, and adaptable to any medical product like blood bags, organs, drugs, test samples, etc. Based on World Health Organization's (WHO) stringent delivery regulations, it is designed for blood and blood components delivery i.e. 4 ± 2 °C for red cells, 25 ± 2 °C for platelets and -25 °C for frozen plasma, respectively.

3.3. 'Appbar', a mobile App

To provide immediate services, a mobile application 'Appbar' has been designed and developed for the above-mentioned emergency cases. Fig.4 (a) shows the steps involved in creating a mobile application. A simple and aesthetically pleasing application is designed in a single solid color with medical transportation details and a logo. Fig. 4 (b) shows the created mobile application called 'Appbar'.
Four accidental cases viz. roadside mishap, heart attack, burns, and electric shocks have been considered, as shown in Figure. The proposed Medkit is modular, highly flexible, and adaptable to any medical product like blood bags, organs, drugs, test samples, etc. Based on World Health Organization's (WHO) stringent delivery regulations, it is designed for blood and blood components delivery i.e. 4 ± 2 °C for red cells, 25 ± 2 °C for platelets and -25 °C for frozen plasma, respectively.

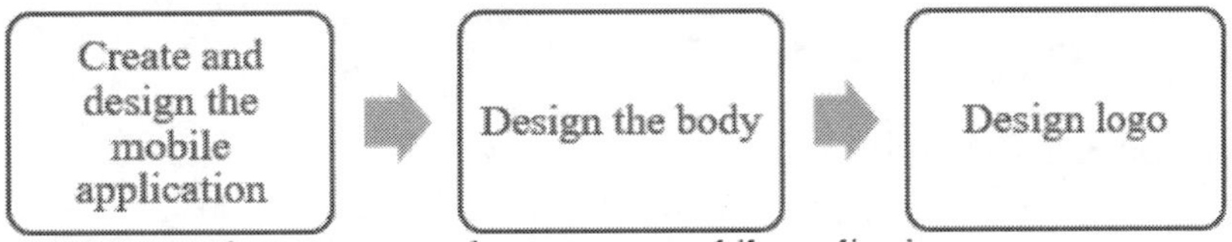

Fig.4(a). Step by step approach to create a mobile application

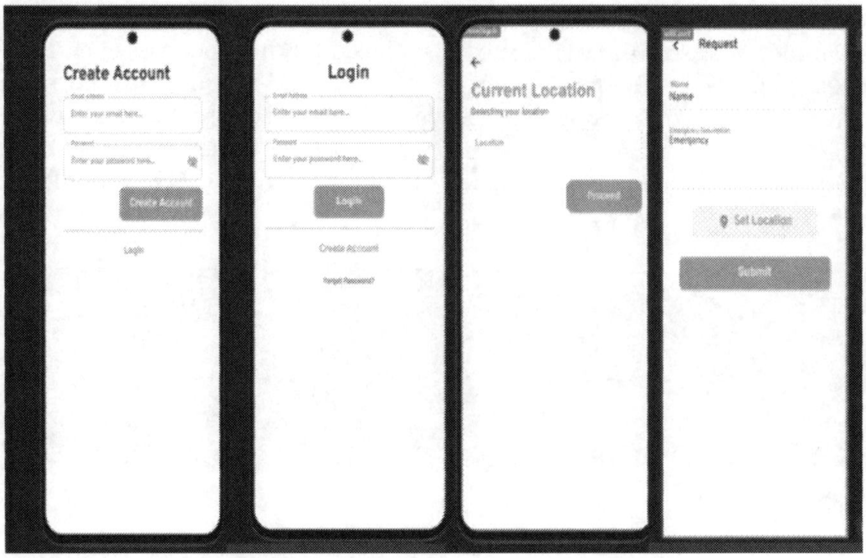

Fig.4(b). Created mobile application "Appbar"

In User Authentication and Profile Management, a customer is logged in with his/her mobile number or email credentials. The mobile GPS is needed to get current location details while providing continuous tracking and updation. The app allows the customer to submit his/her requests for the type of medical help needed with an option to provide additional information on emergencies. The request raised by the customer is then handled and the medical supply is transported using real-time tracking of devices. The app also provides information about the Expected Time of Arrival (ETA) of medical supplies to customers. In the end, feedback on the quality of service is requested by the customers. It helps in upgrading the accountability and reliability of the transportation processes. All the information will be stored in a database for further improvement of services. Step by step approach to generate the mobile application and user identification can be seen in Fig. 5

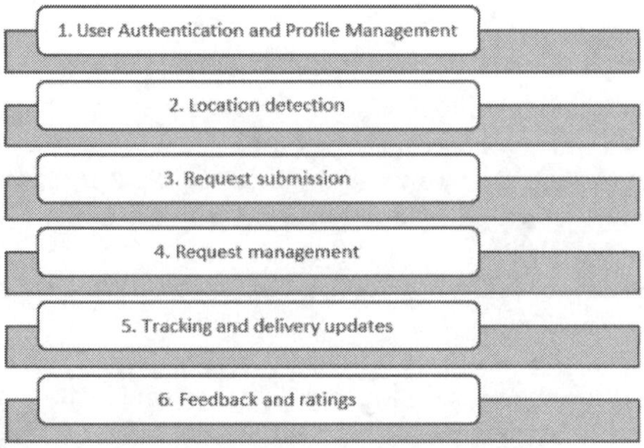

Fig. 5. Steps to create the mobile application with app

3.4. Cost Estimation Analysis

3.4.1 Cost Analysis for Drone

Cost analysis For a drone with the following specifications is presented in table 1 :

- Weight : 2.5 kg

- Power: 170 W/kg
- Dropping mechanism: Servo motor
- Battery: 5200 mAh LiPo
- Flight controller: APM 2.8

The estimated cost breakdown is as follows:

Table 1 : Cost analysis for Medical kit

Component	Cost (in INR)
Frame	5000
Motors	8000
ESCs	4000
Propellers	1500
Battery	3000
Flight controller	2000
GPS	3000
Radio transmitter	3000
Camera	5000
Gimbal	6000
Servo motor	2000
Total	**41500**

3.4.2 Medicine Kit Cost Estimation

Cost analysis For a medicine kit that includes the following items is presented in table 2 :

Table 2 : Cost analysis for Medical kit

Item	Cost (in INR)
One 325 mg uncoated adult aspirin	5
Bandages of various sizes (about 20)	100
5 sterile gauze pads and 5 sterile gauze rolls	150
Cotton balls and cotton-tipped swabs	50
5 antiseptic wipe packets and antiseptic ointment	100
1 bottle of antiseptic solution	50
2 packets of 81 mg aspirin	20
5 antibiotic 1-gram ointment packets	150
2 triangular bandages	50
Adhesive tape, safety pins	50
1 digital thermometer	150
Tweezers and scissors	100
1 flash-light with extra batteries	200
2 insect sting relief pads	50
Simple band-aids	50
CPR breathing mask	300
Disposable gloves (more the merrier)	100
First-aid instruction manual	50
Burns Dressing 20cm x 20cm	200
Burns Dressing 10cm x 10cm	100

Non-Adherent Dressing 10cm x 10cm	100
Non-Adherent Dressing 5cm x 5cm	50
Burn Gel Sachets	150
Eye Wash Pods 20ml	50
Crepe Bandage 10cm x 4.5m	100
Conforming Bandage 7.5cm x 4.5m	50
Gloves (Pair)	50
Microporous Tape 2.5cm x 5m	50
Scissors	50
Gel-soaked burn blanket	500
Water-Jel burn stop dressings	500
Water-Jel Face Mask	100
Burn gel for minor burns	150
H-F Antidote Gel (Hydrofluoric Acid / Calcium Gluconate)	500
Total Cost Estimate	**Approximately 4,225**

4 Methodology

Initially the external fluid flow analysis was performed, which is done by using solidworks flow simulations with certain boundary conditions. Free stream velocity for fluid is about 15m/s for this case (Fig.6). As fluid flows around the body, the smooth streamline and continuous flow came out as desired result. It can be observed that the prototype will sustain and perform well in its actual flight. Also, this provides an idea that, how the aircraft is going to react with such air speed and pressure during its flight phase. The wing horizontal stabiliser is created by importing the coordinates of the RG15 airfoil following fluid simulations (Fig.7). Apart from this, the rudder can be eliminated in the final design as the yaw motion can easily be achieved using the yaw moment of the quad configuration by adjusting the RPM of the diagonal pair of the propellers. Finally for pressure and velocity analysis of wings simulation was performed (Fig. 8).

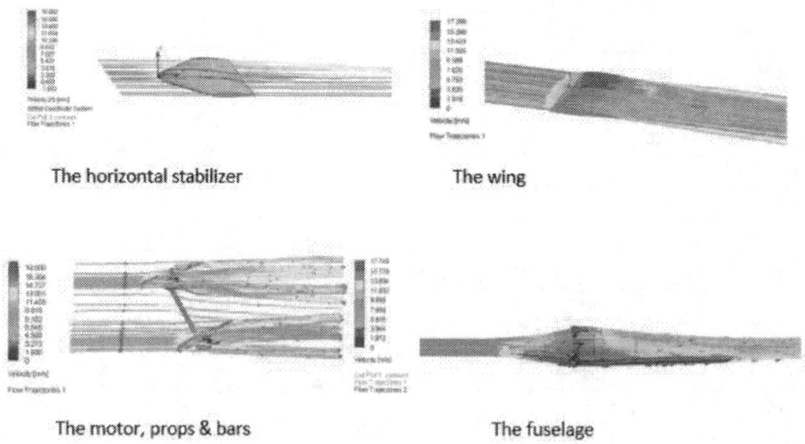

Fig.6. External fluid flow analysis by using solidworks simulation

The 2D cut plots and pressure contour for the airfoil aerodynamic analysis using RG-15 Airfoil using solidworks fluid flow analysis.

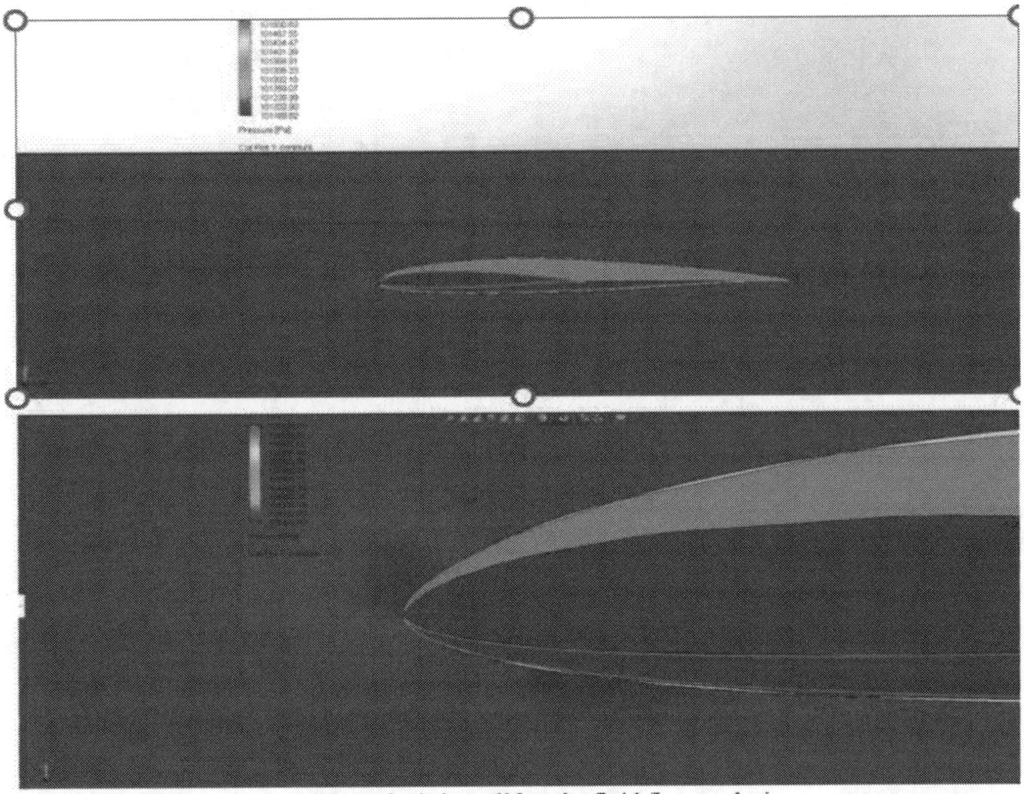

Fig.7 : RG15 Airfoil aerodynamic analysis by solidworks fluid flow analysis

External fluid flow analysis of the wing part for pressure and velocity.

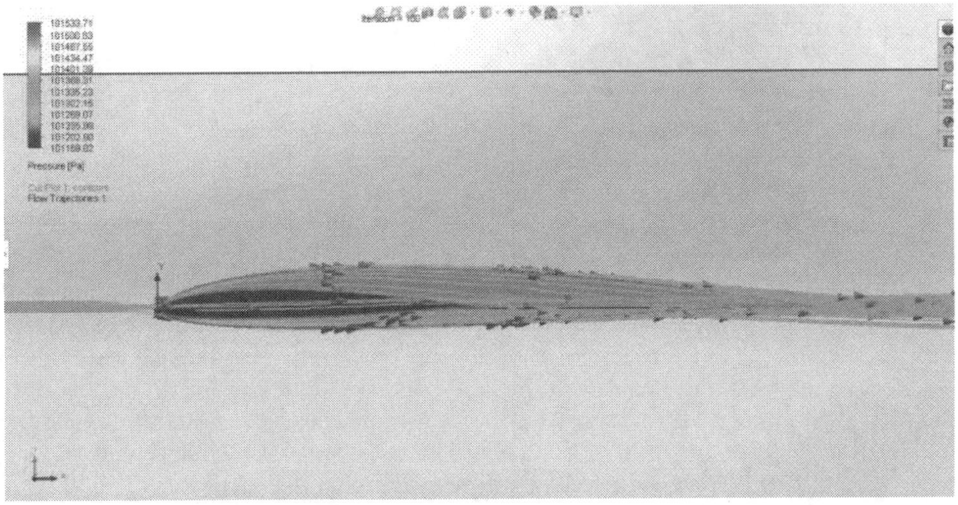

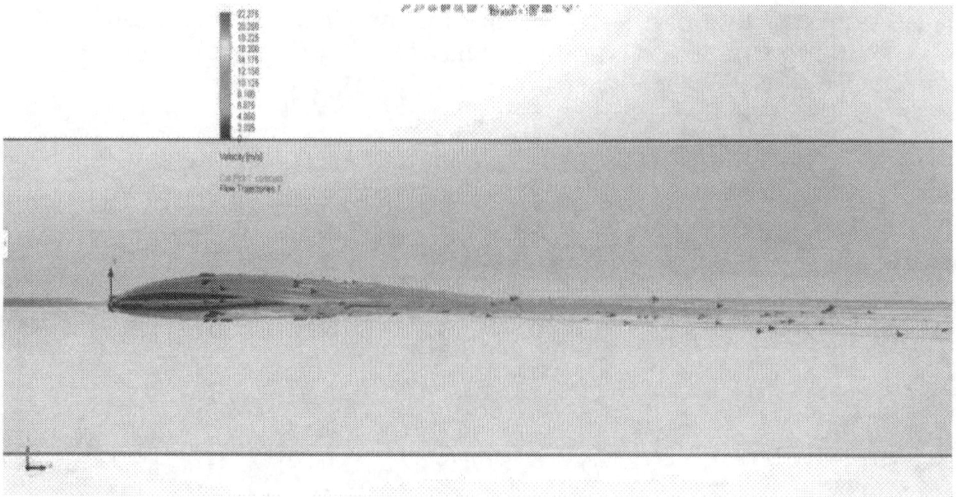

Fig.8 : Simulation for pressure and velocity analysis of wings

It can be observed that if the aircraft is built with this design configuration and approach, accurate test results can be achieved.

Application generation and working : Working of the application can be achieved in following steps

i. The current location of the user would be detected.
ii. The user will generate the request for the emergency stating the kind of emergency that happened.
iii. This information would be stored in the database.
iv. The app also allows user to login to their account.
v. User Interface and Interaction:

The objective of the app is to effectively handle medical supply transportation needs. In addition to offering real-time tracking for better logistics, it streamlines the transportation request procedure. It is possible to talk about the shortcomings and potential areas for improvement, such as the addition of machine learning algorithms for predictive routing.

Expected Results :

1. We anticipate that by employing UAVs to deliver medicine more quickly in emergency situations, patients will receive their medication on time and avoid many scenarios where patients die from not obtaining it in time.
2. Secondly, the UAV can also be used in pandemic situations or contactless delivery, it can also be used for delivering medicine to remote locations where delivering medicine through normal means is difficult.
3. Reduction of the overall Cost, by implementing the cost-saving methods for the drone and medicine kit, it is possible to reduce the overall cost for both significantly. While the exact cost reduction will depend on various factors such as the location, brand, and availability of components. The savings on the drone and medical package might total up to thirty percent. For instance, the initial projected cost of INR 41,500 for the drone may be lowered to INR 29,050, while the initial estimate of INR 35,000 for the medical package might be lowered to INR 24,500. Overall, by implementing cost-saving methods, it is possible to reduce the cost of both the drone and medicine kit significantly without compromising on their functionality and effectiveness.

5 Conclusion

Unmanned aerial vehicles (UAVs) with Vertical Take-Off and Landing (VTOL) capabilities enable the efficient, secure, and rapid transportation of resources from one location to another. Its main goal is to finish its mission quickly, which makes it more helpful in emergency situations like car accidents. The purpose of this work is to design a smartphone app for controlling the operations of an unmanned aerial vehicle (VTOL) and UAV. It will be highly affordable and will also have rotors at the tip of the wing which will be better for yaw controlling. This technology will lead to faster transportation of medical supplies. The UAV will be operated using the app from which we will be getting the appropriate location for the delivery of supplies at the expected time of arrival (ETA) even when there are various obstructions such as heavy traffic on roads. Hereby we have also attached sketches of our designs which is to be accomplished to the highest precision possible. And have set our goal to make a detailed sketch and make a detailed schematic of our app which is to be developed for drone communication and details.

Acknowledgement:

In this research paper, initial attempts were essential. However, without the generous assistance and support of numerous people, it would not have been possible. We would like to sincerely thank Anchal Sharma, Gauri Gupta, Vartika Pal, Ishita Verma, Aagam Jain, Raghvendra Pratap Singh, Soham Nandi, and Tapas Kumar Nag, B.Tech students of VIT Bhopal University, for their help in this research project.

References

1. A. Intwala, Y. Parikh, A review on vertical take-off and landing (vtol) vehicles. International Journal of Innovative Research in Advanced Engineering (IJIRAE). **Feb;2(2):187-91**(2015)
2. Harcus MB. Fused deposition modelling (FDM) to fabricate a transitional vertical take-off and landing (VTOL) unmanned aerial vehicle (UAV) for transportation of medical supplies in underdeveloped areas (Doctoral dissertation) (2020).
3. GJ Ducard, M. Allenspach, Review of designs and flight control techniques of hybrid and convertible VTOL UAVs. Aerospace Science and Technology. Nov 1; **118:107035** (2021).
4. S. Sharma, Economic Survey 2021: More lives lost due to poor medical quality than lack of healthcare access. India Today, January. (2021);29.
5. T. Mesar, A. Lessig, DR King, Use of Drone Technology for Delivery of Medical Supplies During Prolonged Field Care. Journal of special operations medicine: a peer reviewed journal for SOF medical professionals. Jan 1;18(4):**34-5**(2018).
6. S. De Silvestri, M. Pagliarani, F. Tomasello, D. Trojaniello, A. Sanna, Design of a Service for Hospital Internal Transport of Urgent Pharmaceuticals via Drones. Drones. Mar 8;6(3):70 (2022).
7. A. Sikora, R. Czyba, Independent Flight Management System for Unmanned VTOL Aircraft. In2018 23rd International Conference on Methods & Models in Automation & Robotics (MMAR) Aug 27 **(pp. 253-258)**. IEEE (2018).

8. C. Handford, F. Reeves, P. Parker, Prospective use of unmanned aerial vehicles for military medical evacuation in future conflicts. BMJ Military Health. 1;**164(4):293**-6 aug (2018).
9. K. Flemons, B. Baylis, AZ Khan, et. al., The use of drones for the delivery of diagnostic test kits and medical supplies to remote First Nations communities during Covid-19. American Journal of Infection Control. 1;50(8):**849-56** Aug (2022).
10. Rathore, R. Krishna, et al., Response Surface Optimized Robotic Spray-Painting Metamodeling for Fanuc paint Robot, P-250IB/15. (2022).
11. Sambodo F. Abdillah, et al., Preliminary design of dual-system hybrid VTOL UAV for medical delivery purpose. InAIP Conference Proceedings, **Vol. 2366**, No. 1. AIP Publishing (2021). DOI: https://doi.org/10.1063/5.0060064

AUTHOR INDEX